HIGHER

CCEA

GCSE
MATHEMATICS
SECOND EDITION

Practice book

Anne Connolly
Linda Liggett
Robin Liggett

HODDER
EDUCATION
AN HACHETTE UK COMPANY

Although every effort has been made to ensure that website addresses are correct at time of going to press, Hodder Education cannot be held responsible for the content of any website mentioned. It is sometimes possible to find a relocated web page by typing in the address of the home page for a website in the URL window of your browser.

Hachette UK's policy is to use papers that are natural, renewable and recyclable products and made from wood grown in sustainable forests. The logging and manufacturing processes are expected to conform to the environmental regulations of the country of origin.

Orders: please contact Bookpoint Ltd, 130 Milton Park, Abingdon, Oxon OX14 4SB. Telephone: (44) 01235 827720. Fax: (44) 01235 400454. Lines are open 9.00–17.00, Monday to Saturday, with a 24-hour message answering service. Visit our website at www.hoddereducation.co.uk

First published in 2017 by
Hodder Education
An Hachette UK Company,
Carmelite House, 50 Victoria Embankment
London EC4Y 0DZ

Impression number 5	4	3	2	1
Year	2021	2020	2019	2018	2017

Cover photo © Caleb Keiter/Thinkstock

Typeset in 11 on 12 Bembo Regular by Integra Software Services Pvt. Ltd., Pondicherry, India

Printed by CPI Group (UK) Ltd, Croydon, CR0 4YY

A catalogue record for this title is available from the British Library

ISBN 978 1 4718 8992 9

Contents

Introduction

This book contains questions designed for the Higher tier of GCSE Mathematics. It covers the CCEA specification for the modules M3, M4, M7 and M8.

Each chapter matches one in the CCEA Higher GCSE textbook. This book can be used in conjunction with the textbook or separately on its own. Some of the larger chapters have more than one exercise.

On the contents page, the module(s) covered in each chapter is listed in brackets after the chapter title. In some chapters there is an overlap of topics from more than one module.

There are lots of questions in each chapter to provide practice, build confidence and produce satisfaction in being able to do Mathematics. There are problem-solving questions that give opportunities for Mathematics to be used in everyday contexts. The questions usually start out by being quite straightforward. Towards the end of each chapter there are some more challenging questions.

Many of the questions can be completed without a calculator. When a calculator is used, we encourage every step of the working out to be shown in your solution.

At the start of every chapter there is a brief summary of what you will be covering as you work through the questions.

The book has been designed to help prepare students for the Mathematics GCSE exam. Doing every question in this book will certainly do that!

Answers to questions are provided on the Hodder Education website: www.hoddereducation.co.uk/cceagcsemaths

CHAPTER 1 — The language of number

This chapter is about

- being able to find the square, cube, square root, cube root of numbers
- writing any number as a product of primes
- finding the HCF and LCM of pairs of numbers
- interpreting Binary numbers.

1 For the numbers 14, 18, 24, 25 and 60, write down which has a factor of
 a 3
 b 5
 c 12
 d both 6 and 8
 e both 6 and 15.

2 Look at these numbers: 4, 12, 17, 21, 35.
 a Which have a factor of 2?
 b Which have a factor of 5?
 c Which have a factor of 3?
 d Which of these numbers are prime numbers?

3 Look at these numbers: 1, 2, 5, 15, 17, 18. Write down those that are
 a multiples of 2
 b factors of 50
 c multiples of 3
 d prime numbers.

4 Write down the value of
 a 6^2 b 1^2 c 10^2 d 5^2 e 11^2
 f 4^3 g 1^3 h 10^3 i 0.7^2 j 8^1

5 Write down the value of
 a $\sqrt{49}$ b $\sqrt{9}$ c $\sqrt{81}$ d $\sqrt{100}$ e $\sqrt[3]{216}$
 f $\sqrt{1}$ g $\sqrt[3]{1000}$ h $\sqrt[3]{1}$ i $\sqrt[3]{8}$ j $\sqrt{0.36}$

6 Write down the value of
 a $8^2 - 6^2$ b $7^2 + 4^2$ c $5^3 - 2^3$ d $5^2 + 6^3 + 7^2$

7 Write these using index form.
 a $4 \times 4 \times 4$ b $7 \times 7 \times 7 \times 7$
 c $3 \times 3 \times 3 \times 3 \times 3 \times 3 \times 3 \times 3 \times 3$ d $9 \times 9 \times 9 \times 9 \times 9 \times 9$
 e $2 \times 2 \times 2 \times 2 \times 2 \times 2 \times 2$ f $4 \times 4 \times 5 \times 5 \times 5$
 g $8 \times 8 \times 3 \times 3 \times 8 \times 8$ h $2 \times 6 \times 6 \times 2 \times 2 \times 6 \times 6$

8 Find the value of each of these.
 a 2^4 b 1^6 c 4^5 d 7^4 e 2^8

9 Find the product of the first three prime numbers.

10 Find the sum of the 3 largest prime numbers which are less than 100.

11 Which factors of 60 are also multiples of 4?

12 Between which two consecutive whole numbers does the square root of 240 lie?

13 Express each of these numbers as a product of its prime factors. Give your answers using index notation where appropriate.
    a 14   **b** 16   **c** 28   **d** 35
    e 42   **f** 49   **g** 108   **h** 156
    i 225   **j** 424   **k** 864   **l** 6 930

14 a Given that $5\,292 = 2^a \times 3^b \times 7^c$ find a, b, c.
    b Given that $49\,005 = 3^x \times 5^y \times 11^z$ find x, y, z.
    c Given that $274\,400 = 2^m \times 5^n \times 7^p$ find m, n, p.
    d Given that $214\,375 = 5^d \times 7^e$ find d, e.

15 Find the HCF of
    a 14 and 35   **b** 16 and 42   **c** 49 and 108   **d** 156 and 225
    e 240 and 108   **f** 330 and 385   **g** 1 240 and 1 540   **h** 468 and 1 300

16 Find the LCM of
    a 16 and 42   **b** 28 and 35   **c** 108 and 156   **d** 225 and 424
    e 450 and 210   **f** 810 and 735   **g** 1 100 and 450   **h** 374 and 214

17 Work out the prime factorisation of these numbers and use it to write down the HCF and LCM of each pair.
    a 84 and 154   **b** 75 and 135   **c** 150 and 95   **d** 645 and 225

18 Find the HCF and the LCM of each pair of numbers.
    a 17 and 40   **b** 52 and 221   **c** 77 and 98

19 Use prime factorisation to find the cube root of   **a** 21 952   **b** 85 184.

20 What does 5 775 need to be multiplied by to make a square number?

21 Bus A passes a school every 80 minutes. Bus B passes the same school every 48 minutes. If they pass together at 9a.m., when do they next pass together?

22 Write each Binary number as a decimal number.
    a 10101   **b** 11111   **c** 10001
    d 101011   **e** 1010100   **f** 11011011

23 Write each decimal number as a Binary number.
    a 15   **b** 19   **c** 47
    d 80   **e** 90   **f** 119

24 Decide if each of these statements using Binary numbers is true or false.
    a 100 + 100 = 1000
    b 111 + 111 = 1111
    c 1000000 − 1 = 111111
    d A Binary number ending with 1 is always odd.
    e 11110111 + 1000 = 11111111
    f 101 + 1001 + 10001 = 100001

25 Flour is sold in 400g bags at £1.20 per bag. Sugar is sold in 250g bags at 85p per bag. Sally buys an equal mass of flour and sugar. What is the least amount of money she could have spent?

CHAPTER 2

Whole numbers

> ## This chapter is about
>
> - adding and subtracting whole numbers
> - multiplying and dividing by powers, and multiples of the powers of ten
> - multiplying and dividing any whole numbers
> - solving problems involving addition, subtraction, multiplication and division, and understanding the significance of the remainder in the context of division
> - understanding inverse operations and using them as a checking procedure.

Do not use a calculator for any of these questions.

1 Calculate these.
   a $538 + 179$ **b** $347 + 482$ **c** $341 + 5631$ **d** $272 + 61929$
   e $321 + 4560$ **f** $748 + 15709$ **g** $2368 + 492518$ **h** $8236 + 79 + 503$

2 Calculate these.
   a $579 - 432$ **b** $685 - 59$ **c** $967 - 198$ **d** $1468 - 157$
   e $1785 - 532$ **f** $2064 - 262$ **g** $56023 - 1857$ **h** $20307 - 8929$

3 It is 403 miles from San Francisco to Los Angeles and then 127 miles from Los Angeles to San Diego. How far is it from San Francisco to San Diego?

4 Carmel has all of these numbers written on a piece of paper.

   543   123   432   234   321   345

   She claims that she can make exactly 1000 by adding just three of the numbers together. Is this possible?

5 It is 1011 kilometres from Brisbane to Sydney and 957 kilometres from Sydney to Melbourne. How many more kilometres is it from Brisbane to Sydney than from Sydney to Melbourne?

6 Given that $5628 + 3179 = 8807$, write down the answer only to
   a $8807 - 5628$ **b** $8807 - 2179$ **c** $8810 - 5628$ **d** $6628 + 3179$

7 Calculate these.
   a 36×100 **b** $2400 \div 10$ **c** 520×10 **d** $12400 \div 100$ **e** 9×1000

8 Calculate these.
   a 36×500 **b** 18×2000 **c** $36000 \div 90$ **d** $8000 \div 400$ **e** 126×20

9 Calculate these.
   a 47×82 **b** 54×79 **c** 516×65 **d** 3926×48 **e** 4053×917

10 Jill has 15 shelves of books in her study. Each shelf holds 24 books. How many books are there in her study?

11 Work out these.
   a $876 \div 17$ **b** $864 \div 32$ **c** $840 \div 24$ **d** $8775 \div 27$ **e** $27638 \div 40$

12 423 people have booked to go on a special train excursion. Each carriage holds 56 people. How many carriages are needed?

13 In a cinema there are 828 seats. Each row has 36 seats. How many rows are there?

14 Peter opened a 500 g bag of dried pasta. He cooked for four people, using 75 g of pasta per person. How much pasta is left in the bag?

15 One week, Jean used her car for long journeys of 220 miles and 176 miles. At the beginning of the week her milometer showed 17 659. At the end of the week it showed 18 137. For how many miles did she use her car for short journeys that week?

16 At the bakery, Joe bought four jam doughnuts costing 35p each and two Cornish pasties. He gave the assistant £5 and got £1.70 change. How much did each Cornish pasty cost?

17 Given that $83 \times 260 = 21\,580$, write down the answer only to
   a $21\,580 \div 260$ **b** $21\,580 \div 830$

18 Given that $169\,600 \div 320 = 530$, write down the answer only to
   a 530×320 **b** $5\,300 \times 32$

19 Given that $2\,165 + 389 = 2\,554$, write down the answer only to
   a $2\,554 - 389$ **b** $3\,165 + 389$

20 Given that $150 \times 90 = 13\,500$, write down the answer only to
   a $1\,350 \div 9$ **b** 15×9

21 Given that $6\,158 - 1\,749 = 4\,409$, write down the answer only to
   a $4\,409 + 1\,749 + 3\,000$ **b** $(6\,158 - 4\,409) \times 100$

22 Given that $(87 + 53)^2 = 19\,600$, write down the answer only to
   a $\sqrt{19\,600}$ **b** $(87 + 53)^2 - 100^2$

23 Given that $320 \times 15 = 4\,800$, write down the answer to
   a 320×30 **b** $4\,800 \div 160$ **c** 320×14

24 A girl owes the shopkeeper 98p. She has plenty of loose change so, assuming she gives the shopkeeper the exact amount, what is the least number of coins she will need?

CHAPTER 3

Decimals

This chapter is about

- understanding place value in decimals
- ordering decimals
- adding, subtracting, multiplying and dividing decimals, including multiplication and division by powers of ten, and multiples of powers of ten
- understanding the increasing and decreasing effect when multiplying and dividing by decimals
- solving problems involving decimals in context.

Do not use a calculator.

1 Calculate these.
   a $21.43 + 36.31$ b $76.02 + 11.69$
   c $34.986 + 42.37$ d $37.845 - 15.49$
   e $59.26 - 38.164$ f $87.452 - 24.2736$

2 Calculate these.
   a 0.6×8 b 1.8×4
   c 18.7×6 d 13.4×4
   e 3.6×1.4 f 5.8×2.6
   g 8.1×4.32 h 6.51×3.23
   i 74×1.7 j 0.64×3.8

3 Calculate these.
   a $7.2 \div 3$ b $8.5 \div 5$
   c $37.5 \div 3$ d $54 \div 4$
   e $12 \div 0.4$ f $35 \div 0.7$
   g $18.2 \div 0.7$ h $31.2 \div 0.6$
   i $3.614 \div 2.6$ j $9.5846 \div 0.17$

4 A wall is to be 25 m long. So far, the bricklayer has built 18.75 m. How much more has to be built?

5 An antique dealer bought two pictures, one for £35.50 and the other for 85p. He sold the two together for £90. How much money did he make?

6 Two pieces measuring 90 cm each are cut from a length of wood 2.4 m long. What length is left?

7 A paving slab is 59.5 cm long. What is the length of a path made by laying nine of these slabs in a line?

8 A carriage in a train is 29.8 m long. What is the total length of four carriages?

9 A stack of eight textbooks is 18.8 cm high. What is the height of a single textbook?

10 A packet of six large dice weighs 83.4 g. What is the weight of one die?

11 Given that $47 \times 53 = 2491$, write down the answers to these.
   a 4.7×5.3 b 4.7×5300
   c 4.7×0.053 d 0.47×5.3

12 Given that $\dfrac{13\,800}{60} = 230$, write down the answer to

   a 23×6 **b** $13\,800 \div 6\,000$

   c 0.23×600 **d** $\dfrac{138}{60}$

13 Find the cost of six DVDs at £13.34 each.

14 Melissa buys three of these bags of carrots.

WEIGHT	PRICE
0.450 KG	99p

   a What is the total weight?

   b What is the total cost?

15 How many glasses holding 0.3 litre can be filled from a jug containing 1.7 litres?

16 John cycles to and from school each day. He lives 2.4 miles from the school. One weekend he cycled 17.6 miles. How many miles did he cycle altogether that week?

17 At the greengrocers, Harry bought four oranges costing 30p each, some bananas costing £1.08 and three grapefruit. He paid with a £5 note and was given £1.37 in change. How much did each grapefruit cost?

18 Use the correct symbol < or > to compare each of these statements. You should not actually carry out the calculation.

   a 20.6×8 ☐ 20.6 **b** $20.6 \div 0.8$ ☐ 20.6

   c 20.6×1.8 ☐ 20.6 **d** $20.6 \div 8.1$ ☐ 20.6

CHAPTER 4

Negative numbers

This chapter is about

- understanding and being able to use negative numbers in context
- using the rules for addition and subtraction of directed numbers
- using the rules for multiplication and division of directed numbers.

1 The lowest and highest temperatures ever recorded in the countries of the UK are shown in the table.

Country	Lowest temperature (°C)	Highest temperature (°C)
Northern Ireland	−17.5	30.8
England	−26.1	38.5
Wales	−23.3	33.6
Scotland	−27.2	32.9

Which of the countries has the biggest difference between their highest and lowest temperatures?

2 Calculate these.

a $3 - 5$ b $-5 + 2$ c $7 - 15$ d $4 - 60$
e $3 - 4 - 7$ f $-14 + 31$ g $-16 + 25$ h $-13 + 13$
i $-40 + 17$ j $-2 + 10 - 17$ k $-24 - 48$ l $-72 + (-8)$
m $-6 + (-25)$ n $16 - (-38)$ o $-15 - (-7)$ p $16 + (-49)$
q $24 + (-37) - 15$ r $-8 - 19 - 36$ s $-13 + (-29) - (-18)$ t $-6 - 27 - (-82)$
u $5 + (-61) - 29$ v $-42 - 42 - (-6)$ w $-11 - (-19) + (-30)$ x $-16 + (-17) - 28$
y $-19 - (-34) - 67$

3 On Tuesday the temperature at midday was $6\,°C$.
  a By midnight it had dropped by $10\,°C$. What was the temperature at midnight?
  b At 6a.m. the temperature was $-1\,°C$. By how much had it changed since midnight?

4 Calculate these.

a 2×0 b -5×8 c $-6 \times (-2)$ d -4×6
e $5 \times (-7)$ f -3×7 g $-4 \times (-5)$ h $28 \div (-7)$
i $-25 \div 5$ j $-20 \div 4$ k $(-8)^2 \div 4$ l $-15 \div (-3)$
m $-35 \div 7$ n $64 \div (-8)$ o $27 \div (-9)$ p $3 \times 6 \div (-9)$
q $-6 \times (-8) \div (-2)$ r $5 \times 6 \div (-10)$ s $-9 \times 4 \div (-6)$ t $-5 \times 6 \times (-4) \div (-8)$

5 Calculate these.

a $(2 \times -3) + (4 \times -5)$ b $4 - (3 \times -2)$ c $5 - 3 - 2 + 6 - 5$

d $(4 - 1) \times (5 + 3)$ e $\dfrac{4 \times (-6)}{2 + 6}$ f $-2 + (-3) - 4$

g $-6 \times (-3) \div (-9) \times (-2)$ h $(-2 - 3) \div (-7 + 2)$ i $(-1 \times -3 \times -2) + (3 \times -5)$

j $\dfrac{2 - 8 + 3}{-3 \times (-1)}$ k $(-3)^2 + (-2)^2$ l $(-2)^3 \times (-4)$

m $-5 + 6 \times 3$ n $-2 + 20 \div 2$ o $\sqrt{4} - 3^2 - 2^3$

p $\sqrt{25} - (-6)^2$ q $(-12 + 8 \div 2)^2$ r $(-3 + -2 - -1)^3$

6 Write down the two square roots of 64.

CHAPTER 5 Fractions

This chapter is about

- recognising and using equivalent fractions
- converting between improper fractions and mixed numbers
- ordering fractions
- solving problems involving addition, subtraction, multiplication and division of fractions
- calculating a fraction of a quantity.

Exercise A

Do not use a calculator in these exercises.

1 Fill in the missing numbers for each pair of equivalent fractions.

   a $\dfrac{3}{4} = \dfrac{}{20}$ **b** $\dfrac{15}{21} = \dfrac{5}{}$ **c** $\dfrac{1}{2} = \dfrac{}{22}$ **d** $\dfrac{18}{60} = \dfrac{3}{}$ **e** $\dfrac{16}{18} = \dfrac{}{9}$

2 Write each fraction in its lowest terms.

   a $\dfrac{8}{10}$ **b** $\dfrac{15}{18}$ **c** $\dfrac{18}{24}$ **d** $\dfrac{14}{28}$

   e $\dfrac{9}{15}$ **f** $\dfrac{10}{30}$ **g** $\dfrac{25}{40}$ **h** $\dfrac{24}{32}$

   i $\dfrac{20}{55}$ **j** $\dfrac{18}{45}$ **k** $\dfrac{120}{200}$ **l** $\dfrac{54}{162}$

3 Put each set of fractions in order from smallest to largest.

   a $\dfrac{1}{4} \quad \dfrac{3}{10} \quad \dfrac{7}{20} \quad \dfrac{27}{100}$ **b** $\dfrac{3}{5} \quad \dfrac{2}{3} \quad \dfrac{5}{12} \quad \dfrac{13}{20}$

4 Change each of these improper fractions to a mixed number.

   a $\dfrac{7}{2}$ **b** $\dfrac{10}{3}$ **c** $\dfrac{17}{8}$ **d** $\dfrac{14}{9}$ **e** $\dfrac{19}{5}$

   f $\dfrac{10}{7}$ **g** $\dfrac{12}{7}$ **h** $\dfrac{11}{2}$ **i** $\dfrac{13}{4}$ **j** $\dfrac{11}{3}$

5 Change each of these mixed numbers to an improper fraction.

   a $1\dfrac{2}{7}$ **b** $1\dfrac{5}{8}$ **c** $7\dfrac{1}{2}$ **d** $2\dfrac{3}{4}$ **e** $3\dfrac{2}{5}$

   f $2\dfrac{4}{5}$ **g** $3\dfrac{2}{9}$ **h** $4\dfrac{5}{6}$ **i** $5\dfrac{1}{8}$ **j** $4\dfrac{1}{3}$

6 Write each of these in their simplest form (as fractions or mixed numbers).

   a $\dfrac{18}{10}$ **b** $\dfrac{15}{9}$ **c** $\dfrac{28}{24}$ **d** $\dfrac{7}{28}$ **e** $\dfrac{9}{6}$

   f $\dfrac{10}{28}$ **g** $\dfrac{25}{15}$ **h** $\dfrac{24}{18}$ **i** $\dfrac{65}{55}$ **j** $\dfrac{60}{45}$

Exercise B

1 Work out these.

   a $\dfrac{1}{2}+\dfrac{2}{9}$ **b** $\dfrac{11}{12}-\dfrac{2}{3}$ **c** $\dfrac{5}{9}\times\dfrac{1}{10}$ **d** $\dfrac{2}{3}\div\dfrac{5}{6}$ **e** $\dfrac{1}{4}+\dfrac{2}{5}\times\dfrac{1}{3}$

2 Work out these.

   a $4\dfrac{1}{4}+3\dfrac{1}{3}$ **b** $6\dfrac{8}{9}+1\dfrac{2}{3}$ **c** $5\dfrac{3}{8}-\dfrac{1}{4}$ **d** $5\dfrac{11}{16}-2\dfrac{1}{8}$

   e $4\dfrac{3}{4}+3\dfrac{2}{7}$ **f** $6\dfrac{1}{4}-2\dfrac{2}{3}$ **g** $7\dfrac{1}{9}-2\dfrac{2}{3}$ **h** $5\dfrac{3}{10}-4\dfrac{4}{5}$

3 Mary cut two pieces of wood, of lengths $3\dfrac{7}{8}$ inches and $7\dfrac{3}{4}$ inches, from a piece 18 inches long. What length of wood was left?

4 Work out these. Write your answers as simply as possible.

   a $5\dfrac{1}{2}+3\dfrac{5}{9}-4\dfrac{2}{3}$ **b** $6\dfrac{2}{5}-2\dfrac{7}{10}+3\dfrac{1}{2}$ **c** $3\dfrac{2}{5}+4\dfrac{7}{8}-2\dfrac{3}{4}$

5 A shopkeeper had a roll of cloth $16\dfrac{1}{2}$ metres long. Anna bought a piece $2\dfrac{3}{4}$ metres long. How much was left?

6 A shopkeeper sells wire netting by the metre. He buys it in 10-metre rolls. Two customers buy $2\dfrac{2}{5}$ and $3\dfrac{1}{4}$ metres respectively. A third customer wants $4\dfrac{1}{2}$ metres. Will the shopkeeper need to start a new roll? Show your working.

7 Calculate these.

   a $1\dfrac{1}{4}\times\dfrac{3}{5}$ **b** $2\dfrac{1}{2}\times1\dfrac{3}{5}$ **c** $4\dfrac{1}{4}\times2\dfrac{2}{5}$

   d $3\dfrac{2}{3}\times2\dfrac{1}{4}$ **e** $3\dfrac{1}{6}\times\dfrac{3}{5}$ **f** $4\dfrac{1}{5}\times1\dfrac{3}{14}$

8 Calculate these.

   a $3\dfrac{1}{4}\div1\dfrac{1}{8}$ **b** $4\dfrac{1}{6}\div1\dfrac{3}{7}$ **c** $6\dfrac{2}{5}\div\dfrac{8}{15}$

   d $3\dfrac{1}{6}\div1\dfrac{1}{9}$ **e** $2\dfrac{3}{4}\div5\dfrac{1}{2}$ **f** $2\dfrac{2}{7}\div3\dfrac{2}{3}$

9 Calculate these.

   a $4\dfrac{1}{2}\times2\dfrac{1}{6}$ **b** $5\dfrac{1}{4}\times3\dfrac{1}{7}$ **c** $5\dfrac{3}{4}\div\dfrac{5}{8}$ **d** $8\div3\dfrac{1}{3}$ **e** $5\dfrac{1}{7}\times1\dfrac{5}{9}$

   f $1\dfrac{3}{5}\div4\dfrac{1}{10}$ **g** $3\dfrac{5}{9}\times2\dfrac{5}{8}$ **h** $4\dfrac{1}{6}\div1\dfrac{2}{9}$ **i** $4\dfrac{2}{7}\times1\dfrac{5}{16}$ **j** $3\dfrac{3}{4}\div1\dfrac{5}{16}$

   k $4\dfrac{1}{7}\div1\dfrac{3}{14}$ **l** $5\dfrac{1}{3}\times2\dfrac{5}{8}$ **m** $4\dfrac{3}{5}\div5\dfrac{2}{3}$ **n** $2\dfrac{7}{10}\times3\dfrac{2}{9}$ **o** $1\dfrac{4}{5}\div3\dfrac{4}{9}$

10 Calculate these.

   a $1\dfrac{3}{4}\times2\dfrac{1}{2}\times\dfrac{8}{15}$ **b** $1\dfrac{1}{3}\times3\dfrac{2}{5}\div1\dfrac{5}{6}$ **c** $4\dfrac{2}{7}\times2\dfrac{4}{5}\div2\dfrac{1}{4}$

   d $4\dfrac{1}{2}+2\dfrac{1}{3}\times\dfrac{5}{14}$ **e** $5\dfrac{1}{4}-2\dfrac{2}{3}\div2\dfrac{2}{9}$ **f** $5\dfrac{1}{3}+2\dfrac{3}{4}\div3\dfrac{3}{10}$

Exercise C

1 Calculate these.

   a $\frac{3}{7}$ of 196   **b** $\frac{7}{8}$ of £192   **c** $\frac{3}{5}$ of 235 miles   **d** $\frac{4}{7}$ of £357.35

2 Which is larger, a $\frac{3}{8}$ share of £240 or a $\frac{3}{5}$ share of £150? Show your working.

3 Alan spent one-third of his pocket money on sweets and two-fifths on magazines. He saved the rest. What fraction did he save?

4 A farmer owns 400 animals. $\frac{3}{8}$ of them are pigs and $\frac{1}{4}$ of them are cows. The rest are sheep. How many sheep does the farmer own?

5 In 2010 a website had 230 000 hits. In 2011 the number of hits increased by $\frac{2}{5}$. In 2012 the number of hits decreased by $\frac{1}{4}$.
How many hits were there in 2012?

6 A girl spends $\frac{3}{5}$ of her pocket money on clothes. She spends $\frac{1}{4}$ of the rest on make-up. She has just enough money left to buy a book at £4.50.
How much pocket money did she have to start with?

7 Given that $\frac{p}{8}$ and $\frac{3}{q}$ are proper fractions and that $\frac{p}{8} \div \frac{3}{q} = \frac{5}{6}$, find possible values for p and q.

Exercise D

Work out these. Give your answer in its simplest form.

1 $\frac{x}{3} + \frac{x}{5}$ 2 $\frac{m}{2} - \frac{m}{3}$ 3 $\frac{5w}{8} + \frac{w}{5}$ 4 $\frac{9e}{10} - \frac{2e}{3}$

5 $\frac{c}{4} + \frac{d}{5}$ 6 $\frac{4m}{5} - \frac{2e}{7}$ 7 $\frac{x}{2} + \frac{x}{3} + \frac{y}{4}$ 8 $\frac{11y}{12} - \frac{5y}{6} + \frac{x}{3}$

9 $\frac{x}{4} \times \frac{x}{3} \times \frac{10}{y}$ 10 $\frac{x}{4} + \frac{x}{3} \times \frac{x}{2}$ 11 $\frac{x}{3} \times \frac{2}{x}$ 12 $\frac{m}{5} \times \frac{2}{m} \times \frac{x}{6}$

13 $\frac{w}{4} \div \frac{w}{5}$ 14 $\frac{c}{10} \div \frac{d}{3}$ 15 $\frac{x}{4} \div 8$ 16 $\frac{n}{2} \div y$

17 $\frac{x}{2} + \frac{x}{3} \div 7$ 18 $\frac{x}{5} \times \frac{x}{6} \div \frac{x}{4}$ 19 $\frac{5w}{4} \times \frac{2w}{15}$ 20 $\left(\frac{x}{4}\right)^2 \div \frac{m}{8}$

CHAPTER 6 Approximation and estimation

This chapter is about

- rounding a value correct to a given number of decimal places
- knowing the number of significant figures in any value
- rounding a value correct to a given number of significant figures
- estimating answers to practical problems, and using estimation as a check as to the appropriate size of the answer to a calculation.

1 Write each of these numbers
  i correct to 1 decimal place
  ii correct to 2 decimal places.
  a 80.9346   **b** 5.1174   **c** 4.39852
  d 0.034 99   **e** 649.0019   **f** 16.984

2 Round each of these numbers to 1 significant figure.
  a 14.3   **b** 38   **c** 6.54   **d** 308   **e** 1 210
  f 0.78   **g** 0.61   **h** 0.053   **i** 2 413.5 **j** 0.0097
  k 8.4   **l** 18.36   **m** 725   **n** 8 032   **o** 98.3
  p 0.71   **q** 0.0052   **r** 0.019   **s** 407.511 **t** 23 095

3 641 people attended a firework display. How many is this to 1 significant figure?

4 **a** Jenny measured a piece of wood as 2.84 m long. What is this to 1 significant figure?
  b Ben measured the wood as 284 cm. What is this to 1 significant figure?

5 Write each of these numbers to the given degree of accuracy.
  a 851 (1 s.f.)   **b** 568 790 (1 s.f.)   **c** 6.8 (1 s.f.)
  d 0.000 409 (1 s.f.)   **e** 18.6 (2 s.f.)   **f** 0.994 (2 s.f.)
  g 7.389 (3 s.f.)   **h** 2.718 282 (3 s.f.)   **i** 7.994 (2 s.f.)

6 Find an approximate answer to each of the calculations by rounding each number to 1 significant figure. Show your working.

  a $223.7 + 387.2$ **b** $791 \div 81.6$ **c** 18.2×53.7

  d 692^2 **e** $\dfrac{94.6}{37.7 \times 21.7}$ **f** 61.7×5.8

  g 3.7×9.1 **h** 23.127×28.4 **i** 73.4×46.8

  j $\dfrac{17.8 \times 5.7}{39.2}$ **k** $\sqrt{9.7 \times 11.2}$ **l** 0.82×27.3

  m $\dfrac{0.58 \times 73.4}{6.12}$ **n** 21.2^3 **o** 189×0.31

  p $\sqrt{11.1^2 - 8.4^2}$ **q** $\dfrac{51.8 + 39.2}{0.022}$ **r** 71×58

s $\dfrac{614.2}{0.49}$ **t** $\dfrac{5987}{5.1}$ **u** 19.1^2

v 62.7×8316 **w** $\dfrac{5.72}{19.3}$ **x** $\dfrac{32}{49.4}$

y 8152×37 **z** $\dfrac{935 \times 41}{8.5}$

In questions **7** to **16**, show the approximations that you use to get the estimates.

7 Estimate the answers to these calculations.

a 6.32×7.12 **b** $28.7 \div 6.3$
c 48.3×32.1 **d** $7896 \div 189$
e 286×0.32 **f** 18.9^2
g $913 \div 196$ **h** $4.7 \times 6.2 \times 9.8$
i $\dfrac{673 \times 0.76}{3.6 \times 2.38}$ **j** $\dfrac{8.94 \times 6.03}{0.53 - 0.29}$

8 Oisin bought 22 cans of drink at 39p each. Estimate what he paid in total.

9 Tickets for a firework display are £18.50 each. 4125 tickets are sold. Estimate the total value of the sales.

10 A rectangle measures 3.9 cm by 8.1 cm. Estimate its area.

11 A rectangle is 4.8 cm wide and its area is 32.1 cm².
  a Estimate the length of the rectangle.
  b Is the estimate smaller or bigger than the actual length? Explain your answer.

12 Sophie cut 2.9 m of ribbon into pieces 0.18 m long. Estimate how many pieces she had.

13 A café floor is 4.75 metres by 6.35 metres. The owner wants to cover the floor with tiles which cover 0.25 square metres each. Estimate the number of tiles needed.

14 A new computer is priced at £595 excluding VAT. VAT at 20% must be paid on it. Estimate the amount of VAT to be paid.

15 A square paving slab has an area of 6000 cm². Estimate the length of a side of the slab.

16 A circle has radius 4.3 cm. Estimate its area. ($\pi = 3.141\ldots$)

CHAPTER 7

The order of operations

This chapter is about

- applying the BODMAS rule to calculations involving several operations
- inserting the appropriate operations in order to make a calculation correct
- using the various functions on a calculator to evaluate answers
- planning and evaluating a detailed calculation on a calculator
- rounding to various degrees of accuracy appropriate to the context of the question.

Exercise A

Work out these.

1 $4 + 7 \times 2$

2 $3 \times 4 - 2$

3 $(12 - 3) \times 3$

4 $6 + 10 \div 2$

5 $8 \div 2 + 5 \times 3$

6 $3 \times 6 + 8 \div 2 \times 3$

7 $4 + 2(9 - 3 \times 2)$

8 $1.63 + 0.4 \times 1.2$

9 $(12 - 3 \times 2) + 2(18 \div 3 + 1)$

10 $2\frac{1}{4} + 3\frac{1}{2} \times \frac{1}{3}$

11 $\dfrac{16 - 8 \div 4}{5 + 2 \times 1}$

12 $\dfrac{4 + 3(10 + 4 \div 2)}{5 - 1 \times 3}$

Insert the appropriate operations to make each of these calculations correct. You may also insert brackets.

13 $26 \:\square\: 2 \:\square\: 4 = 17$

14 $6 \:\square\: 2 \:\square\: 3 = 15$

15 $18 \:\square\: 6 \:\square\: 2 = 21$

16 $19 \:\square\: 2 \:\square\: 3 = 13$

17 $5 \:\square\: 3 \:\square\: 4 \:\square\: 5 = 35$

18 $6 \:\square\: 5 \:\square\: 1 \:\square\: 2 = 9$

19 $3 \:\square\: 8 \:\square\: 2 \:\square\: 1 = 11$

20 $10 \:\square\: 2 \:\square\: 6 \:\square\: 3 = 5$

Exercise B

Use your calculator to work out each of these.

1 $9.6 - 1.85$

2 3.2^3

3 $4 + (-3) - (-9)$

4 $\dfrac{3.2}{4} + 0.85$

5 $\sqrt{2.56}$

6 $4.25 + 0.95 - 0.05^2$

7 $\dfrac{8 + 2^3}{0.4^2}$

8 $\sqrt[4]{20\,736}$

9 $13\frac{2}{3} - 1\frac{1}{2} \times 4\frac{3}{4}$

10 $15^3 - 6^2 + 2^6$

Exercise C

Use a calculator to work out these. Give your answers either exactly or to 2 decimal places.

1 $-2.7 + 3.8 - 4.9 + 2.1$

2 $-2.1 \times 4.2 + 2.7 \times (-4.6)$

3 $\dfrac{4.6 - 3.7}{9 - 7.4}$

4 $\dfrac{-2.7 \times 3.9}{2.6 + 3.7}$

5 $\dfrac{9.2}{1.3 + 5.4}$

6 $\dfrac{18.4 - 9.1}{3.8}$

7 $\dfrac{10}{0.56 - 0.32}$

8 $\dfrac{6.7 + 9.3}{8.4 \times 4.9}$

9 $(43.7 - 18.4) \div 3.6$

10 $\dfrac{16.7 \times 5.2}{6.1 \times 0.36}$

Work out these on your calculator. If the answers are not exact, give them correct to 3 decimal places.

11 3.2^4

12 $\sqrt{2^2 - 1.8^2}$

13 $2.3 \times 4.7^2 - 4.6 \div 2.89$

14 $\sqrt{17.3 + 16.8}$

15 $\sqrt{68.7 - 2.3^2}$

16 $(7.3 - 2.6)^2$

17 $7.3^2 - 2.6^2$

18 $5.8 \times (1.9 + 7.3)^2$

19 $\sqrt{28.6 - 9.7}$

20 $\sqrt{\dfrac{26.2}{3.8}}$

Work these out on your calculator. Give answers correct to 3 significant figures.

21 $\sqrt[3]{\dfrac{8.6^2 \times 1.3}{2.9}}$

22 $\dfrac{6}{1.3^2} + \dfrac{2}{0.5^2}$

23 $\dfrac{11.6}{9.1} \times 2.3 + 5.4$

24 $\dfrac{6.1 + 9.3 \times 2}{\sqrt{5.6 - 1.3^2}}$

25 $\dfrac{8.21 + \sqrt{2.6^3}}{5.1 + 2.3 \times 1.7^4}$

26 A rectangle is 8.4 cm by 3.2 cm. Work out the area.

27 A square has a side of 6.23 cm. Work out the area.

28 Sarah typed 12 341 words. She is paid 0.64 pence per word. How much does she earn?

CHAPTER 8   Reciprocals

This chapter is about
- finding a reciprocal by writing the value as a fraction in the form $\frac{a}{b}$ and inverting
- knowing the properties of reciprocals
- knowing the calculator function for a reciprocal.

1 Write down the reciprocal of each of these numbers.

   a 3   **b** 6   **c** 640   **d** −10   **e** −317

2 Write down the numbers of which these are the reciprocals.

   a $\dfrac{1}{18}$ **b** $\dfrac{1}{52}$ **c** $\dfrac{1}{1000}$ **d** $\dfrac{2}{5}$ **e** $-\dfrac{1}{2}$

3 Find the reciprocal of each of these numbers, giving your answer as a fraction or a mixed number.

   a $\dfrac{4}{5}$ **b** $\dfrac{3}{8}$ **c** $1\dfrac{3}{5}$ **d** $3\dfrac{1}{3}$ **e** $\dfrac{2}{25}$

4 Calculate the reciprocal of each of these.

   a 2.5   **b** 0.2   **c** 1.25   **d** 0.16   **e** 3.2

5 Write down the reciprocal of each of these.

   a $\dfrac{a}{b}$ **b** y **c** $\dfrac{1}{p^2}$ **d** $\dfrac{-2x}{3}$ **e** $\left(n+\dfrac{1}{n}\right)$

6 What fraction does $1\dfrac{4}{5}$ need to be multiplied by to give an answer of 1?

7 Use the reciprocal function on your calculator to find the reciprocal of

   a 1.6

   b $\dfrac{5}{8}$

   c 4.5^2

   d 2.4^3 (3 s.f.)

   e 0.002

8 **a** Which number has no reciprocal?

   b Find b given that $3\dfrac{3}{4} \times b = 1$

   c If $p > q$ then the reciprocal of p > reciprocal of q. Is this statement true or false? Explain fully.

   d Find e given that $0.4 \times 2\dfrac{1}{2} + 0.3 \times 3\dfrac{1}{3} = e$

   e Find y given that $(1.6y)^2 = 1$

CHAPTER 9 — Ratio

This chapter is about
- simplifying ratios
- dividing in a given ratio
- using ratios in context, for solving problems.

Exercise A

1 Write each of these ratios in its simplest form.
   a $8:6$ b $20:50$ c $35:55$ d $8:24:32$ e $15:25:20$

2 Write each of these ratios in its simplest form.
   a $200\,g:500\,g$ b $60p:£3$ c $1\text{ minute}:25\text{ seconds}$ d $2\,m:80\,cm$
   e $500\,g:3\,kg$ f $£1.60:£2.40$ g $\frac{3}{4}:\frac{7}{8}$ h $2\frac{5}{8}:1\frac{5}{6}$

3 Write each of these ratios in the form $1:n$.
   a $2:10$ b $5:30$ c $2:9$ d $4:9$ e $50\,g:30\,g$
   f $15p:£3$ g $25\,cm:6\,m$ h $20:7$ i $4\,cm:1\,km$ j $\frac{1}{4}\,km:2\,km$

4 Terry, Janine and Abigail receive £200, £350 and £450 respectively as their dividends in a joint investment. Write the ratio of their dividends in its simplest form.

5 Three saucepans hold 500 ml, 1 litre and 2.5 litres respectively. Write the ratio of their capacities in its simplest form.

6 Share £40 between Paula and Tom in the ratio $3:5$.

7 Split £1 950 in the ratio $4:5:6$.

8 Sue, Jane and Christine invest £70 000 between them in the ratio $2:3:5$. How much do they each invest?

9 Grey paint is made with 2 parts white to 3 parts black. How much black is needed to make 2.5 litres of grey?

10 A metal alloy is made up of copper, iron and nickel in the ratio $3:4:2$. How much of each metal is there in 450 g of the alloy?

11 Hollie spends her pocket money on sweets, magazines and clothes in the ratio $2:3:7$. She receives £15 a week. How much does she spend on sweets?

12 In an election the number of votes was shared between the Labour, Conservative and other parties in the ratio $5:4:2$. Labour received 7 500 votes.
   a   How many votes did the Conservatives receive?
   b   How many votes did the other parties receive?

13 A woman shares her inheritance among herself and her two daughters in the ratio $3:1:1$. One of the daughters then shares her money with her son in the ratio $3:2$. If the inheritance was £160 000; how much does the son get?

14 Find the positive value of n such that the ratio $n:8$ is the same as $6:n+8$.

Exercise B

1 On a map a distance of 12 mm represents a distance of 3 km on the ground. What is the scale of the map in the form $1:n$?

2 To make mortar, Fred mixes 1 part cement with 5 parts sand.
 a How much sand does he mix with 500 g of cement?
 b How much cement does he mix with 4.5 kg of sand?

3 Callum uses blue and white tiles in the ratio $1:6$ to decorate his bathroom.
 a He uses twelve blue tiles on the side of the bath. How many white tiles does he use?
 b He uses 240 white tiles in the shower. How many blue tiles does he use?

4 A plan of a forest is drawn using a scale of 2 centimetres to 50 metres.
 a On the plan, the forest is 24 cm wide. What is the actual width of the forest?
 b The actual length of the forest is 950 m. What is the length of the forest on the plan?

5 Two photographs are in the ratio 3 to 4.
 a The smaller photograph is 9 cm long. How long is the larger photograph?
 b The larger photograph is 8.4 cm wide. How wide is the smaller photograph?

6 A model of a theatre set is made on a scale of 1 to 20.
 a A rug on the set is 120 cm long. How long is it on the model?
 b A cupboard on the model is 9 cm high. How high is it on the set?

7 Michael has a recipe for Twinkie Crackle that makes enough for eight people. The recipe uses 200 g of Twinkies.
 a What mass of Twinkies does Michael need to make enough Twinkie Crackle for six people?
 b Michael has 325 g of Twinkies. How many people can he make Twinkie Crackle for?

8 A map is drawn to a scale of $1:40\,000$.
 a Calculate the length of a road which is 6.5 cm on the map.
 b Two villages are 5 km apart. How far apart will they appear on the map?

9 The ratio of pens to pencils in my pencil case is $5:3$. If I have 9 pencils, what is the total number of pens and pencils in my pencil case?

10 $\frac{2}{5}$ of a school are boys. If there are 468 girls, how many students are there in total?

11 A photograph measuring 6 cm by 10 cm is enlarged in the ratio $3:5$.
 Which of these could not be a copy of the original photograph?

  12 cm by 20 cm   9 cm by 15 cm   18 cm by 40 cm

12 A recipe for 12 biscuits uses
 400 ml milk
 150 g butter
 200 g sugar
 Elise has 600 ml of milk, 200 g of butter and 300 g of sugar. What is the maximum number of biscuits she can make?

CHAPTER 10

Percentages and finance

This chapter is about

- knowing the meaning of a percentage
- converting between percentages, fractions, decimals and ratios
- finding a percentage of a quantity, both with and without a calculator
- increasing or decreasing a quantity by a given percentage
- finding one number as a percentage of another
- knowing the formula for percentage change, and applying it in various contexts
- using reverse percentages to calculate an original value
- understanding the concept of simple interest and compound interest and solving a variety of problems
- using percentages and other techniques for solving a variety of finance type problems.

Exercise A

1 Change each of these percentages to **i** a fraction and **ii** a decimal.
   a 45%   **b** 75%   **c** 12%   **d** 110%   **e** 9%   **f** 17.5%

2 Change each of these decimals to **i** a percentage and **ii** a fraction.
   a 0.47   **b** 0.82   **c** 0.04   **d** 0.425   **e** 1.35   **f** $0.\dot{3}$

3 Change each of these fractions to **i** a decimal and **ii** a percentage.
   a $\frac{19}{100}$ **b** $\frac{9}{50}$ **c** $\frac{13}{20}$ **d** $\frac{3}{100}$ **e** $\frac{7}{4}$ **f** $\frac{3}{11}$

4 Put these in order, smallest first.

   a $\frac{29}{100}$, 0.3, 0.32, 31%, $\frac{1}{3}$ **b** 0.25, $\frac{3}{10}$, 0.35, 0.27, 28%

   c 0.46, $\frac{64}{100}$, 0.65, $\frac{9}{20}$, 56%   **d** $\frac{7}{10}$, $\frac{17}{25}$, 72%, 0.84, $\frac{2}{3}$

   e $\frac{1}{3}$, $\frac{7}{25}$, 30%, 0.29, 16.9%   **f** 0.605, $\frac{2}{3}$, 66%, $\frac{7}{9}$, 0.67

5 At Tony's school, $\frac{8}{25}$ of the students support Man. Utd. 27% support Man. City. Which team has the higher support at Tony's school and by how much?

6 Express $\frac{43}{125}$ as an exact decimal.

7 The ratio of boys to girls in a school is $3:5$. What percentage are girls?

8 Find $0.01 \times 0.03 \times 0.05$ as a fraction in its lowest terms.

9 Find
   a 30% of £7.20   **b** 15% of £40   **c** 45% of £240

10 65% of the 840 students in a school eat in the cafeteria. How many students do not?

11 A television normally costs £400. In a sale, the shopkeeper decides to reduce the price by 10% each week until it is sold.
The television is sold during the third week. What price does the television sell for?

12 Work out
   a   37.5% of 6.72 m
   b   17% of £6 300
   c   27% of £12

13 Increase £600 by each of these percentages.
   a   10%   **b**   15%   **c**   2%   **d**   9.5%

14 Decrease £400 by each of these percentages.
   a   25%   **b**   18%   **c**   6%   **d**   11%

15 A shop reduced all its prices by 12%. A coat originally cost £60. What is the new price?

16 Damien earns £15 000 per year. He receives a salary increase of 3.5%. Find his new salary.

17 A shop reduced its prices by 12% in a sale.
On the final day of the sale, there was also a discount of 25% off the sale prices. On the final day, Katie bought a skirt which was originally priced at £36.
How much did she pay for it?

18 A metal bar is 583 mm long. When heated its length increases by 0.12%. Find its new length.

Exercise B

For percentages, give your answer either exactly or to 1 decimal place.

1   Find these.
   a   6 m as a fraction of 75 m   **b**   £3.80 as a fraction of £5   **c**   28 as a percentage of 100
   d   18 as a percentage of 150   **e**   £572 as a percentage of £880   **f**   60 cm as a percentage of 2.5 m

2   For each of these, write the first quantity as a percentage of the second.
   a   14 and 100   **b**   6 and 50   **c**   3 m and 5 m   **d**   27p and £1   **e**   £1.20 and £5
   f   14 and 84   **g**   6 and 35   **h**   2 m and 9 m   **i**   27p and 90p   **j**   £1.47 and £3.75

3   There are 350 people in a cinema audience. 139 of them are under eighteen.
What percentage is this?

4   At a tennis club the membership consists of 85 women and 63 men. What percentage of the members are women?

5   What percentage of 3.5 kg is 450 g?

6   A car was bought for £9 000. A year later it was sold for £7 500. Find the % loss.

7   A girl invests £6 000 in a bank account. One year later, there is £6 240 in the account. What is the % interest rate per annum?

8   A rectangle 24 m by 18 m has both its length and width increased by 5%.
What is the exact % increase in its area?

Exercise C

1 Barbara was given a 3.5% pay rise. Her salary is now £25 378.20 a year. What was her salary before the rise?

2 In a sale, everything is reduced by 30%.
A pair of trousers now costs £31.50. What was the price before the sale?

3 A watch is priced at £1 692 including VAT at 17.5%. What is the cost without VAT?

4 The value of a car has fallen by 30% of its value when new. It is now worth £9 450. How much did it cost when new?

5 The population of a country went up by 15% between 1995 and 2010.
The population in 2010 was 32.2 million. What was the population in 1995?

6 The value of a company's shares fell by 78%.
The value of a share is now £1.32. What was the value before the fall?

7 Electricity prices increased by 66% in 2009. Mr and Mrs Smith's average monthly bill was £69.72 after the increase. What was their average bill before the increase?

8 A train company offers a 40% reduction for travelling at off-peak times.
An off-peak ticket costs £45. What is the full price of a ticket?

9 House prices in a certain area went up by 150% between 1995 and 2005. The price of a house in the area was £180 000 in 2005. What was the price in 1995?

10 A coat is reduced by 15% in the first week of a sale and by a further 10% of the sale price in the second week of the sale. The sale price the second week is £68.85. Calculate the original price of the coat.

Exercise D

1 David is buying a sound system. It will need to be installed. He sees these two adverts for the same sound system.

Sounds Rite	Cheaper Sounds
Cash	**Cash**
£1 199 + VAT at 20%	£1 350
Free Installation	Installation £40
Or easy terms	**Or easy terms**
Pay £500 deposit + 12 monthly payments of £85	20% deposit + 12 monthly payments of £99

Which company is cheaper for David and by how much if
   a he is paying cash?
   b he is paying by easy terms?

2 A colony of rabbits increases by 15% each year. This year there are 250 rabbits. How many will there be in 3 years' time?

3 Mr Brown has £10 000 to invest for 4 years. He can invest it at a simple interest rate of 6.5% p.a. or at a compound interest rate of 5.25% p.a. Which will give him more interest and by how much?

4 The value of a caravan falls by 18% a year. A caravan costs £10 350 new. How much is it worth after 2 years? Give your answer to the nearest pound.

5 The value of a car reduces by 11% per year.
It cost £18 000 new. How many years will it take to reduce in value below £10 000?

6 Is it better to invest £2 000 for 6 years at 4.5% or for 5 years at 5%? Both pay compound interest annually.

7 Naomi has £5 000 to invest. She is looking at these two accounts. They both pay compound interest annually.

Anglo Bank

No Notice Account.
4.75%★★
★★Introductory offer.
Reduces to 4.25%
after 1 year.

Bonus Bank

No Notice Account

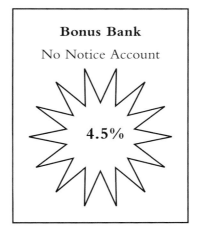

4.5%

Which bank pays more interest over a 3-year investment and by how much?

8 A car depreciates by 12% per year. How many years does it take for the car to halve its value?

9 A motorbike depreciates in value by 5% every year. After 3 years it is worth £5 487.20. Calculate its original value.

10 A girl spends $120 on a handbag whilst on holiday. She sees the same handbag later priced at €115. What was her % saving?
£1 = €1.15
£1 = $1.50

11 What multiplier do you use to increase £4 000 by 8%?

12 What decimal number do you multiply by to decrease £6 000 by 12%?

CHAPTER 11 Bounds

This chapter is about

- understanding the continuous nature of measure
- understanding that measurements given are only approximate to the degree of accuracy used in recording them
- knowing what is meant by the upper and lower bounds
- calculating the maximum and minimum values of expressions given the measurements and their accuracy.

1 Copy and complete the table below.

	Measurement	Accuracy	Lower bound	Upper bound
a	15 cm	nearest cm		
b	5.9 cm	nearest mm		
c	200 cm	nearest 10 cm		
d	23.80 sec	nearest hundredth of a second		
e	464 ml	nearest ml		
f	5.5 m²	1 decimal place		
g	75.0 cm	1 decimal place		
h	6 000 g	Nearest 100 g		

2 Find the maximum value for the sum of each pair of measurements.
   a   43.2 cm and 81.7 cm (both to the nearest millimetre)
   b   10.31 seconds and 19.17 seconds (both to the nearest hundredth of a second)

3 Find the minimum value for the sum of each pair of measurements in question 2.

4 Find the maximum value for the difference between each of these pairs of measurements.
   a   489 m and 526 m (both to the nearest metre)
   b   0.728 kg and 1.026 kg (both to the nearest gram)

5 Find the minimum value for the difference between each pair of measurements in question 4.

6 Find the greatest and least value for the sum of the times 3 hours 23 minutes and 1 hour 37 minutes, each correct to the nearest minute.

7 A length of timber measures 240 mm, correct to the nearest millimetre. Paul cuts off a piece from one end. He intends it to be 90 mm long but his cut is only accurate to the nearest centimetre. Give the greatest and least value of the length of the piece that is left.

8 One day the high tide was 2.3 m above a mark on the harbour wall. The low tide on the same day was 2.0 m below the mark. Both measurements were correct to 1 decimal place. Calculate, in metres, the least value for the difference between the high and low tides.

9 Two parcels weigh 247 g and 252 g, each correct to the nearest gram. The cost of posting parcels increases for those weighing 500 g or more. If these two parcels are fastened together, will they certainly weigh less than 500 g? Explain your answer.

10 Find the greatest value for the area of the floor of rectangular rooms with these dimensions.
  a 4.3 m by 6.2 m (both to 2 significant figures)
  b 4.27 m by 6.24 m (both to the nearest centimetre)

11 Find the least value for the area of the floor of each room in question **10**.

12 Calculate the greatest value for the average speed in each of these situations.
Give your answers to 4 significant figures.
  a 157 km (to the nearest kilometre) in 2.5 hours (to the nearest 0.1 hour)
  b 800.0 cm (to the nearest mm) in 103.47 seconds (to the nearest hundredth of a second)

13 Find the least value for the average speed in each situation in question **12**.

14 Calculate the minimum and maximum widths of these rectangles.
  a Area 400 cm^2 (to the nearest cm^2), length 15 cm (to the nearest cm)
  b Area 24.5 cm^2 (to 3 significant figures), length 5.7 cm (to 2 significant figures)

15 Calculate the maximum and minimum heights of these cuboids.
  a Volume 400 cm^3 (to the nearest cm^3), length 10 cm, width 5 cm (both to the nearest cm)
  b Volume 50.0 m^3 (to 3 significant figures), length 5.0 m, width 2.8 m (both to 2 significant figures)

16 Niall has eight rods, each of length 10 cm, correct to the nearest centimetre.
He places them in the shape of a rectangle as in the diagram.

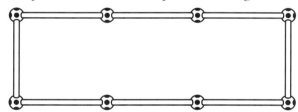

  a What is the minimum length of the rectangle?
  b Calculate the maximum area of the rectangle.

17 The length of Edward's training run is 380 m, correct to the nearest 10 m. Edward completes his run at an average speed of 3.9 m/s, correct to 1 decimal place.
Calculate the greatest possible time Edward takes to complete his run.

18 P and Q are measured as 6.3 and 9.4, each correct to 1 decimal place.
  a Find the maximum value of
    i $P + Q$ **ii** PQ **iii** $\dfrac{Q}{P}$
  b Find the minimum value of
    i $Q - P$ **ii** Q^2 **iii** $\sqrt{P}$

19 K is measured as 420 to the nearest 10. M is measured as 800 to the nearest 100.
  a Find the maximum value of
    i $3K$ **ii** $M - K$ **iii** $\dfrac{M}{K}$
  b Find the minimum value of
    i M^2 **ii** $M + K$ **iii** $\sqrt{K}$

20 C, D and E are measured as 2.64, 1.82 and 4.07, each correct to 3 significant figures.
  a Find the maximum value of $E - C + D$.   **b** Find the minimum value of $\dfrac{3E - 2D}{C}$.

CHAPTER 12

Standard form

This chapter is about

- writing numbers in standard form
- converting numbers from standard form into decimal form
- adding, subtracting, multiplying and dividing numbers in standard form without a calculator
- using standard form on a calculator
- solving a variety of problems using standard form in context.

Exercise A

1 Write these numbers in standard form.
    a 40 000   **b** 4800   **c** 737 000   **d** 25
    e 8 000 000   **f** 12 340   **g** 6 million   **h** 78 900 000

    i 3.6 million   **j** 93 million   **k** $\frac{1}{2}$ million   **l** 2 000 000 000 000

2 Write these numbers in standard form.
    a 0.007   **b** 0.204   **c** 0.000 045   **d** 0.007 07
    e 0.000 000 1   **f** 0.079 36   **g** 0.100 08   **h** 0.000 647

    i $\frac{3}{1000}$ **j** $\frac{29}{10000}$ **k** 0.000 000 000 052   **l** $\frac{73}{1000000}$

3 These numbers are in standard form. Write them as ordinary numbers.
    a 3×10^4 **b** 5.7×10^6 **c** 8.75×10^{-1} **d** 4.02×10^{-3} **e** 7.73×10^2
    f 1.2×10^{-6} **g** 8.03×10^{10} **h** 5.48×10^5 **i** 9.99×10^{-2} **j** 3.86×10^{-4}

4 One trillion is one million million. Write this in standard form.

5 The mass of an oxygen atom is 2.73×10^{-20} mg. If this was written as a decimal, how many zeroes would there be between the decimal point and the 2?

6 Put these numbers in order from smallest to largest.
    1.26×10^{-3} 216×10^{-1} 2.16×10^2 1.62×10^{-1} 6.12×10^0

Exercise B

Work out these, without a calculator. Give your answers in standard form.

1 $(4.12 \times 10^2) + (3.4 \times 10^6)$

2 $(3.7 \times 10^{-3}) + (5.1 \times 10^{-2})$

3 $(8.3 \times 10^5) - (1.75 \times 10^3)$

4 $(6.14 \times 10^{-2}) - (5.3 \times 10^{-4})$

5 $(7.3 \times 10^6) - (6.4 \times 10^5)$

6 $(3.7 \times 10^{-2}) + (4.2 \times 10^{-3})$

7 $(8 \times 10^4) \times (4 \times 10^6)$

8 $(6 \times 10^8) \div (3 \times 10^3)$

9 $(1.5 \times 10^{-2}) \times (2 \times 10^{-3})$

10 $(1.5 \times 10^2) \div (5 \times 10^{-2})$

11 $(6.8 \times 10^{-5}) \div (2 \times 10^{-2})$

12 $(4.96 \times 10^{-3}) \times (6.1 \times 10^6)$

13 Calculate 8 million times 400 000.

14 A sheet of paper is 9.6×10^{-3} cm thick. How thick will a ream of paper be in mm? (ream = 500 sheets)

15 A particle has a mass 6.2×10^{-24} g. Find the total mass of 4 million particles.

16 A rectangle is 2.3×10^3 mm long and 1.4×10^2 mm wide. Calculate its perimeter.

Exercise C

You may use a calculator for these questions.

1 Work out these. Give your answers in standard form.
  a $(2.7 \times 10^2) \times (1.6 \times 10^6)$ **b** $(9.87 \times 10^6) \div (4.7 \times 10^4)$
  c $(6.8 \times 10^{-5}) \times (4.3 \times 10^3)$ **d** $(6.5 \times 10^{-2}) \times (8.9 \times 10^{-4})$
  e $(4.088 \times 10^5) \div (5.6 \times 10^{-2})$ **f** $(4.324 \times 10^{-4}) \div (4.7 \times 10^6)$
  g $(1.7 \times 10^{-2})^2$ **h** $(8.8 \times 10^6)^2$
  i $(3.27 \times 10^{-2}) - (8.16 \times 10^{-4})$ **j** $(7.26 \times 10^8) + (4.81 \times 10^6)$
  k $(1.44 \times 10^{-5}) + (3.75 \times 10^{-6})$ **l** $(3.7 \times 10^{-4}) - (4.81 \times 10^{-3})$

2 Work out these. Give your answers in standard form correct to 3 significant figures.
  a $(7.26 \times 10^4) \times (8.92 \times 10^6)$ **b** $(2.73 \times 10^8) \div (7.89 \times 10^3)$
  c $(9.15 \times 10^{-3}) \div (5.28 \times 10^4)$ **d** $(4.832 \times 10^{-6}) \times (7.021 \times 10^5)$
  e $(1.79 \times 10^5)^2$ **f** $\sqrt{5.2 \times 10^{12}}$

3 Work out these. Give your answers in standard form
  a $320\,000 \times 460\,000$ **b** $37\,200\,000 + 1\,800\,000$
  c $0.000\,006\,2 - 0.000\,005\,12$ **d** $46\,000\,000 \div 200\,000$
  e $0.000\,32 \times 41\,000$ **f** $(234\,000)^2$

4 There are 8.64×10^4 seconds in a day. There are 3.6525×10^4 days in a century.
  a How many seconds are there in a century? Give your answer in standard form correct to 3 significant figures.
  b What fraction of a century is a second? Give your answer as a decimal in standard form correct to 3 significant figures.

5 6 000 hydrogen atoms weigh 1.002×10^{-23} kg. Find the mass of one atom in grams.

6 Light travels at a speed of 186 000 miles per second. How far will it travel in a year (365 days)? Give your answer in standard form, correct to 2 significant figures.

CHAPTER 13   Rational and irrational numbers

This chapter is about

- distinguishing between a rational and an irrational number
- writing all rational numbers as fractions of the form $\frac{a}{b}$
- solving problems involving rational and irrational numbers in context
- changing a recurring decimal to a fraction.

1 Decide whether each of these values is rational or irrational, giving reasons for your answers.

   a 0.1   **b** 0.123   **c** $\dfrac{5\pi}{2}$

   d 0.54   **e** $\sqrt{144}$ **f** $\sqrt{66}$

   g $5 \div 2\sqrt{3}$ **h** $\dfrac{5}{6}$ **i** $0.\dot{8}$

2 Decide whether each of these values is rational or irrational, giving reasons for your answers. For those that are rational, write them in the form $\dfrac{a}{b}$ in their lowest terms.

   a 0.56   **b** $0.\dot{5}\dot{6}$ **c** $\dfrac{3}{11}$

   d $\sqrt{24}$ **e** $\sqrt{25}$ **f** $\dfrac{4}{91}$

   g $\sqrt{121}$ **h** $\sqrt{8}$ **i** $0.\dot{1}0\dot{3}$

   j $3 + 7\pi$ **k** $16 - \sqrt{12}$

3 Convert these recurring decimals to fractions or mixed numbers in their lowest terms.

   a $0.\dot{7}\dot{2}$ **b** $0.4\dot{8}$ **c** $0.3\dot{0}\dot{6}$

   d $0.\dot{1}2\dot{3}$ **e** $1.\dot{2}\dot{7}$ **f** $1.3\dot{8}$

   g $0.6\dot{1}\dot{2}$ **h** $0.5\dot{6}$ **i** $0.3\dot{1}6\dot{2}$

4 **a** Write down three different irrational numbers.

    b Write down two irrational numbers that have a difference of 5.

5 For each of these numbers, find an irrational number which can be multiplied by it to give a rational result.

   a $\dfrac{5}{\sqrt{3}}$

   b $\dfrac{\pi}{6}$

   c $5\sqrt{7}$

6 Find two different irrational numbers whose product is rational.

7 Find an irrational number N such that N^2 is rational.

8 For the isosceles, right-angled triangle shown, state whether each of these are rational or irrational.

   a sin 45
   b cos 45
   c tan 45

9 Write down two irrational numbers between 6 and 7.

10 Write down a rational number between $\sqrt{14}$ and $\sqrt{18}$.

CHAPTER 14

Surds

This chapter is about

- simplifying surds using square numbers
- knowing and applying the rules of surds

$$\sqrt{a} \times \sqrt{a} = a$$

$$\sqrt{a} \times \sqrt{b} = \sqrt{ab}$$

$$\frac{\sqrt{a}}{\sqrt{b}} = \sqrt{\frac{a}{b}}$$

$$a\sqrt{b} \pm c\sqrt{b} = (a \pm c)\sqrt{b}$$

- rationalising the denominator of a surd.

Simplify each of these.

1 $\sqrt{28}$

2 $\sqrt{63}$

3 $\sqrt{125}$

4 $\sqrt{600}$

5 $3\sqrt{40}$

6 $4\sqrt{50}$

7 $2\sqrt{216}$

8 $3\sqrt{112}$

9 $\dfrac{\sqrt{108}}{3}$

10 $\dfrac{\sqrt{50}}{5}$

11 Simplify each of these.

  a $\sqrt{8} \times \sqrt{50}$

  b $\sqrt{75} \div \sqrt{27}$

  c $\sqrt{32} \times \sqrt{8}$

  d $\sqrt{15} \times \sqrt{27}$

  e $\dfrac{\sqrt{15}}{\sqrt{27}}$

  f $\dfrac{\sqrt{10} \times \sqrt{12}}{\sqrt{15}}$

12 Simplify each of these as far as possible.

  a $\sqrt{18} + \sqrt{72}$

  b $4\sqrt{3} + \sqrt{12}$

  c $5\sqrt{50} + 2\sqrt{8}$

  d $\sqrt{80} - \sqrt{5}$

  e $\sqrt{75} + \sqrt{27}$

  f $6\sqrt{8} - 5\sqrt{2}$

13 If $x = 5 + \sqrt{2}$ and $y = 5 - \sqrt{2}$, simplify these.

  a $x + y$

  b $x - y$

  c xy

14 If $x = 7 + \sqrt{3}$ and $y = 5 - 2\sqrt{3}$, simplify these.

  a $x + y$

  b $x - y$

  c x^2

  d xy

15 Expand and simplify these.

  a $(\sqrt{3} + 5)^2$ ✳

  b $(2 - \sqrt{3})(2 + \sqrt{3})$

  c $(6 - 2\sqrt{3})^2$

  d $(2\sqrt{7} + 5)(3\sqrt{7} - 2)$

16 Find the exact value of x, expressing your answer as simply as possible.

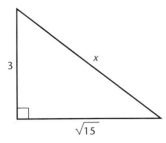

17 A square has sides of $\sqrt{8}$ cm.
   a Calculate the exact length of the diagonal, in its simplest form.
   b Calculate the exact perimeter, in its simplest form.
   c Calculate the exact area, in its simplest form.

18 Simplify each of these by rationalising the denominator.

   a $\dfrac{1}{\sqrt{3}}$ **b** $\dfrac{3}{\sqrt{5}}$

   c $\dfrac{7}{\sqrt{10}}$ **d** $\dfrac{10}{\sqrt{5}}$

   e $\dfrac{5}{2\sqrt{3}}$ **f** $\dfrac{1}{3\sqrt{2}}$

   g $\dfrac{\sqrt{3}}{3\sqrt{2}}$ **h** $\dfrac{7\sqrt{5}}{5\sqrt{7}}$

19 Simplify as far as possible $\dfrac{\sqrt{6}}{10} + \dfrac{2}{\sqrt{6}}$

20 Simplify as far as possible $\sqrt{50} - \dfrac{1}{\sqrt{2}}$

21 Which of the following are true and which are false?

   a $\sqrt{6} + \sqrt{6} = \sqrt{12}$

   b $\dfrac{\sqrt{8}}{2} = \sqrt{2}$

   c $\sqrt{7} \times \sqrt{7} = 7$

   d $3\sqrt{18} = \dfrac{18}{\sqrt{2}}$

   e $(3\sqrt{2} + 2\sqrt{3})^2 = 6(5+\sqrt{6})$

CHAPTER 15 Basic algebra

This chapter is about

- forming an algebraic expression for given information
- simplifying algebraic expressions by gathering like terms
- evaluating algebraic expressions by substituting in various numerical values.

Exercise A

Simplify the expressions in questions **1** to **30**.

1 $a + a + a + a$

2 $2 \times a$

3 $a \times b \times c$

4 $y \times y$

5 $p + p + p + r + r + r + r$

6 $c + c + c - c$

7 $x + x + y + x + y + x + y + y + x$

8 $2a + 3b + 5a - b$

9 $4a - 2b + 3b$

10 $2 \times y + 3 \times s$

11 $a \times a \times a \times b \times b$

12 $4x \times 3y$

13 $8x + 3y + 2x - 3y$

14 $5ab + 3ac - 2ab$

15 $2a \times 3b$

16 $5x + 3y - 2x - y$

17 $3a + 2b + 3c + b + 3a - 3c$

18 $3a + 5b - 2a - 3b$

19 $5a^2 + 3a - 3a^2 + 6a$

20 $6p + 2q - 3 - 2p - 5q + 7$

21 $7p + 3q - 2p - 2q$

22 $3x^2 - x + 4x^2 + 5x - 2x^2 - 7x$

23 $5x^3 - 3x^2 + 2x$

24 $3ab - 2ac - ab + 5ac$

25 $3x - 2y + 4z - 2x - 3y + 5z + 6x + 2y - 3z$

26 $3a^2 - 3ab - 2ab + 4b^2$

27 $3a + 2b + 3c + b + 3a - 3c$

28 $1 + a + 2 - b + 3 + c$

29 $6ab + 3ac - 2ab + 5ab + 3ac$

30 $a^2 - 3ab - 3a^2 + 6ab - 2ab$

31 Thomas bought 6 packets of sweets at a pence each and 2 chocolate bars at $2a$ pence each. Find and simplify an expression for the total amount he spent in pence.

32 A shape is made up of a square of side a, a rectangle of sides $2a$ and $3a$ and a rectangle of sides $3a$ and b. All measurements are in cm. Find and simplify an expression for the total area of the complete shape.

33 If x is the largest of three consecutive even numbers, write an expression for the total of the three numbers in terms of x.

34 What must be added to each of these expressions to make $2a - b$?
   i　$a - 4b$
   ii　$5a + 2b$
   iii　$-a - b + 6$

Exercise B

1 Find the value of these expressions when
   $a = 4$ and $b = 2$.

   a $a + b$ **b** $a - b$

   c $2a$ **d** $7a$

   e $5b$ **f** $8b$

   g ab **h** a^2

   i $3ab$ **j** $2a + b$

   k $3a - b$ **l** $5a + 4b$

   m $3a - 4b$ **n** $a^2 - b^2$

   o $3a^2$ **p** $6b^2$

   q $a^2 b$ **r** b^3

   s $\dfrac{a}{b}$ **t** $\dfrac{10b}{a}$

   u $b - a$ **v** $\dfrac{4b}{2a}$

   w $\dfrac{a^2}{b^2}$ **x** $\dfrac{b}{a}$

   y $2a^2$ **z** $(2a)^2$

2 Work out the value of $5t + 2$ when

   a $t = 2$

   b $t = 8$

   c $t = -12$

   d $t = 20$

   e $t = \dfrac{1}{2}$

3 Work out the value of $2C - F$ when

   a $C = 4, F = 4$

   b $C = 3, F = -5$

   c $C = -8, F = -3$

   d $C = 0, F = 7$

4 Work out the value of $w + nd$ when

   a $w = 1, n = 2, d = 1$

   b $w = 4, n = 2, d = -6$

   c $w = 12, n = 10, d = 0$

   d $w = 50, n = -30, d = -10$

   e $w = -11, n = 9, d = \dfrac{1}{2}$

5 Find the value of these expressions when
   $k = 3$ and $m = 2$.

   a $2k + 5m$

   b $3k - 10m$

   c $3km - 4k^2$

   d $2m^2 k - 3km^2$

   e $(2m + 3k)^2 - 2m^2 k^2$

6 Evaluate each of these expressions for the
   values given.

   a $4y - 5x$ when $y = 7$ and $x = -2$

   b $fh + \dfrac{g}{h}$ when $f = 1.2$, $g = 3.5$ and
      $h = 0.5$

   c $a^2 - c^3$ when $a = -5$ and $c = -3$

   d $\dfrac{a}{2bc}$ when $a = 6$, $b = \dfrac{3}{4}$ and $c = -2$

   e $mx^2 + c$ when $m = 3$, $x = -12$ and
      $c = 6$

7 Given that $p = \dfrac{1}{2}$, $q = 10$, $r = -2$ and $s = 5$

   find the value of $\dfrac{pq(s - 5r)}{pqs}$.

8 Given that $a = -4$, $b = 2$, $c = -1$ and

   $d = -10$ find the value of $\dfrac{a^2 c(d - 2c)}{-4(5b + 2c)}$.

CHAPTER 16 Brackets

This chapter is about

- expanding a single bracket
- expanding and simplifying an expression with brackets
- multiplying two sets of brackets
- squaring brackets
- solving problems involving brackets (including proof of identities).

Exercise A

Expand these brackets.

1 $7(3a + 6b)$

2 $-5(2c + 3d)$

3 $-4(3e + 5f)$

4 $3(7g - 2h)$

5 $3(4i + 2j - 3k)$

6 $-3(5m - 2n + 3p)$

7 $6(4r - 3s - 2t)$

8 $8(4r + 2s + t)$

9 $4(3u + 5v)$

10 $-6(4w + 3x)$

11 $2(5y + z)$

12 $4(3y + 2z)$

13 $-5(3v + 2)$

14 $3(7 + 4w)$

15 $-5(1 - 3a)$

16 $3(8g - 5)$

17 $x(2x + y)$

18 $5p(2p - 1)$

19 $ab(a - b^2)$

20 $-2k(5 - 3k)$

In questions **21** to **28** expand and simplify.

21 $4 + 3(2a - 1) - 4a$

22 $3(2p + 1) - (p + 7)$

23 $x + 2(3x - 1) - 5$

24 $5t(2t - 1) - t(6t + 5)$

25 $8 - (3y + 5) + 2(6 - y)$

26 $x(3x + 4) + 2x(x - 1)$

27 $4b - 2(3b + 4) + 5 - b$

28 $5t + 3(r - 2) - 6t + r$

29 Find the area of this shape in terms of x.
All the lengths are in cm.

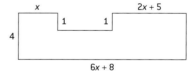

30

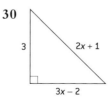

All the lengths are in cm.
Find the expression in terms of x for
  i   the perimeter
  ii   the area.

Exercise B

In questions **1** to **33** expand the brackets.

1 $(x + 1)(x + 2)$

2 $(x + 3)(x + 4)$

3 $(x + 2)(x - 1)$

4 $(x + 5)(x - 3)$

5 $(x - 1)(x - 2)$

6 $(x + 1)(x - 5)$

7 $(x + 3)(x - 3)$

8 $(x + 2)^2$

9 $(x - 7)^2$

10 $(x - 12)(x - 9)$

11 $(x - 4)(x - 5)$

12 $(x + 3)(x - 7)$

13 $(x - 8)(4x - 1)$

14 $(2x + 4)(3x + 2)$

15 $(3x + 5)(x - 2)$

16 $(4x + 3)(2x - 4)$

17 $(7x - 2)(2x - 7)$

18 $(3x - 5)(2x + 3)$

19 $(3x + 7)^2$

20 $(3a + 4b)(5a + 2b)$

21 $(2m - 3n)(3m + 2n)$

22 $(5p - 3q)(2p - q)$

23 $(a + 3b)(3a - 2b)$

24 $(3x - 4y)(2x + 3y)$

25 $(5a + 2b)(5a - 2b)$

26 $(a + b)(a - 2b)$

27 $3(2p - 1)(3p + 4)$

28 $(x - 2)^2 + (x + 2)^2$

29 $(3y - 1)^3$

30 $10 - (2x + 3)^2$

31 $5(2x - 1)(3x + 2)$

32 $4(2y - 3)^2$

33 $(2x + 1)^2 - (2x - 1)^2$

34 Find an expression for the area of this trapezium.
Simplify your answer.
All the lengths are in centimetres.

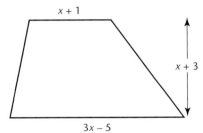

35 Find the values of a and b for which $p^2 - 8p + a \equiv (p + b)^2$.

36 Prove that $(2x + 1)^2 - 3x - (x + 1)$ is always a multiple of 4.

37 Prove that $(2x + 1)^2 - 3x^2 - (1 - x) \equiv x(x + 5)$.

38 Find the value of c for which $(x + c)^2 \equiv x^2 + 12x + c^2$.

39 Find d given that $(x + d)^2 \equiv x^2 + 2dx + 16$.

40 Given that $(mx + 3)^2 - n \equiv 16x^2 + px + 1$ find the values of m, n, and p.

CHAPTER 17 Linear equations

This chapter is about

- solving a basic linear equation
- solving an equation with variables on both sides
- solving an equation with brackets
- solving an equation with fractions
- solving a variety of problems by forming and solving a suitable equation.

Exercise A

Solve these equations.

1 $2x + 5 = 19$

2 $3x - 2 = 16$

3 $11 = 3x + 5$

4 $2 - 5x = 6$

5 $4x + 7 = 13$

6 $4x + 9 = 1$

7 $11 - 2x = 1$

8 $-6 - 4x = 3$

9 $40 = 25 - 2p$

10 $6y - 2 = 13$

11 $8a - 3 = -5$

12 $0 = 1 - 2k$

13 $4 - 3y = -2$

14 $4 = 10 + 2d$

15 $12x - 1 = 29$

16 $5 - 4x = -2$

Exercise B

Solve these equations.

1 $5x - 1 = 3x + 5$

2 $5x + 1 = 2x + 13$

3 $7x - 2 = 2x + 8$

4 $6x + 1 = 4x + 21$

5 $9x - 10 = 4x + 5$

6 $5x - 8 = 3x - 6$

7 $6x + 2 = 10 - 2x$

8 $2x - 10 = 5 - 3x$

9 $15 + 3x = 2x + 18$

10 $2x - 5 = 4 - x$

11 $3x - 2 = x + 7$

12 $x - 1 = 2x - 6$

13 $2x - 4 = 2 - x$

14 $9 - x = x + 5$

15 $3x - 2 = x - 8$

16 $6y = 5 - 2y + 3$

17 $4p - 3 = 9p - 2$

18 $3a + 6 + 5a = 9 - 2a + 4$

19 $3t - 1 + t = -5 + 4t - t$

20 $8 - 6y + 3 = 2y - 1$

Exercise C

Solve these equations.

1 $3(x - 2) = 18$

2 $4 + 2(1 + x) = 8$

3 $3(x - 5) = 6$

4 $10 - 2(x + 3) = 4$

5 $5(2x - 3) = 15$

6 $3 + 2(8 - 3y) = 5y + 1$

7 $4 - 3(2a + 3) - 5 = 2a - 1$

8 $2 - 3(t + 2) + 4(t - 3) = 9t$

9 $6(3p - 2) - 4 = 1 - 2(3p + 1)$

10 $4 - (h + 6) = 5h - 2(3 - h)$

11 $2(x - 3) + 3x = 6(x - 2)$

12 $3(x + 2) + 2(x - 5) - 4(x - 6) = 25$

13 $4(3y + 2) - 3(2y + 1) = 7$

14 $3(x + 2) - 2(x + 1) + 5(x - 2) = 15$

15 $4(x - 1) - (3x + 2) = 2(x + 7)$

16 $3 - 2(2x - 5) + 3(2 - x) = 1 - x$

Exercise D

Solve these equations.

1 $\frac{x}{2} = 5x - 18$

2 $\frac{2x}{5} = x - 12$

3 $\frac{2x}{3} = 10 - x$

4 $\frac{3x}{2} = 5x - 1$

5 $\frac{x}{5} + 4 = 11$

6 $\frac{40}{x} = 5$

7 $\frac{300}{x} = 6$

8 $\frac{20}{3x} = 2$

9 $\frac{1}{5}(3x - 2) = 5$

10 $\frac{3x - 10}{4} = 2x$

11 $\frac{12}{x} + 5 = 13$

12 $\frac{7}{3x} - 4 = 1$

13 $\frac{x}{3} - \frac{x}{5} = 2$

14 $\frac{2x}{3} - \frac{x}{4} = 5$

15 $\frac{2x - 1}{3} = \frac{x + 5}{2}$

16 $\frac{5 + x}{x - 3} = 2$

Exercise E

Solve these equations.

1 $\frac{x}{3} + \frac{2x}{5} = 1$

2 $\frac{x}{7} + \frac{1}{3} - \frac{2}{7} = -\frac{2}{3}$

3 $\frac{x + 1}{4} + \frac{x - 1}{3} = \frac{17}{12}$

4 $\frac{2(x - 1)}{5} + \frac{3(1 - x)}{2} = \frac{19}{10}$

5 $\frac{x}{3} - \frac{x - 1}{4} = 7$

6 $\frac{x + 1}{5} + \frac{x - 3}{3} = 4$

7 $\dfrac{2(x-4)}{3} - \dfrac{x+2}{5} - 3 = 0$

8 $\dfrac{3x-1}{2} - \dfrac{x-2}{5} = 9$

9 $\dfrac{1}{2}(x+3) - \dfrac{1}{3}(x+2) = 4$

10 $\dfrac{2}{3}(x+1) + \dfrac{3}{4}(x-2) = 1$

11 $\dfrac{x+4}{3} + \dfrac{2x+3}{2} - 2 = 0$

12 $\dfrac{1}{6}(2x+5) - \dfrac{1}{4}(x-3) = \dfrac{2}{3}(8-4x) + 1$

Exercise F

1 A triangle has a base of b cm and a perpendicular height of 7 cm. The area is 21 cm². Write down an equation in b and solve it to find the length of the base.

2 Emma thinks of a number. Her number divided by 3 gives the same answer as dividing her number by 4 and adding 2. Let the number be n. Write down an equation in n and solve it to find the number Emma thought of.

3 James had a rope x metres long. He cut off 7 metres and there was still three quarters of the rope left.
  a Write down an equation in x.
  b Solve your equation to find how much rope James started with.

4 Grace had £80. She bought a pair of jeans for £x. The amount she has left is one third of what she has spent.
  a Write down an equation in x.   **b** Solve the equation to find the amount spent.

5 The angles of a triangle are as shown in the diagram.

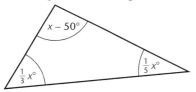

  a Write down an equation in x.
  b Solve the equation to find the angles of the triangle.

6 The lengths of the sides in this quadrilateral are in centimetres. The perimeter of the quadrilateral is 68 cm.

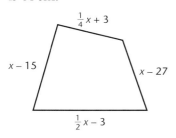

  a Write down an equation in x.
  b Solve the equation to find the lengths of the sides in this quadrilateral.

7 Three consecutive odd numbers have a sum of 381. What are the three numbers?

8 Finn buys 8 packets of crisps at 42 pence each and k ice lollies at 24 pence each. He pays with a £5 note and gets 44 pence change. Write an equation in k and solve it to find how many ice lollies he buys.

9 Calculate the perimeter of the equilateral triangle.

$2x - 4$ $x + 2$

$3x - 10$

CHAPTER 18

Trial and improvement

> **This chapter is about**
> - solving a polynomial equation by trial and improvement, correct to 1 decimal place
> - solving a polynomial equation by trial and improvement, correct to 2 decimal places
> - solving a problem, in context, by trial and improvement, to a suitable degree of accuracy.

1 Show that a solution of $x^3 - 2x - 1 = 0$ lies between 1 and 2.
Use trial and improvement to find the solution correct to 1 decimal place.

2 Show that a solution of $x^3 + 5x - 3 = 0$ lies between 0 and 1.
Use trial and improvement to find the solution correct to 1 decimal place.

3 Show that a solution of $x^3 - 5x + 2 = 0$ lies between -3 and -2.
Use trial and improvement to find the solution correct to 1 decimal place.

4 Show that the equation $x^3 - 5x + 2 = 0$ has another solution between 0 and 1.
Use trial and improvement to find this solution correct to 1 decimal place.

5 Show that the equation $\dfrac{2}{x} = x^2 - 2$ has a solution between $x = 1$ and $x = 2$.
Then find this solution, correct to 2 decimal places.

6 Solve the equation $x^3 - 2x = 180$ by trial and improvement.
Give your solution correct to 1 decimal place.

7 Solve the equation $4x^2 - 2x + 1 = 85$ by trial and improvement.
Give your solution correct to 1 decimal place.

8 Solve the equation $x^3 = 2x^2 + 5$ by trial and improvement.
Give your solution correct to 2 decimal places.

9 The cube of a number subtracted from eight times the original number squared is 35.
Find the original number, correct to 1 decimal place.

10 The hypotenuse of a right-angled triangle is 25 m.
The base of the triangle is 4 m less than the height.
Calculate the height of the triangle correct to 1 decimal place.

11 The distance s metres travelled by a particle in t seconds is given by $s = 20t + 5t^2$.
Find the time taken by the particle to travel a distance of 120 m.
Give your answer correct to 2 decimal places.

CHAPTER 19

Factorisation

This chapter is about

- factorising using common factors
- factorising using grouping
- factorising using the difference of two squares
- factorising quadratics
- simplifying expressions involving several types of factorisation.

Exercise A

Factorise these expressions fully.

1 $8x + 20$

2 $3x + 6$

3 $9x - 12$

4 $5x - 30$

5 $16 + 8x$

6 $9 + 15x$

7 $12 - 16x$

8 $8 - 12x$

9 $4x^2 + 16x$

10 $6x^2 + 30x$

11 $8x^2 - 20x$

12 $9x^2 - 15x$

13 $6x + 3y$

14 $4a - 10b$

15 $5a + 7a^2$

16 $6x^2 - 4x$

17 $3a^2 - ab + ac$

18 $4xy + 2y^2 - y$

19 $2y + 8xy$

20 $9a^2 + 3ab$

21 $8ab - 4ab^2$

22 $a^3 + a^2b$

23 $2x^2 - x^2y$

24 $6x^3 - 15xy$

25 $5a^2b + 15ab^2$

26 $12abc + 15ab^2 - 3a^2c$

27 $8x^2yz + 4xy^2z + 12xyz$

28 $2\pi rh + 2\pi r^2$

29 $3x^2 - x^4$

30 $6p^2q + 10pq^3 + 8pq^2$

Exercise B

Factorise each of these expressions.

1 $ak + bk + am + bm$

2 $p^3 + 3p^2 - 6 - 2p$

3 $ab - ac + bc^2 - c^3$

4 $x^2 + x - xz - z$

5 $6a^3 + 2a^2t - t^2 - 3at$

6 $m^2 - nt + mn - mt$

7 $r^3 + r^2 + r + 1$

8 $ac - 8 + a - 8c$

9 $6c^2 - cd - 48c + 8d$

10 $12ac - 2ab + 18cd - 3bd$

Exercise C

Factorise the expressions in questions **1** to **20**.

1 $x^2 - 9$

2 $x^2 - 64$

3 $x^2 - 196$

4 $x^2 - y^2$

5 $25 - y^2$

6 $16 - c^2$

7 $a^2 - 10\,000$

8 $81 - s^2$

9 $x^2 - 36$

10 $49 - y^2$

11 $9x^2 - 25$

12 $4y^2 - 9$

13 $1 - 64t^2$

14 $x^2 - 121y^2$

15 $81a^2 - 16b^2$

16 $36a^2 - 25b^2$

17 $100 - 49y^2$

18 $2x^2 - 50y^2$

19 $p^2 - \dfrac{1}{100}$

20 $0.25a^2 - 0.36b^2$

21 Evaluate each of these without a calculator.

a $59^2 - 58^2$

b $(7\frac{1}{2})^2 - (6\frac{1}{2})^2$

c $20\,001^2 - 19\,999^2$

22 Use the difference of two squares to simplify

$$(4x + 3)^2 - (3x + 3)^2$$

Exercise D

Factorise these expressions.

1 $x^2 + 4x + 3$

2 $x^2 + 7x + 12$

3 $x^2 - 5x + 4$

4 $x^2 + 21x + 20$

5 $x^2 + 10x + 9$

6 $x^2 + 9x + 20$

7 $a^2 - 9a + 8$

8 $a^2 - 10a + 16$

9 $y^2 - 11y + 30$

10 $x^2 - 17x + 16$

11 $x^2 + 2x - 15$

12 $x^2 + 6x - 7$

13 $x^2 - 5x - 14$

14 $x^2 + 5x - 36$

15 $x^2 + 9x - 10$

16 $a^2 + 4a - 12$

17 $b^2 + 5b - 24$

18 $c^2 - 5c - 36$

19 $x^2 + 13x - 14$

20 $a^2 - 4a - 21$

21 $x^2 + 8x + 12$

22 $2x^2 - 5x + 2$

23 $4x^2 - 8x + 3$

24 $4x^2 + 21x + 5$

25 $5x^2 - 16x + 3$

26 $x^2 - x - 12$

27 $2x^2 - 5x - 12$

28 $6x^2 + x - 1$

29 $8x^2 - 2x - 3$

30 $6x^2 - 7x - 10$

Exercise E

Factorise fully each of these expressions.

1 $p^3 - p$

2 $4x^2 + 2x - 42$

3 $12x^2y^3 + 20xy^2$

4 $10b^2 - 10$

5 $3xy + 6x + 3y + 6$

6 $40 - 6k - 4k^2$

7 $6x^2 + 6xy + y^2$

8 $6c^2 - cd - d^2$

9 $8m^2 + 26mn + 15n^2$

10 $10w^2 + 19ew - 15e^2$

CHAPTER 20

Algebraic fractions

This chapter is about

- simplifying an algebraic fraction, using factorisation and the rules of indices
- adding and subtracting algebraic fractions
- multiplying and dividing algebraic fractions.

1 Simplify each of these fractions.

a $\dfrac{16}{20}$

b $\dfrac{24}{56}$

c $\dfrac{27}{45}$

2 Simplify each of these fractions.

a $\dfrac{4p^2q}{12p}$

b $\dfrac{36y}{(6y)^2}$

c $\dfrac{24pq^2}{40p^2q}$

3 Simplify each of these.

a $\dfrac{5(x+10)}{x(x+10)}$

b $\dfrac{3x(x-1)}{12(x-1)}$

c $\dfrac{5x(x+2)}{10x^2(x+2)}$

d $\dfrac{24x^2(x-3)}{30x^3(x-3)}$

4 Simplify each of these.

a $\dfrac{p^2-p}{p}$

b $\dfrac{18x^2y^3}{6x^2y-12xy^2}$

c $\dfrac{2a-1}{6a^2-3a}$

d $\dfrac{8-4t}{6-3t}$

e $\dfrac{x^2-2x}{x^2+x-6}$

f $\dfrac{t^2-1}{t+1}$

5 Simplify each of these.

a $\dfrac{7x}{x^2-5x}$

b $\dfrac{2x^2-6x}{x^2-x-6}$

c $\dfrac{x^2+2x-3}{x^2+8x+15}$

d $\dfrac{x^2+2x-8}{x^3-2x^2}$

e $\dfrac{x^2-7x+12}{x^2-9}$

f $\dfrac{x^2-5x}{2x^2-11x+5}$

g $\dfrac{3x^2+14x-5}{x^2+7x+10}$

h $\dfrac{9x^2-4}{3x^2-x-2}$

i $\dfrac{x^2-ax+bx-ab}{x^2-a^2}$

j $\dfrac{k^2-9r^2}{k^2-9kr+18r^2}$

6 Simplify each of these.

a $\dfrac{a}{2} + \dfrac{a+1}{3}$

b $\dfrac{2x-1}{5} + \dfrac{x+3}{2}$

c $\dfrac{x-3}{3} + \dfrac{x-2}{5}$

d $\dfrac{1}{b} + \dfrac{2}{b+1}$

e $\dfrac{5}{x-2} + \dfrac{3}{x+3}$

f $\dfrac{p-1}{3} + \dfrac{p+2}{4}$

g $\dfrac{p+1}{3} - \dfrac{2p+1}{4}$

h $\dfrac{2x+1}{7} - \dfrac{x+2}{3}$

i $\dfrac{3}{x-2} + \dfrac{4}{x}$

j $\dfrac{7}{m+1} - \dfrac{3}{m+2}$

k $\dfrac{x+1}{x+2} + \dfrac{x+2}{x+1}$

l $\dfrac{1}{x} - \dfrac{1}{x-1} + \dfrac{1}{x+1}$

m $\dfrac{5}{a^2+a} + \dfrac{2}{a^2-a}$

n $\dfrac{5p}{p^2-1} - \dfrac{2}{p+1}$

o $\dfrac{3}{2x-6} + \dfrac{2x}{x-3}$

7 Simplify each of these.

a $\dfrac{20a^2b^3}{5b} \times \dfrac{b^4}{2a}$

b $\dfrac{2x^2y-1}{6y^2} \times \dfrac{3y^2x}{2x^2}$

c $\dfrac{a-3}{15} \div \dfrac{2a-6}{5}$

d $\dfrac{4p^2}{9q} \div \dfrac{2p}{3q^2}$

e $\dfrac{m^2-q}{15} \times \dfrac{5}{m+q}$

f $\dfrac{4x-2}{15-3x} \div \dfrac{2x-1}{x-5}$

g $\dfrac{5+5t}{t-3} \times \dfrac{6-2t}{8t+8}$

h $\dfrac{k^2-t^2}{(k+t)^2} \times \dfrac{k+t}{k-t}$

i $\dfrac{a^2+5a+6}{a^2-5a+6} \div \dfrac{a^2+6a+8}{a^2-6a+8}$

j $\dfrac{10-2p}{p^2} \div \dfrac{p-5}{p}$

k $\left(\dfrac{3}{x+6} - \dfrac{2}{x}\right) \div \dfrac{4x-48}{x^2-36}$

l $\dfrac{16a^2b}{8a+4b} \times \dfrac{a}{2b} \div \dfrac{24a+12b}{(2a+b)^2}$

m $\dfrac{2x^2-18}{x^2+5x+6} \div \dfrac{x^2+2x-15}{2x^2+9x-5}$

CHAPTER 21

Solving quadratic equations

This chapter is about

- solving a quadratic equation by factorisation
- solving a quadratic equation using the quadratic formula
- solving algebraic fractional equations (including those with algebraic denominators)
- solving a variety of problems involving quadratic equations, by choosing an appropriate form of solution.

Exercise A

Solve these equations by factorisation.

1 **a** $x(x - 3) = 0$
   b $(x - 2)(x + 6) = 0$
   c $(4x + 1)^2 = 0$
   d $(2p - 1)(3p - 4) = 0$
   e $2y(3y - 1) = 0$
   f $5(2x - 1)(3x + 2) = 0$

2 **a** $x^2 + 4x + 3 = 0$
   b $x^2 - x - 30 = 0$
   c $3x^2 - 2x - 8 = 0$
   d $5x^2 - 23x + 12 = 0$
   e $6x^2 + 19x - 20 = 0$
   f $6x^2 + 7x - 5 = 0$

3 **a** $x^2 + 4x = 0$
   b $x^2 - 49 = 0$
   c $4y^2 - 36 = 0$
   d $6p^2 + p = 0$
   e $h^2 - \dfrac{1}{36} = 0$
   f $9x^2 - 25 = 0$

4 **a** $x^2 = 3x$
   b $x^2 = 4x$
   c $x^2 = 35 - 2x$
   d $3x^2 = 24x$
   e $(x + 3)(x - 2) = 2(2x - 1)$
   f $5x(x + 2) = 2x^2 - 3$

5 **a** $x^2 - 19x + 18 = 0$
   b $3x^2 - 15x = 0$
   c $4x^2 + 5x = 6$
   d $x(5x + 13) = 6$
   e $\dfrac{p}{2} = 6p^2$
   f $\dfrac{2m - 1}{3} = m(m + 2)$

6 Solve each of these equations, giving your answers correct to 2 decimal places.
   a $x^2 + 7x + 3 = 0$
   b $5x^2 + 11x + 1 = 0$
   c $3x^2 - x - 8 = 0$
   d $9x^2 + 6x - 19 = 0$
   e $3x^2 + 8x - 9 = 0$
   f $9x^2 - 8x + 1 = 0$

7 Solve each of these equations, giving the exact answers in surd form.
   a $x^2 + 4x - 10 = 0$
   b $x^2 - 3x - 7 = 0$
   c $x^2 + 18x - 35 = 0$
   d $9x^2 + 6x - 19 = 0$
   e $2x^2 - 4x + 1 = 0$
   f $5x^2 - 4x - 8 = 0$

8 Solve each of these equations, giving your answers correct to 3 significant figures.
   a $x^2 + 12x = 5$
   b $\dfrac{x^2 + 2}{x} = 4$
   c $(t - 3)^2 = 6$
   d $(x - 5)(x + 1) = 12$
   e $x(x - 6) = -1$
   f $x(1 - 3x) = 5x - 2$

Exercise B

Solve each of these equations. Give your answer either exact or correct to 2 decimal places.

1 $\dfrac{4}{x} + \dfrac{3}{2x} = \dfrac{11}{4}$

2 $x + 2 = \dfrac{x + 2}{x - 3}$

3 $\dfrac{2}{3x} - \dfrac{6}{5x} = -8$

4 $\dfrac{1}{x + 1} + \dfrac{1}{x - 1} = 4$

5 $(x - 2)(x + 5) = 6x + 8$

6 $\dfrac{2}{x - 1} - \dfrac{1}{x + 2} = \dfrac{1}{2}$

7 $\dfrac{3}{n - 2} - \dfrac{1}{2} = \dfrac{1}{n + 3}$

8 $\dfrac{6}{x} - 4 = \dfrac{-3}{x + 1}$

9 $\dfrac{10}{x - 1} + \dfrac{12}{x + 2} = \dfrac{7}{2}$

10 $\dfrac{5}{2p + 1} + \dfrac{4}{p + 1} = 3$

11 $(5 - 3x)(1 + 2x) = (x + 1)^2$

12 $\dfrac{5x}{x + 1} - \dfrac{3}{1 - 2x} + \dfrac{10}{3} = 0$

Exercise C

Solve each of these problems by forming and solving a suitable quadratic equation.

1 The length of a rectangle is 6 m greater than its width. The area of the rectangle is 160 m². Find the width.

2 The sum of the squares of two consecutive odd numbers is 650. Find the numbers.

3 The width of a rectangle is 7 cm less than its length. The diagonal is 1 cm more than the length. Calculate the length of the rectangle.

4 A driver has a journey of 300 km to complete. On one particular journey she travelled 10 km/h faster than her usual average speed and her journey took 1 hour less. What is her usual average speed?

5 Ellie has £3.20 to spend on pens. However, each pen is 2p dearer than she expected, so she has to buy 8 fewer than planned. What is the cost of each pen?

6 A 400 g packet of biscuits contains b biscuits. A 450 g packet has 5 extra biscuits, but each biscuit in the second packet is 1 g lighter than those in the first. Find the number of biscuits in the 400 g packet.

7

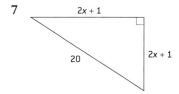

Calculate the perimeter of the triangle, correct to 2 decimal places. All the lengths are given in cm.

8 The sum of a number and its own reciprocal is 3.5125. Find the number and its reciprocal.

CHAPTER 22

Straight lines and linear graphs

This chapter is about

- recognising and drawing a linear graph
- finding the length of a line joining two points
- finding the midpoint of a line joining two points
- finding the gradient of a line
- knowing the properties of gradients, including the gradients of parallel and perpendicular lines
- understanding and using the equation $y = mx + c$
- drawing a straight line using the gradient and crossing point (including lines not given in the form $y = mx + c$)
- finding the equation of a straight line given the graph
- finding the equation of a straight line given the gradient and one point on the line
- finding the equation of a straight line given two points on the line
- interpreting gradients and intercepts of linear graphs.

Exercise A

1 **a** Copy and complete this table for the equation $y = 6x + 1$.

x	−3	−2	−1	0	1	2	3
$y = 6x + 1$		−11					

   b Draw the graph.

2 **a** Copy and complete this table for the equation $y = 3x − 5$.

x	−2	−1	0	1	2	3	4
$y = 3x − 5$							

   b Draw the graph.

3 **a** Copy and complete this table for the equation $y = 4 − 5x$.

x	−3	−2	−1	0	1	2	3
$y = 4 − 5x$							

   b Draw the graph.

4 **a** Copy and complete this table for the equation $y = 3x − 3$.

x	−2	0	4
$y = 3x − 3$			

   b Draw the graph of the equation $y = 3x − 3$ for values of x from −2 to 4.
   c From the graph find the value of x when $y = 4$.
      Give your answer correct to 1 decimal place.

5 Draw the graph of $3x + 4y = 12$.

6 Draw the graph of $2y − 5x = 15$.

7 a Does the point (2, 5) lie on the line $y = 2x + 3$?
 b Does the point (3, 1) lie on the line $y = 2x - 5$?
 c Does the point (−7, 0) lie on the line $y = 7 - x$?
 d Does the point (4, −2) lie on the line $2x - y = 10$?

8 (−2, k) lies on the line $3x + 2y = 4$. Find k.

9 For each of the lines in the diagram, find the co-ordinates of the midpoint.

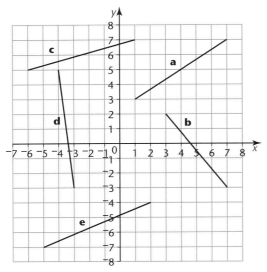

10 Find the co-ordinates of the midpoint of the line joining each of these pairs of points.
 a A(3, 7) and B(−5, 7)
 b C(2, 1) and D(8, 5)
 c E(3, 7) and F(8, 2)
 d G(1, 6) and H(9, 3)
 e I(−7, 1) and J(3, 6)
 f K(−5, −6) and L(−7, −3)

11 B is the point (5, −3). The midpoint of AB is (1, 2). Find the co-ordinates of A.

12 Find the length of each of the lines in the diagram.
 When the answer is not exact, give your answer correct to 2 decimal places.

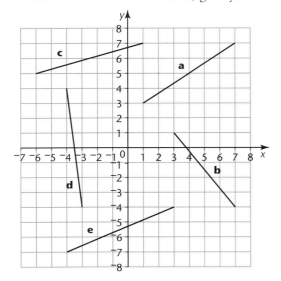

13 Find the length of the line joining each of these pairs of points.
When the answer is not exact, give your answer correct to 2 decimal places.

   a A(2, 2) and B(4, 7)   **b** C(2, 9) and D(7, 2)

   c E(−2, −4) and F(−3, 5)   **d** G(6, 3) and H(3, −1)

   e I(0, −6) and J(−5, 6)   **f** K(5, −7) and L(−3, 10)

14 A is (1, 3). B is (2, 5). C is (5, 1).
By finding the lengths of the three sides, show that triangle ABC is a right-angled triangle.

Exercise B

1 Find the gradient of each line.

   a

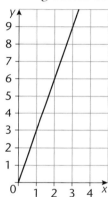

   b

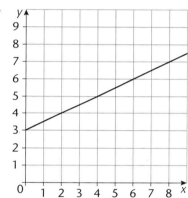

   c

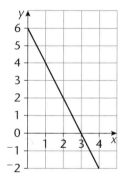

   d

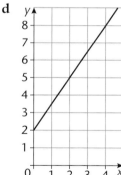

   e

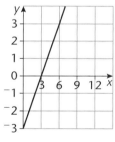

2 Work out the gradient of the line joining each of these pairs of points.

   a (1, 5) and (2, 8)   **b** (3, 4) and (5, 8)

   c (−1, 4) and (9, 6)   **d** (−2, 1) and (0, −7)

   e (−3, −4) and (−7, −3)   **f** (2, 1) and (−1, $\frac{1}{2}$)

3 Write down the gradient of each line.

   a $y = 7x + 2$ **b** $y = 5 - 4x$

   c $y = x + 8$ **d** $y = 4 - x$

   e $2y = 4x + 8$ **f** $3y + 12x = 5$

   g $7x - 2y = 3$ **h** $x + 3y = 6$

4 For each of these, which two lines are parallel?

  a $y = 2x + 4$ $y = 4x + 3$ $y = 4x + 5$

  b $y = 3x - 2$ $y = 5 + 3x$ $y = 5x - 5$

  c $y = 6 - x$ $y = 6 + 3x$ $y = -x + 3$

  d $y = x + 4$ $y = 2x + 1$ $2y = 2x + 3$

  e $3x + y = 2$ $y = 5 - 3x$ $y = 3x - 2$

5 Write down the gradient of a line which is parallel to $y = 6 - 4x$.

6 Write down the gradient of a line which is perpendicular to $y = 2x + 7$.

7 Write down the gradient of a line which is perpendicular to $y + 4x = 8$.

8 Write down the y-intercept of each of these lines.

  a $y = 3x + 5$ **b** $y = 6 - 2x$ **c** $y = 8x$

  d $2y = 3x + 4$ **e** $3y - x + 6 = 0$ **f** $8y - x + 2 = 0$

9 For each pair of lines, write down which line is steeper.

  a $y = 5x + 1$ $y = 2x + 7$ **b** $y = 3x - 8$ $y = x + 3$

  c $y = \dfrac{1}{2}x + 4$ $y = x + 2$ **d** $y = 5 + 6x$ $y = 10x + 4$

10 Match each line to its equation.

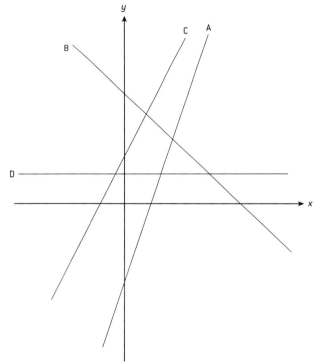

  i $y = 2x + 3$ **ii** $y = 2$ **iii** $y = 3x - 5$ **iv** $y = 7 - x$

11 Find the equation of the line with gradient 3 which passes through the point $(0, 7)$.

12 Find the equation of the line with gradient 2 which passes through the point $(0, -4)$.

13 Find the equation of each line.

a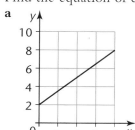

b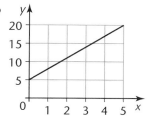

c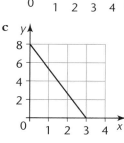

d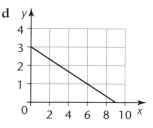

14 Find the equation of the line parallel to $y = 4x + 5$ which passes through $(0, 1)$.

15 Find the equation of the line parallel to $2y = 6x - 3$ which passes through $(0, -5)$.

16 Find the equation of the straight lines which pass through these pairs of points.

    a $(0, 2)$ and $(1, 6)$ **b** $(0, -4)$ and $(2, 4)$

    c $(0, 5)$ and $(3, -4)$ **d** $(3, 5)$ and $(4, 7)$

    e $(2, 5)$ and $(6, -3)$ **f** $(-2, -7)$ and $(3, -3)$

17 Find the equation of the straight line which passes through $(0, 10)$ and is perpendicular to $y = 4x + 3$.

18 Find the equation of the straight line which passes through $(-1, 5)$ and is parallel to $y - 3x = 4$.

19 Find the equation of the straight line which passes through $(5, 4)$ and is perpendicular to $2y + 5x = 3$.

20 A straight line has x-intercept 3 and y-intercept 6. Write down its equation.

21 This graph shows what a plumber charges for his labour.
The charge is made up of a call-out fee and a cost per hour.

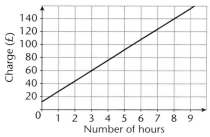

    a What is the plumber's charge for 3 hours?

    b **i** What is the intercept on the vertical axis?

       ii What does this value represent in this situation?

    c **i** What is the gradient of the line?

       ii What does this value represent in this situation?

CHAPTER 23 Indices

This chapter is about

- knowing the rules for indices
- simplifying algebraic indices
- evaluating numerical indices
- solving index equations.

Exercise A

1 Write these in a simpler form, using indices.
  a $2 \times 2 \times 2 \times 2$
  b $2 \times 2 \times 3 \times 3 \times 5 \times 5 \times 5$
  c $a \times a \times a \times a \times a$

2 Write these in a simpler form, using indices.
  a $2^6 \times 2^5$ **b** $3^6 \times 3^2$
  c $4^2 \times 4^3$ **d** $5^6 \times 5$

3 Write these in a simpler form, using indices.
  a $5^5 \div 5^2$ **b** $7^8 \div 7^2$
  c $2^6 \div 2^4$ **d** $3^7 \div 3^3$

4 Write each of these expressions as a power of a single base number.

  a $\dfrac{2^5 \times 2^4}{2^3}$ **b** $\dfrac{3^7}{3^5 \times 3^2}$

  c $\dfrac{5^5 \times 5^4}{5^2 \times 5^3}$ **d** $\dfrac{7^5 \times 7^2}{7^2 \times 7^4}$

  e $\dfrac{125 \times 5^2}{5^3 \times 625}$ **f** $\dfrac{3^6 \times 27}{9 \times 3^4}$

5 Simplify these.
  a $a^4 \times 2a^2$ **b** $a^5 \div a^3$
  c $3a^3 \times 4a^2$ **d** $12a^3 \div 4a^2$
  e $p^{-6} \times p^2$ **f** $y^{1.5} \div y^{0.5}$
  g $k^{\frac{1}{5}} \times k^{\frac{-1}{2}}$ **h** $\sqrt{t^4}$

  i $\dfrac{x}{\sqrt{x} \times \sqrt{x}}$ **j** $6y^3 \times y^{-5}$

  k $(4a^2)^3$ **l** $(6p^2q^3)^2$

6 Simplify where possible.
  a $2a^3 \times 3a^2$ **b** $6x^2 \times 3x^4$
  c $5a^2b \times 4ab^3$ **d** $4a^3b^4 - 2a^2b^3$

  e $5a^4b^5 - 3a^4b^5$ **f** $\dfrac{4c^3 \times 7c^5}{2c^2 \times c^3}$

  g $\dfrac{\left(8y^6\right)^2 \times 3y}{12y^3}$ **h** $\dfrac{4p^3q^2 \times 3p^4q^3}{6p^5q^4}$

  i $\dfrac{3pq \times 16p^3q^2}{12p^7q^3}$ **j** $\dfrac{16p^4}{(2p^2)^3}$

  k $\dfrac{\left(2m^2n^3\right) \times \left(3m^3n^2\right)}{4m^4n^4}$ **l** $\dfrac{10p^8}{5p^{-5}}$

  m $4p^2q^3 \times 3p^3q^3 \div (3p^2q^2)^2$

  n $10a^2b^3 \times 4a^3b^2 - 4a^5b^5$

  o $\dfrac{2b^2 \times 3b^3 \times 4b^4 \times 5b^5}{10b^{10}}$

7 Simplify.
  a $x^5 \times x^3 \times x^4$ **b** $\sqrt{(x^3)^4}$

  c $\dfrac{x^2 \times x^{\frac{1}{2}}}{4\sqrt{x}}$ **d** $4\sqrt{n^3}$

  e $m^6n^2 \div m^5n$ **f** $4\sqrt{81k^8}$

  g $\dfrac{p^6 \times p^{-4}}{p^5}$ **h** $(49t^2)^{\frac{1}{2}}$

  i $(3p^{-2})^2$ **j** $\sqrt{36m^6n^4}$

8 Write each of these as a power of 2.
  a 64 **b** $16^{\frac{3}{4}}$
  c 0.125 **d** 1
  e $2^3 \times \sqrt{2}$ **f** $2^n \times 2^{n+1}$

9 Write this expression as a power of a single base.

$$\frac{8^{3a} \times 4^{b}}{2^{3} \times 16^{2b}}$$

10 If $5^{x} = 8$ and $5^{y} = 6$ find the value of

a 5^{x+y} **b** $(5^{x})^{2}$
c 5^{3y} **d** 5^{x-y}
e 25^{y} **f** 125^{x}

Exercise B

In questions **1** to **5**, work out the exact value without a calculator.

1 a 5^{-1} **b** 5^{0}
  c $25^{\frac{1}{2}}$ **d** $125^{\frac{2}{3}}$
  e $8^{\frac{5}{3}}$ **f** $16^{-0.25}$

2 a $27^{\frac{5}{3}}$ **b** $1000^{\frac{2}{3}}$
  c $32^{\frac{4}{5}}$ **d** 8^{-2}
  e $4^{\frac{5}{2}}$ **f** $0.01^{-1.5}$

3 a $36^{-\frac{1}{2}}$ **b** $6^{-2} \div 6^{-3}$
  c $64^{\frac{5}{6}}$ **d** $9^{1\frac{1}{2}}$
  e $(25^{2})^{-\frac{1}{2}}$ **f** $\sqrt{1\frac{11}{25}}$

4 a $3^{2} \times 16^{\frac{1}{2}}$
  b $3^{-1} \times 5^{-2}$

c $2^{-2} + 6^{0} + 5^{-1}$
d $64^{\frac{2}{3}} \times 8^{\frac{-1}{3}}$
e $4^{3} - 81^{\frac{1}{4}} + \left(\frac{1}{8}\right)^{\frac{1}{3}}$

5 Evaluate.

a $\dfrac{36^{2} \times \left(6^{3}\right)^{2}}{\sqrt{36} \times 6^{4}}$ **b** $9^{\frac{3}{2}} \div 3^{2} \times \sqrt[3]{27^{2}}$

6 What is half of 4^{6}? Give your answer in index notation.

7 Find the missing index.

a $\dfrac{1}{\sqrt{10}} = 10^{\square}$

b $\sqrt[3]{6^{2}} = 6^{\square}$

c $8^{-3} \div 2 = 2^{\square}$

Exercise C

Find the value of x in each of these equations.

1 $3^{x} = 81$

2 $2^{x+2} = 64$

3 $4^{-x} = \dfrac{1}{16}$

4 $5^{2x} = 625$

5 $10^{2x} = 0.01$

6 $3^{-2x} = 27$

7 $\dfrac{1}{6^{x}} = \sqrt{36^{2}}$

8 $5^{x+3} = 25^{2}$

9 $81^{x} = 9^{4}$

10 $100^{3x} = 10 \times \dfrac{1}{100}$

11 $3^{3} \times 9^{4x} = 3^{2}$

12 $4^{3x} = \dfrac{1}{\sqrt{2^{8}}}$

13 $\sqrt{100} = 10^{-x}$

14 $\left(\dfrac{1}{3}\right)^{2x} = 81^{-2}$

15 $5^{3} \times 25^{2x} = (\sqrt[4]{125})^{6}$

16 Given that $\dfrac{16^{p}}{4} = 8^{q}$, write p in terms of q.

CHAPTER 24 — Sequences

This chapter is about

- recognising special number sequences
- describing the pattern for a sequence in words
- continuing a sequence (including use of the difference method)
- writing a sequence from given information
- finding the nth term of a linear sequence
- finding the nth term of a quadratic sequence
- solving a variety of problems using sequences
- recognising the role of counterexamples in the context of sequences.

1 Write down the next three terms in each of these sequences.

   a 7, 10, 13, 16, …, …, …

   b 87, 80, 73, 66, …, …, …

   c 0.2, 0.4, 0.6, 0.8, …, …, …

   d 1, 8, 15, 22, 29, 36, …, …, …

   e 7, 8, 10, 13, 17, …, …, …

   f 9, 17, 25, 33, 41, 49, …, …, …

   g 4, 3, 5, 2, 6, …, …, …

   h 16, 21, 26, 31, 36, 41, …, …, …

   i 20, 18, 14, 8, …, …, …

   j 100, 50, 25, 12.5, …, …, …

2 Write down the first four terms of each of these sequences.

   a Starts with 3 and 3 is added each time.

   b Starts with 26 and 2 is subtracted each time.

   c Starts with 1 and it is multiplied by 10 each time.

   d Starts with x and 4 is added each time.

   e Starts with 1 000 and it is divided by 10 each time.

   f Starts with n and it is halved each time.

3 Write down a rule in words which generates each of these sequences and hence write down the next three terms.

   a 24, 12, 6, 3, …, …, …

   b 100, 99, 98, 97, …, …, …

   c 1, 2, 4, 8, …, …, …

   d 6, 5.5, 5, 4.5, …, …, …

   e 29, 23, 17, 11, …, …, …

   f 25, 18, 11, 4, …, …, …

   g 16, 11, 6, 1, …, …, …

   h 2, 4, 8, 16, 32, …, …, …

   i 27, 19, 11, 3, …, …, …

   j 42, 29, 16, 3, …, …, …

   k 1, 1, 2, 3, 5, …, …, …

   l 22, 17, 12, 7, …, …, …

4 Find the missing numbers in each of these sequences. For each sequence, write down the term-to-term rule that connects the numbers.

   a 1, 7, …, 19, 25, …

   b 11, …, 35, 47, …, 71

   c 85, …, 67, …, 49, 40

   d …, 87, 83, 79, …, 71

   e 7, 22, …, …, 67, 82

   f 34, …, 20, 13, 6, …

5 Use the difference method to find the next 3 terms in each of these sequences.

   a 0, 1, 4, 9, 16, …, …, …

   b −1, 4, 15, 32, 55, …, …, …

   c 2, 9, 18, 29, 42, …, …, …

6 Find the rule for the nth term for each of these sequences.

a 2, 4, 6, 8, ...

b 4, 7, 10, 14, ...

c 0, 3, 6, 9, ...

d 21, 22, 23, 24, ...

e 1, 5, 9, 13, ...

f 10, 13, 16, 19, ...

g −3, −1, 1, 3, ...

h 25, 20, 15, 10, ...

i 4, 2, 0, −2, ...

j 3, 2, 1, 0, ...

k −3, −6, −9, −12, ...

l $\frac{1}{6}, \frac{2}{7}, \frac{3}{8}, \frac{4}{9}, ...$

7 Find the first four terms of each sequence given by the nth term.

a $n + 6$

b $6n$

c $n - 2$

d $3n + 2$

e $2n - 7$

f $3n - 2$

g $4n + 4$

h $-n$

i $5n + 2$

j $1 - 2n$

k $3 - n^2$

l $5 + 2n^2$

8 The 4th term of a sequence is 30, the 5th term is 39 and the 6th term is 48. Find the nth term of this sequence.

9 The 4th term of a sequence is 8, the 6th term is 14 and the 8th term is 20. Find the nth term of this sequence and hence find the 100th term.

10 Which term in the sequence 1, 7, 13, 19, ... will be 79?

11 Here are four patterns made using sticks.

Pattern 1   Pattern 2   Pattern 3   Pattern 4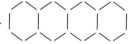

a How many sticks will be needed for pattern 10?

b Find a rule connecting the number of sticks (s) and the pattern number (p).

c How many sticks are needed for pattern 15?

d Which pattern number can be created with 81 sticks?

12 a Find a rule connecting n and T.

n	1	2	3	4
T	4	7	10	13

b Find a rule connecting r and K.

r	2	3	5	6
K	7	11	19	23

c Find a rule connecting m and P.

m	3	6	8	10
P	8	23	33	43

d Find a rule connecting d and W.

d	2	5	8	10
W	9	15	21	25

13 Here are three patterns of tiles.

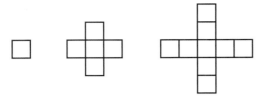

   a Copy and complete this table.

Pattern	1	2	3	4	5	n
Squares						

   b Which pattern has 57 squares?

14 Here are three patterns of squares and crosses.

 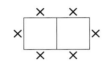 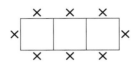

   a Copy and complete this table.

Squares	1	2	3	4	5	n
Crosses						

   b How many crosses are needed for 18 squares?
   c If there are 78 crosses, how many squares are there?
   d Is it possible to have a pattern with 53 crosses? Explain your answer.

15 Prove that every term in this sequence is 6.

   $(3 \times 4 - 1 \times 6), (4 \times 5 - 2 \times 7), (5 \times 6 - 3 \times 8), (6 \times 7 - 4 \times 9), \ldots$

16 How many terms of the sequence with nth term $12n - 3$ can be listed before a term greater than 180 occurs?

17 A sequence is given by nth term $\dfrac{4n - 3}{n^2}$.

   Two terms of the sequence are 1. Which two terms?

CHAPTER 25 Formulae

This chapter is about

- finding the value of a defined quantity using a formula
- changing the subject of a formula involving brackets, fractions, powe...
- changing the subject of a formula where the subject appears more tha...
- solving problems requiring formulae and changing the subject.

Exercise A

1 For the formula $F = a^2 - b^2$, find F when $a = 1.2$, $b = 0.5$.

2 For the formula $C = 5e^2 + 6f$, find C when $e = 1.5$, $f = 2.6$.

3 A regular polygon has n sides. The size of each angle A can be found from the formula

$A = 180 - \dfrac{360}{n}$. Calculate the size of each angle A in

   a a regular nonagon.   **b** a regular 20-sided polygon.

4 The formula for converting a temperature F degrees Fahrenheit to C degrees Celsius is
$C = \dfrac{5}{9}(F - 32)$.

   a Calculate C when F is 86.   **b** Calculate F when C is 120.

5 $T = 2\pi\sqrt{\dfrac{L}{g}}$. Find T when $L = 40$ and $g = 9.8$. Give your answer correct to 3 significant figures.

6 $v^2 = u^2 + 2as$. Calculate v when $u = 54$, $a = 9.8$ and $s = 0.6$.

7 The volume of a container V is given by $V = \dfrac{1}{3}\pi h(2R^2 - r^2)$ where r is the radius of the
base, R is the radius of the top and h is the height of the container. Calculate V when
$\pi = \dfrac{22}{7}$, $h = 14$, $R = \dfrac{2}{3}$ and $r = \dfrac{1}{2}$. Give your answer as a mixed number.

8 A man travels x km to work and then home again each day. On a particular day his mileometer
shows m miles. After p days his mileometer now shows k miles.

   a Assuming his only journeys are to and from work, write a formula for k in terms of x, p and m.
   b Hence calculate k when $x = 32$, $p = 12$ and $m = 82$.

Exercise B

1 Rearrange each formula to make the letter in brackets the subject.
   a $a + b + c = 180$ (b)   **b** $x = 5y + c$ (y)
   c $c = p - 3t$ (t)   **d** $p = 2g - 2f$ (g)

2 Make h the subject in $S = \pi r^2 + 2\pi rh$.

3 Make a the subject in $v - at = u$.

n $C = 2\pi r$.

bject in $dx + e = f$.

e subject in $p - 3t - r = 5t + 1$.

rrange each formula to make the letter in brackets the subject.

a $A = \dfrac{bh}{2}$ (h)

b $a = 180(n - 2)$ (n)

c $A = p(q + r)$ (q)

d $F = \dfrac{m + 4n}{t}$ (n)

8 Make x the subject in $4(x - y) = 8y + x$.

9 Make x the subject in $y = \dfrac{8}{x} - 3$.

10 Make m the subject in $E = \dfrac{1}{2}mv^2$.

11 Make h the subject in $V = \dfrac{1}{3}\pi r^2 h$.

12 Make R the subject in $I = \dfrac{PRT}{100}$.

13 Rearrange each formula to make the letter in brackets the subject.

a $A = 2\pi r^2 h$ (r)

b $P = 1 + a^2$ (a)

c $a^2 = b^2 + c^2$ (c)

d $y = 3x^2 - 4$ (x)

14 Make v the subject in $v^2 - u^2 = 2as$.

15 Make t the subject in $S = 15 - \dfrac{1}{2}at^2$.

16 Make x the subject in $\sqrt{x} = tv$.

17 Make x the subject in $p^2 + q^2 = (x + y)(x - y)$.

18 Make v the subject in $r = \sqrt[3]{\dfrac{3v}{4\pi}}$.

19 Rearrange each formula to make the letter in brackets the subject.

a $f = \dfrac{uv}{u - v}$ (u)

b $pq + r = rq - p$ (r)

c $y = x + \dfrac{px}{q}$ (x)

d $s = ut + \dfrac{at}{2}$ (t)

e $pq - rs = rt$ (r)

f $y = \dfrac{x + 4}{x - 3}$ (x)

g $K = \dfrac{2y}{3 + y}$ (y)

h $\dfrac{1}{uv} + \dfrac{1}{v} = \dfrac{1}{f^2}$ (v)

20 Decide if each of these is an expression, equation, formula or identity.

a $2x + 3 = 1$

b $5x + 4 = 4 + 5x$

c $3x^2 + 8$

d $d = 5x^2 - 7$

e $6x + 3 = 2x$

f $x(x + 4) = x^2 + 4x$

CHAPTER 26

Inequalities

This chapter is about

- knowing how to read and write inequalities
- displaying and interpreting inequalities on a number line
- solving inequalities
- solving problems involving inequalities
- identifying a region satisfied by a number of inequalities
- finding the inequalities that define a region
- finding the maximum or minimum value of a function subject to the constraints of a given set of inequalities.

Exercise A

1 Write down the integers which satisfy the following.
   a $2 < x < 5$ **b** $3 \leqslant x < 7$ **c** $-2 \leqslant x \leqslant 1$
   d $-8 < x \leqslant -6$ **e** $-4 < x < -1$ **f** $-5 \leqslant x < -4$

2 Write down the integers which satisfy the following.
   a $-3\frac{1}{2} \leqslant x \leqslant -1$ **b** $\frac{7}{2} \leqslant x \leqslant \frac{11}{2}$ **c** $\frac{5}{3} < x < \frac{20}{3}$

   d $-\frac{11}{4} < x \leqslant \frac{7}{6}$ **e** $-\frac{20}{7} \leqslant x \leqslant -\frac{6}{5}$ **f** $-4\frac{1}{4} \leqslant x < 2\frac{2}{5}$

3 Given that n is an integer write down the smallest value that n can have given that $n > 2$.

4 Given that n is an integer write down the largest value that n can have given that $n \leqslant -3$.

5 Write down an inequality in x that is represented by each number line.
   a
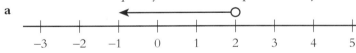
   b
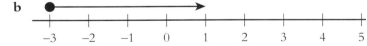
   c
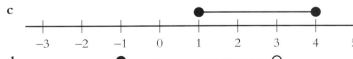
   d

6 Show each of these inequalities on a number line.

   a $x > 4$ **b** $x \leqslant -1$ **c** $-3 < x \leqslant 2$ **d** $-5 < x < -2$

7 Solve each inequality.

   a $2x + 1 \leqslant 6$ **k** $5x + 2 < 7x - 4$

   b $3x - 6 \geqslant 0$ **l** $3(3x + 2) \geqslant 2(x + 10)$

   c $7 \leqslant 2x - 1$ **m** $5 - 3x > 8$

   d $5x < x + 12$ **n** $2(3 - x) \leqslant x$

   e $4x \geqslant x + 9$ **o** $3 - 6x > 2x + 14$

   f $4 + x < 0$ **p** $-2 < 2x < 14$

   g $3x + 1 \leqslant 2x + 6$ **q** $1 < 3x + 1 \leqslant 10$

   h $2(x - 3) > x$ **r** $-8 \leqslant 4(2x - 3) \leqslant 12$

   i $5(x + 1) > 3x + 10$ **s** $-1 < 5(3x + 1) < 5$

   j $7x + 5 \leqslant 2x + 30$ **t** $2x + 4 \leqslant 3x - 2 < x + 18$

Exercise B

In questions **1** to **10**, label the required region using the letter R.

1 Draw a set of axes and label them from -5 to 5 for x and y.
Shade the region $x \geqslant -3$.

2 Draw a set of axes and label them from -5 to 5 for x and y.
Shade the region $y > x$.

3 Draw a set of axes and label them from -5 to 5 for x and y.
Shade the region $x + y \leqslant 3$.

4 Draw a set of axes and label them from -3 to 3 for x and -5 to 5 for y.
Shade the region $y \leqslant 2x - 1$.

5 Draw a set of axes and label them from -2 to 6 for x and y.
Shade the region $4x + 3y < 12$.

6 Draw a set of axes and label them from -3 to 7 for x and y.
Show, by shading, the region where $x > -2$, $y < 4$ and $y > 2x + 1$.

7 Draw a set of axes and label them from -5 to 5 for x and y.
Show, by shading, the region where $x < 4$, $y < 3$ and $y \geqslant -x$.

8 Draw a set of axes and label them from 0 to 7 for x and y.
Show, by shading, the region where $x \geqslant 0$, $y \geqslant 0$ and $3x + y \leqslant 6$.

9 Draw a set of axes and label them from -3 to 5 for x and y.
Show, by shading, the region where $x \geqslant -1$, $y \geqslant 0$ and $x + 2y \leqslant 4$.

10 Draw a set of axes and label them from -2 to 8 for x and y.
Show, by shading, the region where $x \geqslant 1$, $x + y > 5$ and $y > 2x + 1$.

11 In each part of this question, write down the inequality that describes the shaded region R.

a

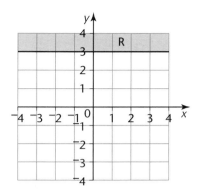

b

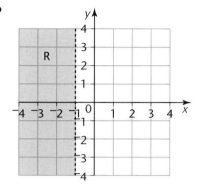

c

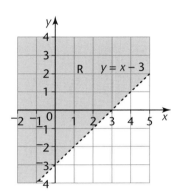

d

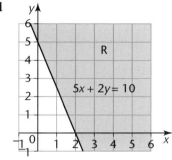

12 Find the set of inequalities which defines each of the regions R.

a

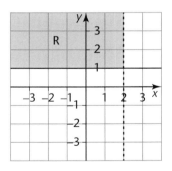

b

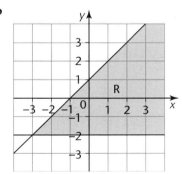

c

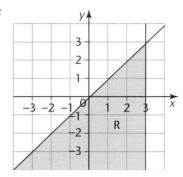

d

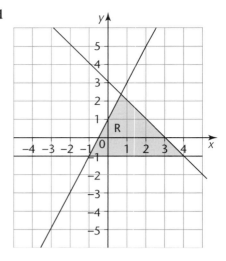

Exercise C

1 Find the co-ordinates of the point with integer values which
   a maximises the function $3x - 2y$ for the region R
   b minimises the function $x + 4y$ for the region R.

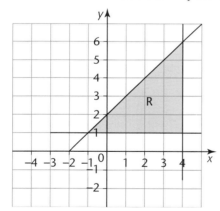

2 Find the point with integer co-ordinates in the region R defined by the three inequalities
   $$y \leqslant 2x + 1 \qquad x + y \leqslant 4 \qquad y \geqslant -1$$

   which
   a maximises the function $4x + 3y$
   b minimises the function $2x - y$.

CHAPTER 27 Proportion and variation

This chapter is about

- solving numerical problems involving direct or inverse proportion
- solving problems involving direct or inverse variation algebraically by setting up a formula involving the constant of proportionality
- using the formula derived for direct or inverse variation to find further unknown quantities
- recognising graphs which represent variables which are directly or inversely proportional.

1 Eight men build a wall in 20 days. How long would it take if there were 10 men?

2 Twelve pens cost £4.44. How much would 5 of these pens cost?

3 A recipe for 6 people uses 150 g of butter and 210 g of flour.
How much of each will be needed if the recipe is adapted for 10 people?

4 A car travelling at 80 km/h can complete a journey in 4 hours.
How long would it take to complete the same journey if the car was travelling at 60 km/h?

5 Five photocopiers print a total of 3 800 pages in 20 minutes.
If one photocopier breaks down, how many pages will be printed in the next half hour?

6 A team of 24 factory workers produce x items per day.
If production is to be increased by 25%, how many workers will be needed?

7 y is directly proportional to w. $y = 20$ when $w = 5$.
 a Find a formula connecting y and w.
 b Find y when $w = 8$.
 c Find w when $y = 6$.

8 m is inversely proportional to h. $m = 2$ when $h = 8$.
 a Find a formula connecting m and h.
 b Find m when $h = 20$.
 c Find h when $m = 100$.

9 $y \propto x^2$ and $y = 13$ when $x = 2$. Find y when $x = 5$.

10 As part of their training, astronauts have to experience the effects of large accelerations, called 'G forces'. The training consists of being placed in a pod at the end of a rotor arm which is spun at speed. The force experienced, F, is proportional to the square of the speed, s, of the pod. This relationship can be written as $F \propto s^2$. When the pod is spinning with a speed of 6 m/s the force experienced is 2G. How many Gs will be experienced when the speed is 18 m/s?

11 y is proportional to the square root of x. When $x = 25$, $y = 15$. Calculate x when $y = 24$.

12 The weight of a spherical particle varies directly as the cube of its radius. A particle of radius 4 mm weighs 115.2 g.
  a What is the weight of a particle of radius 8 mm?
  b A particle has weight 77.175 g. What is its radius?

13 Given that $t \propto r^3$, what is the effect on t when
  a r is doubled? **b** r is divided by 3?

14

r	2	3	6
P	1.6	3.6	14.4

Which law connects the values in the table?

$$p \propto r \qquad p \propto r^2 \qquad p \propto r^3$$

15 Boyle's law states that the pressure of a gas, P pascals, varies inversely with its volume, $V\,m^3$, at constant temperature.
  a If $P = 10$ pascals when $V = 100\,m^3$, find the formula connecting P and V.
  b Find P when $V = 500\,m^3$.

16 $y \propto \dfrac{1}{x^2}$ and $y = 5$ when $x = 2$. Find y when $x = 5$.

17 The gravitational force, F, between a satellite and the Earth is inversely proportional to the square of its distance, d, from the centre of the Earth. The rule can be written as $F \propto \dfrac{1}{d^2}$.

  a The radius of the Earth is 6400 km. So, at 12800 km above the Earth's surface, d is three times its value on the Earth's surface. What is the effect on the gravitational force at this height?
  b How far above the Earth's surface is the satellite when the gravitational force is one quarter that on the Earth's surface?

18 Given that $p \propto \dfrac{1}{\sqrt{r}}$, copy and complete the table of values.

r	100	4	
p	1.2		3

19 The intensity of the illumination from a light bulb is inversely proportional to the square of the distance from the bulb. What needs to happen to the distance for the illumination to be four times greater?

20 The temperature t (°C) of sea water is inversely proportional to the depth below sea level d (m). At a depth of 400 m the temperature was 6 °C. How far below sea level will the temperature be 1.5 °C?

21 Given that $a \propto \dfrac{1}{b^2}$, what is the effect on a when:
  a b is trebled? **b** b is halved?

22 Match the correct graph with each statement.
  a $y \propto x^2$ **b** $y \propto \dfrac{1}{x}$ **c** $y \propto x$

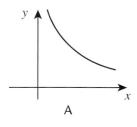

A

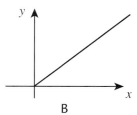

B

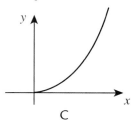

C

CHAPTER 28

Non-linear graphs

This chapter is about

- recognising and drawing quadratic and cubic graphs
- recognising and drawing a reciprocal graph
- recognising and drawing an exponential graph
- recognising and drawing graphs of the trig functions sin, cos and tan.

Exercise A

1 Copy and complete this table for $y = x^2 + 3x - 5$. Draw the graph.

x	−2	−1	0	1	2	3	4	5
x^2								
+3x								
−5								
$y = x^2 + 3x - 5$								

2 Draw the graph of $y = 2 + x - x^2$, for values of x from −3 to 3.

3 Draw the graph of $y = 2x^2 - 3x - 1$, for values of x from −2 to 3.

4 Draw the graph of $y = (3 - 2x)(1 - x)$, for values of x from −2 to 4.

5 a Copy and complete the table for $y = x^3 - 5x$.

x	−3	−2	−1	0	1	2	3
x^3	−27						
−5x	15						
y	−12						

b Draw the graph of $y = x^3 - 5x$.

6 a Make a table of values for $y = x^3 + 4x^2 - 6$ for values of x from −4 to 2.
b Draw the graph of $y = x^3 + 4x^2 - 6$.

7 Draw the graph of $y = x^3 + 3x^2$ for values of x from −3 to 2.

8 Draw the graph of $y = 5x - x^3$ for values of x from −3 to 3.

Exercise B

1 Draw the graph of $y = \dfrac{3}{x}$ for $-3 \leqslant x \leqslant 3$.

2 Draw the graph of $y = \dfrac{-10}{x}$ for $-5 \leqslant x \leqslant 5$.

3 Draw the graph of $y = \dfrac{1}{2x}$ for $-4 \leqslant x \leqslant 4$.

4 Draw the graph of $y = \dfrac{8}{2 - x}$.

Write down the equation of the vertical asymptote.

5 a Draw the graph of $y = 2.5^x$ for values of x from -2 to 4.
   b Use your graph to estimate
     i the value of y when $x = 3.4$ **ii** the value of x when $y = 11$.

6 a Copy and complete this table of values for the equation $y = 0.8^x$.

x	-8	-6	-4	-2	0	2	4	6	8
y	5.96			1.56	1				0.17

   b Draw the graph of $y = 0.8^x$ for values of x from -8 to 8.
   c Use your graph to solve these equations.
     i $0.8^x = 0.5$ **ii** $0.8^x = 5$

7 Draw the graph of $y = 3^{-x}$ for values of x from -3 to 3.

8 Draw the graph of $y = 2^x + 1$ for values of x from -3 to 3. Write down the equation of the horizontal asymptote of this graph.

Exercise C

1 a Sketch the graph of $y = \cos x$ for values of x from $0°$ to $360°$.
   b Use your graph and your calculator to find the solutions of $\cos x = 0.5$ between $0°$ and $360°$.

2 a Sketch the graph of $y = \sin x$ for values of x from $0°$ to $360°$.
   b Use your graph and your calculator to find the solutions of $\sin x = 0.7071$ between $0°$ and $360°$.

3 Given that $\cos 40° = 0.766$, use the symmetry of the graph of $y = \cos x$ to find the solutions of $\cos x = -0.766$ between $0°$ and $360°$.

4 Give one other angle that has a sine value equal to each of these. Use the graph of $y = \sin x$.
   a $\sin 30°$ **b** $\sin 120°$ **c** $\sin 45°$ **d** $\sin 200°$

5 Give one other angle that has a cosine value equal to each of these. Use the graph of $y = \cos x$.
   a $\cos 120°$ **b** $\cos 30°$ **c** $\cos 210°$ **d** $\cos 90°$

6 For $0° \leqslant x \leqslant 360°$, solve these equations. Use the graphs and your calculator.
   a $\sin x = 0.1$ **b** $2 \cos x = 1$ **c** $\tan x = -3$ **d** $2 \sin x = \sqrt{3}$
   e $\cos x = -\sin x$

CHAPTER 29

Real-life graphs

This chapter is about

- using graphs to represent real-life functions
- understanding and using the terms exponential growth, exponential decay and half-life
- interpreting the gradient at a point on a curve as the instantaneous rate of change.

1 This graph shows Michael's journey from Dorton to Canburn on a bicycle.

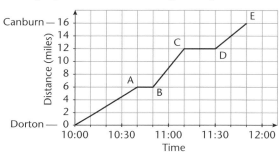

  a Which parts of the graph show Michael not moving?
  b At what time did Michael get to Canburn?
  c How far did he travel altogether?
  d During which part of the journey was Michael cycling fastest?
  e Calculate Michael's average speed from D to E.

2 John and Imran live in the same block of flats and go to the same school. The graph represents their journeys home from school.
  a Describe Imran's journey home.
  b After how many minutes did Imran overtake John?
  c How many minutes before John did Imran arrive home?
  d Calculate John's speed in kilometres per hour.

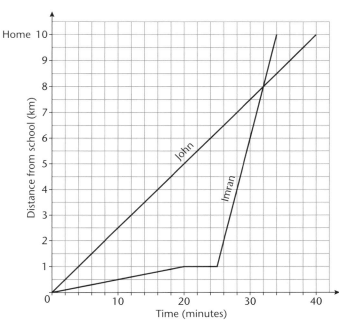

3 Copy the axes below onto graph paper.

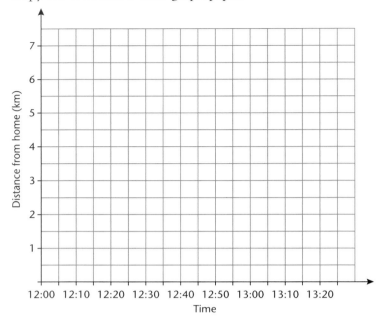

Aaron leaves his house at 12.00 and cycles into town. After 2.5 km he stops at 12.10 to answer his phone. At 12.15 he proceeds into town, cycling the remaining 3.5 km at an average speed of 10 km/h. He spends 20 minutes in town and then cycles home without stopping. He arrives home at 13.20.

a Show this journey on a distance–time graph.

b What was Aaron's average speed on the journey home from the town?

4 The height of a ball above ground level is given by the formula $H = 4t - t^2$.

a Copy and complete the table of values for t and H.

t (seconds)	0	1	2	3	4
H (metres)					

b Draw the graph of H against t.

c What was the maximum height reached by the ball?

d From the graph find the two times when the ball is 3 metres above the ground.

e i Draw a tangent to the curve at $t = 1$.

ii Calculate the gradient of the tangent.

iii What does the value of the gradient represent?

f i Draw a tangent to the curve at $t = 3$.

ii Calculate the gradient of the tangent.

iii What does the value of the gradient represent?

5 The average temperature of a family-size frozen meal is given by the formula $T = x^2 - 16$ where x $(0 < x \leqslant 8)$ is the number of minutes the meal is placed in the microwave and T is the temperature in °C.

   a Copy and complete the table of values for x and T.

x (minutes)	0	1	2	3	4	5	6	7	8
T (temperature °C)									

   b Draw the graph of T against x.
   c What was the average temperature of the meal when it was taken from the freezer?
   d Mary wants the meal heated to a temperature of 40 °C. Estimate how long she will need to place the meal in the microwave.
   e i Draw a tangent to the curve at $x = 4$.
   ii Work out the gradient of the tangent.
   iii What does the value of the gradient represent?

6 A radioactive substance is decaying. Its half-life is 12 years. In 2010 the mass is 160 g.

   a Draw a graph to show the mass of the substance over the next 60 years.
   b Use your graph to estimate the time when the mass will be 50 g.

7 A population of rabbits doubles every 3 months. The initial population is 10 rabbits.

   a Draw a graph to show the population of rabbits over the next 12 months.
   b Use your graph to estimate the population of rabbits after 5 months.
   c Use your graph to work out when the population passes 50 rabbits. Give your answer to the nearest month.

8 £800 is invested for 6 years and earns compound interest. At the end of the first year the value of this investment is £880.
   a Work out the compound interest rate per annum.
   b Draw a graph to show the value of the investment over the 6-year period.
   c Write a formula for the value V, in £, of the investment at any time t years.
   d Use trial and improvement to work out the number of complete years that it will take for the investment to exceed £2 000.

CHAPTER 30

Simultaneous equations

This chapter is about

- solving two simultaneous linear equations graphically
- solving two simultaneous linear equations algebraically
- solving problems by forming and solving simultaneous equations
- solving two simultaneous equations where one is linear and the other is not linear
- understanding the geometrical significance of simultaneous equations.

Exercise A

1

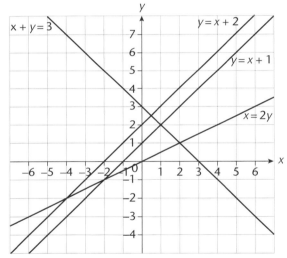

a Use the graph to write down the solutions of the simultaneous equations.

   i $y = x + 1$ **ii** $x = 2y$ **iii** $y = x + 1$ **iv** $y = x + 2$
       $x + y = 3$ $x + y = 3$ $x = 2y$ $x = 2y$

b Explain why there are no solutions for the simultaneous equations.
   $y = x + 1$
   $y = x + 2$

2 On graph paper, draw a set of axes with values of x from 0 to 5 and values of y from −2 to 14.
   Draw $y = x + 2$ and $y = 3x − 2$. At what point do the lines intersect?

3 On graph paper, draw a set of axes with values of x from 0 to 5 and values of y from −2 to 20.
   Solve the simultaneous equations $y = 4x − 1$ and $x + y = 4$ using their graphs.

4 On graph paper, draw a set of axes with values of x from 0 to 5 and values of y from -3 to 7. Solve the simultaneous equations $2x + y = 7$ and $2y = x + 3$ graphically.

5 The equations for the sides of a triangle are $x + 2y = 8$, $2y = x$ and $x + 4y = 18$. Find the vertices of the triangle using graphs.

Exercise B

Solve these simultaneous equations.

1 $2x + 5y = 48$

$2x + y = 8$

2 $2x + y = 8$

$3x - y = 17$

3 $x + 3y = 13$

$2x + y = 11$

4 $4x - 3y = 1$

$x + y = -5$

5 $5x + y = 10$

$3x - 5y = 10$

6 $2x + 3y = 17$

$3x + 4y = 24$

7 $5x - 2y = 34$

$3x + 5y = 8$

8 $x - 3y = -5$

$3x - 2y = 6$

9 $4x - 5y = 7$

$4x + y = -11$

10 $2y = 4 - 3x$

$4y + x = 13$

11 $4x = 11 + 3y$

$y = 5x - 22$

12 $4x - 3y + 3 = 0$

$2x + 5y = 18$

13 $x = 2y - 1$

$3x - 4y = 6$

14 $4x + 2y = 2$

$3y = 19 + 2x$

15 $2x - 7y = 43$

$5x - 4y = 13$

16 $6x = 7y + 71$

$2(5x - 4y) = 89$

Exercise C

1 Two numbers, x and y, have a sum of 40 and a difference of 14. Write down two equations in x and y and solve them to find the numbers.

2 The line $y = mx + c$ passes through the points $(-2, 2)$ and $(4, 1)$. Write down two equations and solve them to find m and c. Hence write down the equation of the line.

3 Tickets for the cinema cost £a for adults and £c for children. Mr and Mrs Kennedy go to the cinema with their four children. The total cost is £27. Mrs Jones takes her three children. The total cost is £17. Write down two equations in a and c and solve them to find the cost of an adult ticket and of a child ticket.

4 Pens cost p pence and rulers cost r pence. Jack bought three pens and two rulers. The total cost was £1.35. Jenny bought four pens and three rulers. The total cost was £1.90. Write down two equations in p and r and solve them to find the cost of a pen and of a ruler.

5 For a coach trip the adult fare is twice the child fare. The total fare for two adults and three children is £84. Write down two equations and solve them to find the cost of each fare.

6 a The total of Lan's age and his mother's age is 46. If Lan's age is x years and his mother's age is y years, write down an equation in x and y.
  b Next year Lan's mother will be three times as old as Lan. Write down a second equation in x and y. Simplify this equation.
  c Solve your two equations to find both ages.

7 The time T (minutes) taken to cook a turkey is given by the formula $T = a + bM$ where M is the mass of the turkey in kg. A 10 kg turkey takes 3 hours and 50 minutes to cook, whilst an 8 kg turkey takes 3 hours and 10 minutes to cook. How long will it take to cook a 12 kg turkey?

8 A taxi fare has a fixed charge and an additional cost per mile. A girl who travelled 12 miles was charged £13.20, whilst a boy who travelled 15 miles was charged £15.90. What will the fare be for a journey of 24 miles?

Exercise D

1 a Draw the graph of $y = x^2 - 6x + 8$ for $x = 0$ to 6.
  b On the same grid, draw the line $y = 2x - 5$.
  c Write down the co-ordinates of the points where the curve and the line intersect.

2 Use the graph of $y = x^2 - 6x + 8$ from question **1**. By drawing another line on the graph, solve the simultaneous equations $y = x^2 - 6x + 8$ and $2y + x = 8$ graphically.

3 Solve each pair of simultaneous equations algebraically. Where necessary give your solutions correct to 3 significant figures.

  a $y + x - 3 = 0$
   $y = x^2 + 1$

  b $y = x^2 + 3x - 1$
   $x - 2y - 4 = 0$

  c $x + y = 3$
   $4x - y^2 = 0$

  d $y - x = 5$
   $y = x^2 + 1$

  e $y(x + 1) = 10$
   $y = 2x + 3$

  f $2y + 3x = 5$
   $x^2 + 2y^2 = 11$

  g $xy = 8$
   $3y - 2x = -2$

  h $3x + 4y = 10$
   $2x^2 - 5xy - 3 = 0$

  i $x^2 + y^2 = 2$
   $y - 2x - 1 = 0$

  j $2x^2 + 3y = 12$
   $x - y + 1 = 0$

CHAPTER 31

Graphical solutions of equations

This chapter is about

- understanding the significance of points of intersection of graphs in the context of solving equations
- determining the lines that need to be drawn on a graph in order to solve a particular equation
- solving a variety of equations using a particular graph
- determining the equation of a function whose solution is a particular point of intersection.

1   **a**   Draw the graph of $y = x^2 - 2x - 8$ for values of x from -3 to 5.
  b   Use your graph to solve

     i $x^2 - 2x - 8 = 0$
     ii $x^2 - 2x - 8 = 5$
     iii $x^2 - 2x - 8 = -3$

2   **a**   Draw the graph of $y = 3x^2 - 2x$ for values of x from -3 to 3.
  b   Use your graph to find
     i   the minimum value of y
     ii   the solutions of the equation $3x^2 - 2x = 10$.

3   **a**   Draw the graph of $y = \frac{1}{2}x^2 - 2x + 5$ for values of x from -2 to 6.

  b   Use your graph to solve these equations.

     i $\frac{1}{2}x^2 - 2x + 5 = 9$

     ii $x^2 - 4x + 10 = 8$

4   **a**   Draw the graph of $y = 5x - 2x^2$ for values of x from -2 to 4.
  b   Use your graph to find
     i   the maximum value of y and the value of x where it occurs
     ii   the solutions of the equation $5x - 2x^2 = 0$
     iii the solutions of the equation $3x - 2x^2 = 0$.

5   **a**   Draw the graph of $y = x^2 - 5x + 6$ for values of x from 0 to 5.
  b   Use your graph to solve these equations.
     i $x^2 - 5x + 6 = 0$
     ii $x^2 - 5x + 3 = 0$

6   The table of values below is for the equation $y = x^2 - x - 4$.

x	-2	-1	0	1	2	3
y	2	-2	-4	-4	-2	2

  a   Draw the graph for values of x from -2 to 3.
  b   Use your graph to solve these equations.
     i $x^2 - x - 4 = 0$
     ii $x^2 - x - 1.5 = 0$

7 a Draw the graphs of $y = x^2$ and $y = 5 - \frac{1}{2}x$ for $x = -3$ to 3 on the same grid.

  b What is the equation whose solution is found at the intersection of the two graphs?
  c Use your graphs to solve this equation.

8 a Draw the graph of $y = 2x^2 + 3x - 9$ for values of x from -3 to 2.
  b Use your graph to solve these equations.
    i $2x^2 + 3x - 9 = -1$
    ii $2x^2 + 3x - 4 = 0$

9 a Draw the graph of $y = x^2 - 4x$ for $x = -3$ to 3.
  b Draw a straight-line graph on the same grid so that you can solve the equation $x^2 - 5x + 3 = 0$.
  c Use your graphs to solve the equation.

10 a Draw the graph of $y = x^2 - 5x + 3$ for values of x from -2 to 8.
  b Use your graph to solve these equations.
    i $x^2 - 5x + 3 = 0$
    ii $x^2 - 5x - 2 = 0$
    iii $x^2 - 7x + 3 = 0$

11 a Draw the graph of $y = x^2 - 3x$ from $x = -2$ to 5.
  b Use your graph to solve these equations.
    i $x^2 - 3x - 2 = 0$
    ii $x^2 - 4x + 1 = 0$

Do not draw the graphs in questions **12** to **14**.

12 The intersection of two graphs is the solution to the equation $x^2 - 2x - 3 = 0$.

One of the graphs is $y = x^2 - x - 1$. What is the other graph?

13 The graphs of $y = x^2 + 2x$ and $y = 2x + 1$ are drawn on the same grid.

What is the equation whose solution is found at the intersection of the two graphs?

14 The graph of $y = x^2 - 3x - 7$ is drawn on a grid.

What is the equation of the straight-line graph that you would draw to solve each of these using the graph of $y = x^2 - 3x - 7$?
  a $x^2 - 5x + 1 = 0$ **b** $x^2 + x - 5 = 0$ **c** $x^2 - 2x = 6$

15 The curve $y = x^2 - 2x + 3$ is drawn on a graph. What is the equation of the straight line that you should draw on the graph to solve each of these equations?
  a $x^2 - 3x + 1 = 0$
  b $2x^2 - 3x + 5 = 0$

CHAPTER 32 Co-ordinate geometry of the circle

This chapter is about:

- writing the equation of a circle, centre the origin and radius r
- finding the equation of a tangent to a circle at a given point
- solving various problems in the context of a circle.

Exercise A

1 Write down the centre and radius for the circle with equation
  a $x^2 + y^2 = 100$
  b $x^2 + y^2 = 1$
  c $x^2 + y^2 - 64 = 0$
  d $3x^2 + 3y^2 = 27$
  e $\dfrac{x^2}{2} + \dfrac{y^2}{2} = 72$

2 Which of the following equations do represent circles and which do not?
  a $x^2 + y^2 = 25$
  b $x^2 + y^2 - 9 = 0$

  c $5x^2 + 5y^2 = 30$
  d $9x^2 + 9y^2 = 49$
  e $x^2 + 2y^2 = 16$

3 A circle has centre $(0, 0)$ and radius $3\,\text{cm}$. Write down the equation of the circle.

4 A circle has centre $(0, 0)$ and diameter $7\,\text{cm}$. Write down the equation of the circle.

5 Show that the point $(6, -4)$ lies on a circle with equation $3x^2 + 3y^2 = 156$. Write down the exact radius of this circle.

Exercise B

1 Find the gradient of the tangent to the circle $x^2 + y^2 = 68$ at the point $(-2, 8)$.

2 Find the equation of the tangent to the circle $2x^2 + 2y^2 = 17$ at the point $(1\frac{1}{2}, 2\frac{1}{2})$.

3 A circle has the equation $x^2 + y^2 = 52$. The point P$(-6, 4)$ lies on the circumference. The tangent at P meets the y-axis at K. Find the co-ordinates of K.

4 A circle with centre $(0, 0)$ has radius $2\sqrt{10}$. The point P$(-6, -2)$ and the point Q$(6, -2)$ lie on the circumference of the circle. Tangents are drawn to the circle at P and at Q. Find the co-ordinates of the point T where the two tangents intersect.

5 A$(-2, 5)$ and M$(2, -5)$ are the ends of the diameter of a circle. Find the equation of the circle.

Angles in circles

Exercise A

1   Draw a circle with centre O. Mark four points A, B, C, D on the circumference of the circle as shown in the diagram.

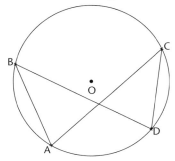

   Measure angle ABD and angle ACD. Write a sentence relating your findings.

2   Draw a circle with centre O. Join O to a point C on the circumference. Draw a tangent to the circle at point C. Label it MN. Measure angle MCO. Write a sentence relating your findings.

3   Draw a circle with centre O. Mark four points P, Q, R, S on the circumference of the circle. Join them to form a quadrilateral. Measure angle PQR. Measure angle PSR. Write a sentence relating your findings.

4   Draw a circle with centre O. Draw a diameter in your circle FG. Join F and G to a third point P on the circumference of your circle. Measure angle FPG. Write a sentence relating your findings.

5   Prove that POR = 2PQR. O is the centre of the circle. P, Q and R are points on the circumference of the circle. Let PQO = x and RQO = y.

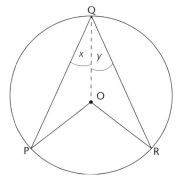

In each of the following diagrams, O is the centre of the circle.
Find the size of each of the lettered angles.
Give reasons for your answers.

6

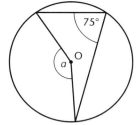

10

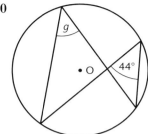

7

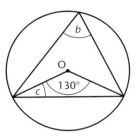

11

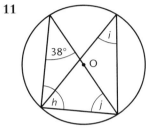

8

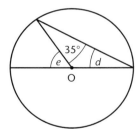

12

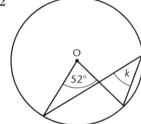

9

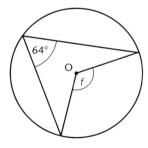

13

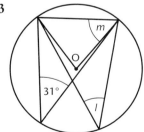

Exercise B

In each of the following diagrams, O is the centre of the circle.
Calculate the size of each of the lettered angles.
Give reasons for your answers.

1

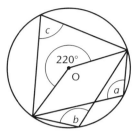

4

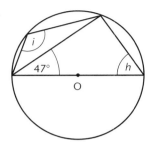

2

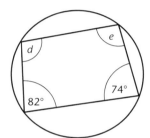

5

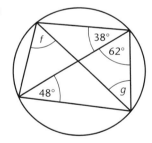

3

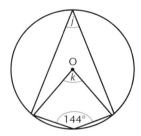

6

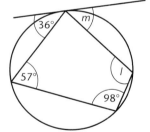

Exercise C

In questions **1** to **4**, calculate the size of the angles marked with letters. O is the centre of each circle. X and Y are the points of contact of the tangents to each circle.

1

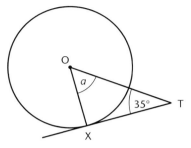

2

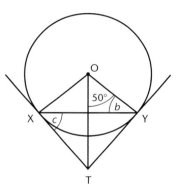

3

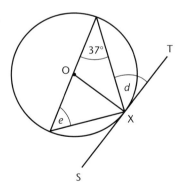

4

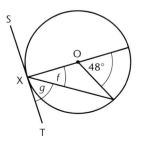

In questions **5** and **6**, find the lengths marked with letters.

5

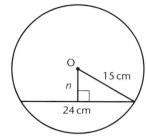

6

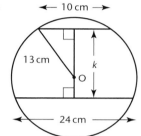

7 TB and TD are tangents to the circle, which has centre O.
Angle APC = $x°$.
Find the size of angle ATC in terms of x.

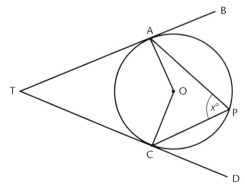

Exercise D

In these questions, ST is a tangent to the circle, which has centre O.
Find the size of the lettered angles, giving reasons for your answers.

1

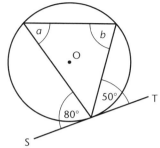

4

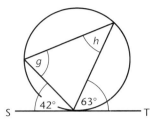

2

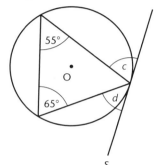

5

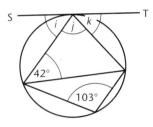

3

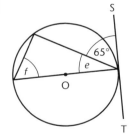

6

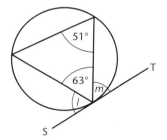

Exercise E

In each of the following diagrams, O is the centre of the circle.
Find the size of each of the lettered angles, giving reasons for your answers.

1

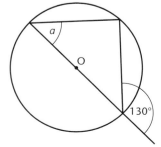

2

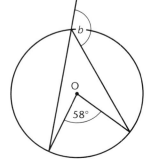

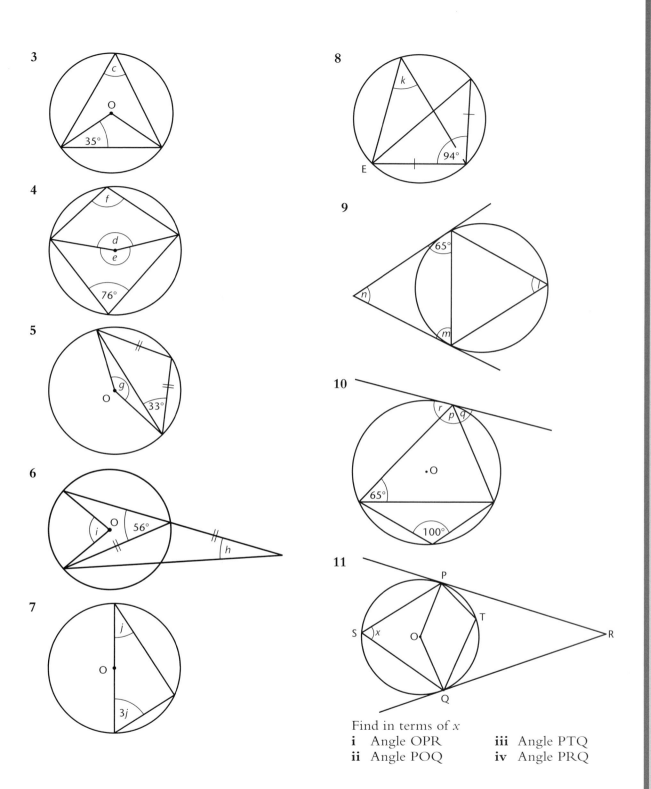

3

c

O

35°

4

f

d
e

76°

5

O

g

33°

6

i   O   56°

h

7

O

j

3j

8

k

E   94°

9

65°

n

m

l

10

r p q

•O

65°

100°

11

P

S   x   O   T   R

Q

Find in terms of x
i   Angle OPR   **iii**   Angle PTQ
ii   Angle POQ   **iv**   Angle PRQ

This chapter is about

- knowing the names of polygons with 5, 6, 7, 8, 9 and 10 sides
- knowing that all the sides are equal in a regular polygon and that all the angles are equal
- knowing that an interior angle and an exterior angle at a vertex add up to 180°
- knowing that the sum of the exterior angles is 360°
- knowing that the sum of the interior angles of an n-sided polygon is 180° × (n − 2)
- finding the number of sides in a regular polygon if you know each angle
- finding the angles in a regular polygon if you know the number of sides
- finding the angle at the centre of a regular polygon
- finding the missing angles in an irregular polygon.

Find the missing angles in each of the polygons in questions **1** to **3**.

1

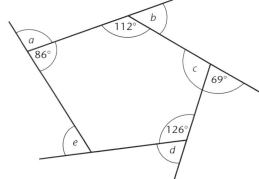

2

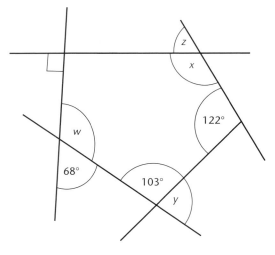

3

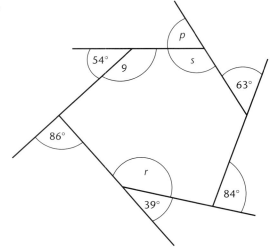

4 A regular polygon has 18 sides. Find the size of each of its exterior and interior angles.

5 A regular polygon has 24 sides. Find the size of each of its exterior and interior angles.

6 A regular polygon has an exterior angle of 12°. How many sides does it have?

7 A regular polygon has an interior angle of 172°. How many sides does it have?

8 Name the regular polygon which has interior angles of 140°.

9 Two sides of a regular pentagon PQRST are produced to meet at M. Calculate the size of the angle M.

10 A polygon has nine sides. Work out the sum of the interior angles of this polygon.

11 A polygon has 13 sides. Work out the sum of the interior angles of this polygon.

12 Four of the exterior angles of a hexagon are 93°, 50°, 37° and 85°. The other two angles are equal.
   a Find the size of these equal exterior angles.
   b Find the size of each interior angle.

13 Four of the exterior angles of a pentagon are 170°, 80°, 157° and 75°.
   a Find the size of the other interior angle.
   b Find the size of each exterior angle.

14 The angles in a heptagon are $2x$, $2x + 10$, $2x$, 140, 135, $2x + 30$ and $2x - 5$. Calculate the value of x.

15 The interior angle sum of a polygon is 2520°. How many sides does the polygon have?

16 Each angle at the centre of a regular polygon is 24°. Calculate the sum of the interior angles of the polygon.

17 Is it possible to have a regular polygon with interior angle 164°? Explain your answer.

18 ABCDEF is a regular hexagon. Calculate each of the angles in the triangle ACD.

19 The angle at the centre of a regular polygon is x. Write an expression in terms of x for the number of sides that the polygon has.

20 Joseph measured all the interior angles of a polygon. His total was 1 901° but he had missed an angle.
   a How many sides does the polygon have?
   b What was the missing angle?

21 A square and two other identical regular polygons meet at a point. What type of polygons are these? Justify your answer.

CHAPTER 35

Pythagoras and trigonometry

This chapter is about

- using Pythagoras' theorem to find missing sides
- applying Pythagoras' theorem to solve problems in 2D
- using trigonometry to find missing sides
- using trigonometry to find missing angles
- applying trigonometry to solve problems in 2D
- applying Pythagoras' theorem and trigonometry to solve problems in 3D.

Exercise A

Write an equation relating the sides of each of these triangles.

1

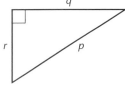

2

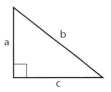

3

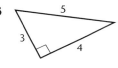

Calculate the missing area of the squares shown in each of these diagrams.

4

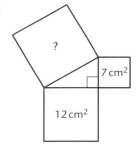

5

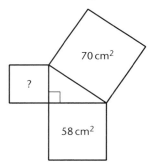

Find the length of the hypotenuse, x, in each of these triangles. Give your answers either exactly or correct to 2 decimal places.

6

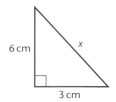

7

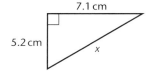

8

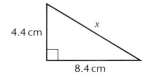

9 PQR is a right-angled triangle. PQ = 13 cm, QR = 7 cm and angle PQR = 90°. Calculate the length of PR.

10 WXY is a right-angled triangle. WX = 12 cm, XY = 9 cm and angle WXY = 90°. Calculate the length of WY.

Find the length marked x in each of these triangles. Give your answers either exactly or correct to 2 decimal places.

11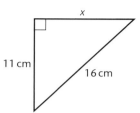

12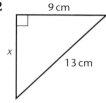

13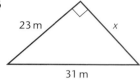

14 ABC is a right-angled triangle. AB = 4.2 cm, AC = 14.81 cm and angle ABC = 90°. Calculate the length of BC.

15 KLM is a right-angled triangle. KL = 19.14 cm, KM = 18 cm and angle KML = 90°. Calculate the length of LM.

16 Ann can walk home from school along two roads or along a path across a field.

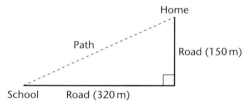

How much shorter is her journey if she takes the path across the field?

17 This network is made of wire. What is the total length of wire?

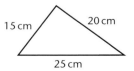

18 A ladder 10 m long has its base 3 m from the foot of a vertical wall. How far up the wall will the ladder reach?

19 The sides of a rectangle are 5 cm and 6 cm long. Find the length of the diagonal of the rectangle.

20 An aircraft flies 120 km South, then 150 km East, then 200 km North. How far is the aircraft from its starting position?

21 State whether or not each of these triangles is right-angled. Explain your reasoning.

a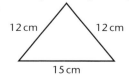

b

Exercise B

Copy and label each of the three sides in these triangles with the letters O, A and H.

1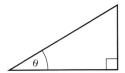

2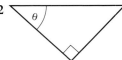

Find the trig ratio connecting the given sides and angle θ in these triangles.

3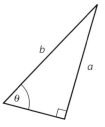

4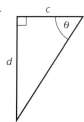

5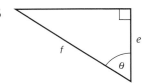

6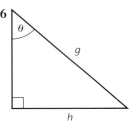

In these diagrams find the lengths marked with a letter. Give your answers to 3 significant figures.

7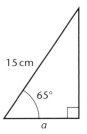

8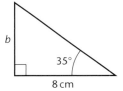

9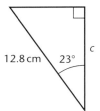

10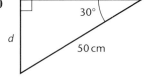

11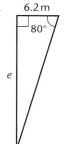
6.2 m
80°
e

12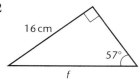
16 cm
57°
f

13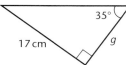
35°
17 cm
g

14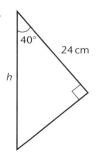
40°
24 cm
h

In these diagrams find the size of the angles marked with a letter. Give all answers to 3 significant figures.

15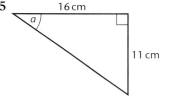
16 cm
a
11 cm

16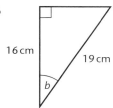
16 cm
19 cm
b

17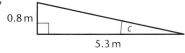
0.8 m
c
5.3 m

18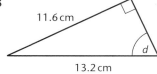
11.6 cm
13.2 cm
d

In these diagrams find the lengths marked with a letter. Give your answers to 3 significant figures.

19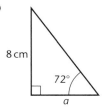
8 cm
72°
a

20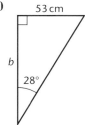
53 cm
b
28°

21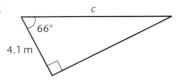

22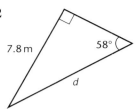

In these diagrams find the size of the angles marked with a letter. Give all answers to 3 significant figures.

23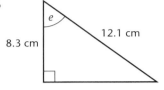

24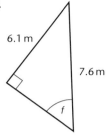

25 ABC is a right-angled triangle in which angle CAB = 90°. AB = 28 cm. Angle ACB = 41°. Calculate the length of BC.

26 PQR is a right-angled triangle in which angle PQR = 90°. PQ = 65 m, QR = 102 m. Calculate angle PRQ.

Exercise C

1 The diagram shows the side view of a bin.

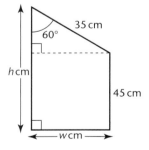

    a Find the width, w cm, of the bin.
    b Find the height, h cm, of the bin.

2 Look at this triangle.

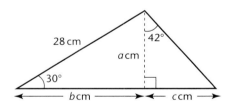

    a Find the height, a cm, of the triangle.
    b Find the length, b cm.
    c Find the length, c cm.
    d Find the area of the triangle.

3 A ship sails on a bearing of 125° for 220 km.
  a Find how far east the ship has travelled.
  b Find how far south the ship has travelled.

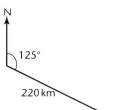

4 Calculate the area of this trapezium.

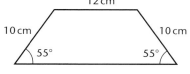

5 The diagram shows a step ladder. The two sections of the ladder are opened to 30°. The feet of the two sections are 1.2 m apart.

Calculate the length, *l*, of each section of the ladder.

6 a Find the length of the base, *a* cm, of the triangle.

  Give your answer to 2 d.p.

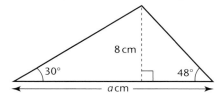

  b Find the area of the triangle.

7 The bearing of A from B is 230°. A is 20 km west of B.

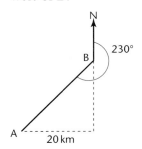

Calculate how far A is south of B.

8 Look at this diagram.

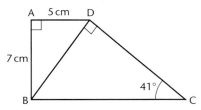

Calculate the length of BC.

9 A rectangle has length 12 cm and width 8 cm. Calculate the size of the angle that the diagonal makes with the longer side.

10 A yacht sails from A to B. B is 43 km east of A and 25 km south of A.

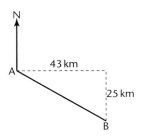

Calculate the bearing of B from A.

11 The diagram shows an isosceles triangle.

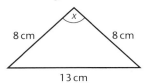

Calculate the size of the angle x.

12 A ship sails for 200 miles from A to B on a bearing of 080°. How far north and east of A is the ship when it reaches B?

Exercise D

1 ABCDEFGH is a cuboid with dimensions as shown.

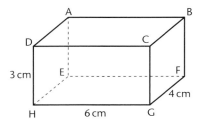

Calculate these.
a AC
b Angle BAC
c AG
d Angle CAG

2 PABCD is a square-based pyramid with P vertically above the midpoint of the square base.

AB = 12 cm, AP = BP = CP = DP = 15 cm.

E is the midpoint of BC. F is the midpoint of AD.

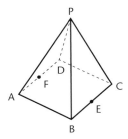

Calculate these.
a DB
b Angle PBD
c PE
d Angle PEF

3 ABCDEF is a triangular prism with angle BCF = 90°.

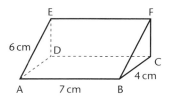

Calculate these.
a FC
b Angle FBC
c EB
d Angle EBD

4 PABCD is a square-based pyramid with P vertically above the midpoint of the square base.

AP = BP = CP = DP = 9.5 cm. BC = 5.8 cm.

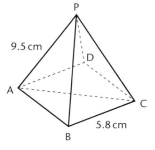

Calculate angle PAC.

5 A pencil case is a cuboid with a base measuring 6 cm by 15 cm. A pencil 17 cm long just fits in the box.

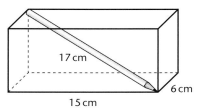

Calculate the height of the pencil case.

6 The points A, B and C are in the same horizontal plane. The angle of elevation of a vertical mast MC from A is 24.7°. AC is 34 m and BC is 57 m.

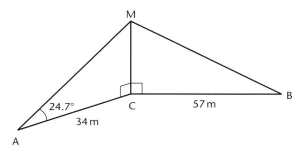

Find the angle of elevation of M from B.

7 A cuboid has a base with sides 5.6 cm by 8.2 cm. Its height is 4.3 cm.

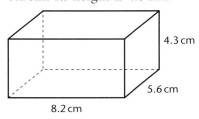

Calculate the angle between a space diagonal of the cuboid and
a the base
b a 5.6 cm by 4.3 cm face.

8 A pyramid is 8 cm high and has a square base with sides of 6 cm. Its sloping edges are all of equal length. Calculate the angle between a sloping edge and the base.

9 A square-based pyramid has a height of 8.3 cm. Its sloping faces each make an angle of 55° with the base.
a Find the length of the sides of the base.
b Find the length of the sloping edges of the pyramid.

10 ABCDEF is a triangular wedge. The base ABFE is a horizontal rectangle. C is vertically above B.

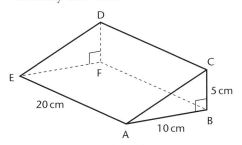

a Calculate the length AD.
b Calculate the angle which AD makes with
i the base ABFE
ii the face ABC.

CHAPTER 36

Similarity

This chapter is about

- knowing the definition of similar figures
- calculating missing lengths in similar figures
- proving that two triangles are similar
- knowing the relationship between the ratios of length, area and volume of similar figures
- using the relationship between the ratios of length, area and volume of similar figures to calculate lengths, areas and volumes.

Exercise A

1 Explain why these trapeziums are not similar.

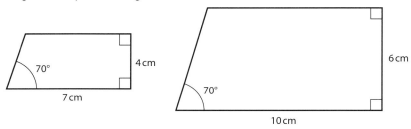

2 Look at these two triangles.

Explain why the triangles are similar.

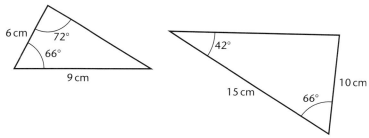

3 State, with reasons, which of these rectangles are similar.

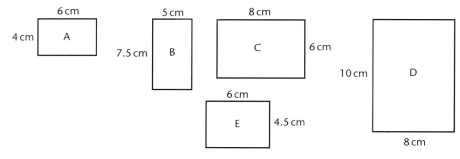

4 Any two squares must be similar. Write down the names of three other types of shapes for which this statement is true.

5 These two rectangles are similar. Calculate x.

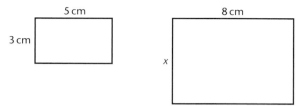

6 These two parallelograms are similar. Calculate the length marked x.

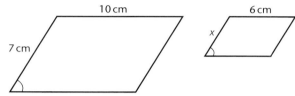

7 The triangles PQR and STU are similar. Calculate the lengths of ST and SU.

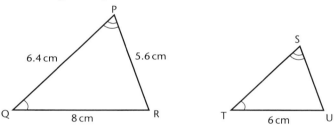

8 The triangles ABC and DEF are similar. Calculate the lengths of AB and EF.

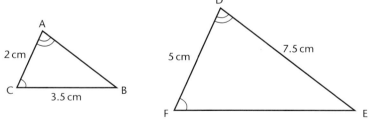

9 The quadrilaterals ABCD and PQRS are similar.

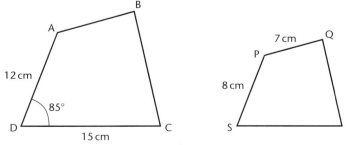

   a Write down the size of angle PSR.
   b Calculate the lengths of AB and SR.

10 Decide if each pair of triangles is similar or not. You must explain your answer fully, commenting on the angles or the ratios of the corresponding sides.

a

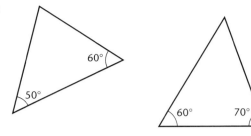

b

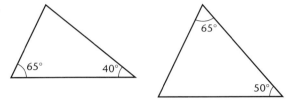

c

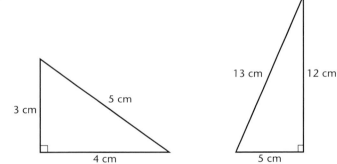

d

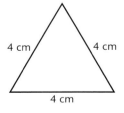

e

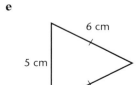

Exercise B

Remember to give reasons when stating that two angles are equal.

1 Prove that triangle ABC is similar to triangle ADE.

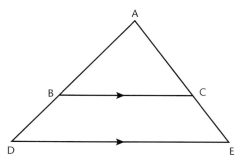

2 Prove that triangle MON is similar to triangle POQ.

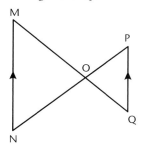

3 Prove that triangle FGH is similar to triangle GHI.

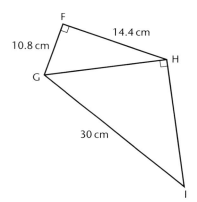

4 Prove that triangle JKL is similar to triangle KTL.

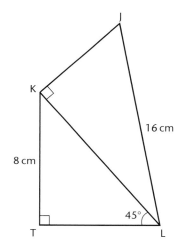

Find each missing length.

5
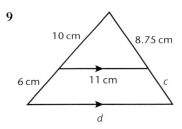
2 cm
8 cm
6 cm
x

6
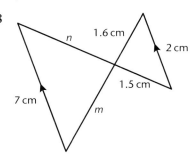
6 cm
3 cm
2.5 cm
y

7
6 cm
q
8 cm
p
12.5 cm
5 cm

8
n
1.6 cm
2 cm
1.5 cm
7 cm
m

9
10 cm
8.75 cm
6 cm
11 cm
c
d

10 An oval mirror has an area of $162\,\text{cm}^2$.
What is the area of a similar mirror which is one and a half times as long?

11 The three tables in a nesting set of tables are similar, with heights in the ratio $1:1.2:1.5$.
The area of the smallest table top is $120\,\text{cm}^2$. What is the area of the middle-sized table top?

12 A model of a theatre set is made to a scale of $1:20$.
A cupboard on the model has a volume of $50\,\text{cm}^3$.
Find the volume of the cupboard on the actual set, giving your answer in m^3.

13 Two jugs are similar, with capacities of 1 litre and 2 litres respectively. The height of the larger jug is $14.8\,\text{cm}$.
What is the height of the smaller jug, to the nearest millimetre?

14 Julia is making a model of a wooden statue. She uses the same kind of wood as the original.
The model is on a scale of $1:20$.
  a The statue is $15\,\text{m}$ high. How high will Julia's model be?
  b 500 litres of varnish were needed for the statue. How many litres will be needed for the model?
  c Julia's model weighs $3\,\text{kg}$. Estimate the weight of the statue.

15 LEGO® and DUPLO® are types of building bricks for children.
Duplo bricks are designed for younger children and are twice as large as Lego bricks.

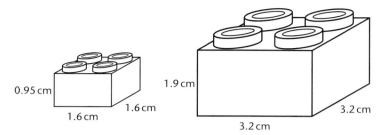

a Janet makes a shape using Duplo bricks. Her shape is 9.5 cm high. Elaine makes a similar shape from Lego. How high is Elaine's shape?

b The area of the base of Elaine's shape is 64 cm². What is the area of the base of Janet's shape?

c The volume of Elaine's shape is 304 cm³. What is the volume of Janet's shape?

16 This is a set of Russian dolls. The larger dolls are enlargements of the smallest doll.
The heights of the dolls are 3 cm, 4.5 cm and 8 cm respectively.

a The width of the smallest doll is 1.2 cm.
What is the width of the largest doll?

b The surface area of the middle-sized doll is 8.4 cm².
What is the surface area of the smallest doll?

c The volume of the smallest doll is 4.2 cm³.
What is the volume of the middle-sized doll?

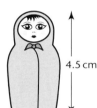
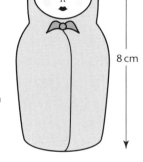

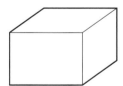

17 These two cuboids are similar.
The volume of the smaller cuboid is 2 000 cm³.
The volume of the larger cuboid is 8 000 cm³.
The surface area of the smaller cuboid is 1 200 cm².
Calculate the surface area of the larger cuboid.

CHAPTER 37

Trigonometry rules

This chapter is about

- knowing when to use the sine rule and the cosine rule
- calculating missing sides and angles using either rule
- knowing how to use the area rule of a triangle.

Exercise A

1 Find x.

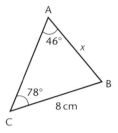

2 Find x.

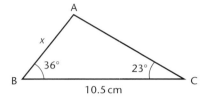

3 ABC is a triangle in which BAC = 58°, ABC = 42° and AC = 6.6 cm. Calculate BC.

4 PQR is a triangle in which QPR = 55°, PQR = 30° and QR = 14 cm. Calculate PQ.

5 Find angle ACB.

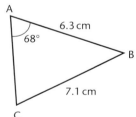

6 Find angle ACB.

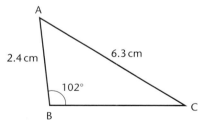

7 PQR is a triangle in which QR = 7 cm, PR = 15 cm and angle PQR = 124°. Calculate angle QPR.

8 ABC is a triangle in which BC = 5.9 cm, AC = 4.7 cm and angle BAC = 106°. Calculate angle BCA.

9 Find length AC.

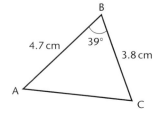

10 Find length AB.

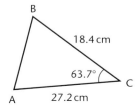

11 ABC is a triangle in which ABC = 120°, BC = 8 cm and AB = 6 cm. Calculate AC.

12 WXY is a triangle in which WXY = 76.5°, XY = 13.4 cm and WX = 19.2 cm. Calculate WY.

13 Find angle CAB.

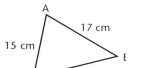

14 Find angle CAB.

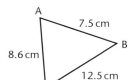

15 PQR is a triangle in which QR = 7.3 cm, PR = 9.2 cm and PQ = 6.5 cm. Calculate the size of angle R.

16 ABC is a triangle in which BC = 5 cm, AC = 9 cm and AB = 6 cm. Calculate the size of the largest angle.

17 Find the area of each of these triangles.

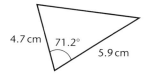

18 In triangle ABC, AB = 15.3 cm, AC = 9.6 cm and angle BAC = 53.6°. Find the area of triangle ABC.

19 The area of triangle WXY = 14.1 cm². WX = 6.5 cm and angle WXY = 62°. Calculate XY.

20 The area of triangle JKL = 9.37 m². JK = 5.7 m and KL = 3.6 m. Calculate angle JKL.

Exercise B

1 Find　　　　**a** Angle A　　　**b** Angle C　　　**c** Side AB

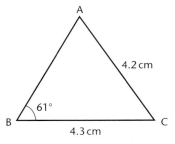

2 In triangle PQR, PQ = 7.8 cm, angle R = 79° and angle P = 51°. Find these.
　a Side QR
　b Angle Q
　c Side PR

3 The angles of elevation of the bottom (B) and top (T) of a flagpole on top of a tower (AB) are 37° and 42°. These are measured 200 m from the base of the tower at a point C.

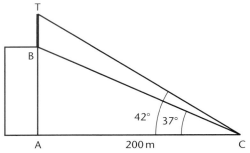

a Find the length BC.
b Find the height of the flagpole, BT.

4 In triangle PQR, PR = 7.2 cm, QR = 6.3 cm and angle PRQ = 37°. Find these.
a The length of PQ
b Angle PQR

5 The lengths of the sides of triangle ABC are AB = 4.6 cm, BC = 11.5 cm and CA = 7.8 cm. Find the sizes of the angles of the triangle.

6 Find   **a** BD   **b** AB   **c** the area of ABCD.

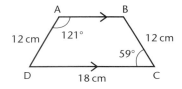

7 A, B and C are points on an orienteering course. The bearing of B from A is 040°. The bearing of C from B is 125°. AB is 3.7 km and BC is 2.3 km.

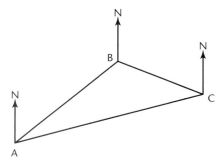

8 ABCD is a kite. Find its area.

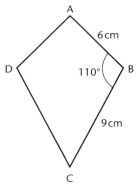

What is the distance AC?

9 In triangle EFG,
EF = 7 m, FG = 9 m and angle G = 39°.
Find these.

   a Angle E

   b Angle F

   c Side EG

10 ABC is a flower bed in a garden.
AB is 3.56 m long.

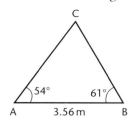

Find the lengths of the other two sides.

11 P, Q and R are three buoys marking a
sailing course. The bearing of P from Q is
035°. The bearing of P from R is 310°. The
bearing of R from Q is 075°. The length of
QR is 5.3 km.

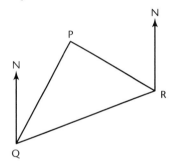

Find the total length of the three stages of
the course from P back to P.

12 Find   **a** BC   **b** AD

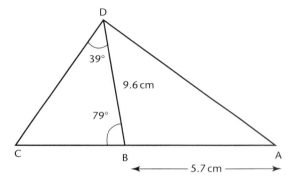

13 ABCD is a children's playground.
OA = 83 m, OB = 122 m, OC = 106 m,
OD = 78 m.

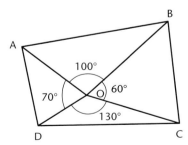

Calculate the area and the perimeter of the
playground.

14 Find angle BAC.

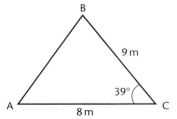

CHAPTER 38

Perimeter, area and volume

This chapter is about

- Calculating the perimeter and area of a rectangle, parallelogram and trapezium
- Calculating the circumference and area of a circle
- Calculating the length of an arc of a circle
- Calculating the area of a sector of a circle
- Calculating the surface area and volume of solids.

Exercise A

For questions **1** to **6**, find   **a** the perimeter   **b** the area.

1

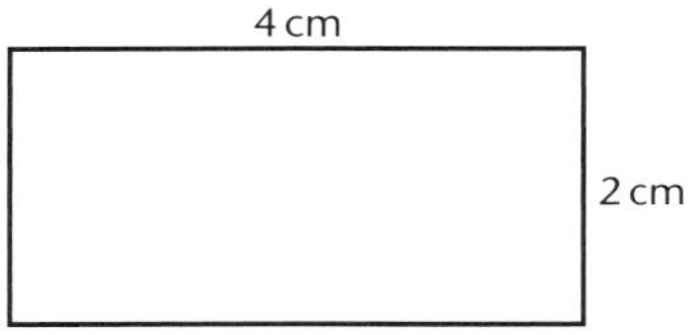

4 cm

2 cm

2

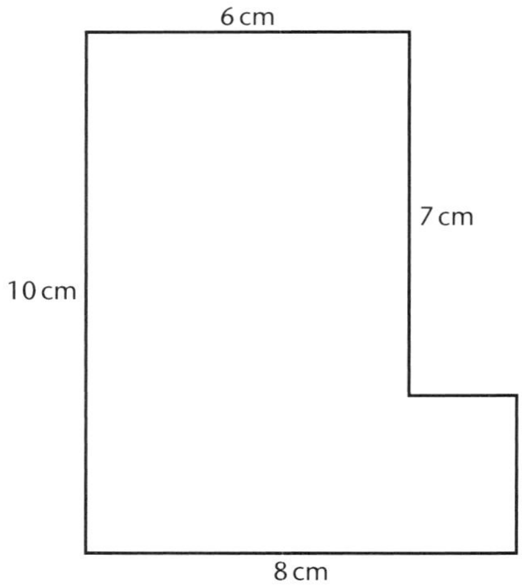

6 cm

7 cm

10 cm

8 cm

3

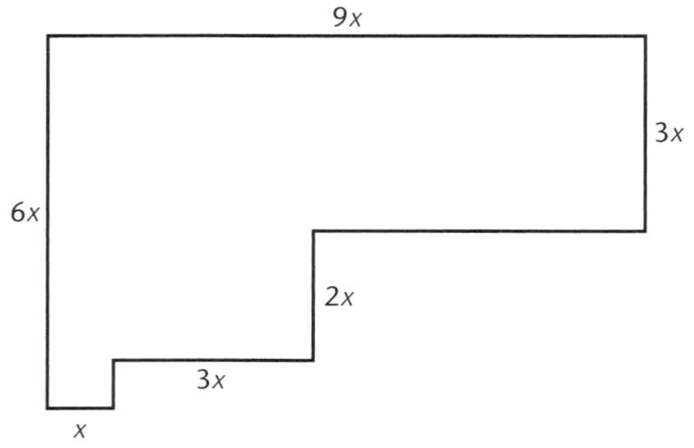

9x

3x

6x

2x

3x

x

All lengths are in cm. Give answers in terms of x in their simplest form.

4

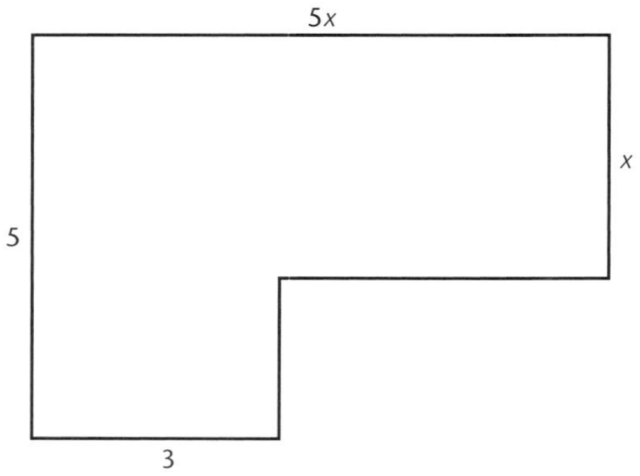

5x

x

5

3

All lengths are in cm. Give answers in terms of x in their simplest form.

5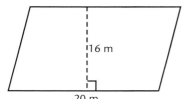

All lengths are in cm. Give answers in terms of x and y in their simplest form.

6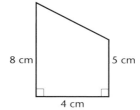

7 Find the area of the parallelogram.

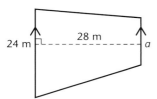

8 Find the area of the trapezium.

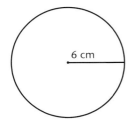

9 The area of a parallelogram is $7.56\,\text{m}^2$. Its base is $5.4\,\text{m}$. Calculate the perpendicular height of the parallelogram.

10 The parallel sides of a trapezium are $8\,\text{cm}$ and $11\,\text{cm}$. The area of the trapezium is $42.75\,\text{cm}^2$. Calculate the perpendicular height of the trapezium.

11 Calculate a given that the area is $360\,\text{m}^2$.

Exercise B

1 Find the circumference of each of the following circles. Give your answers in terms of π.

a

6 cm

b

50 m

2 Find the diameter of the circle with circumference **a** 42π cm **b** 0.8π m

3 Find the radius of the circle with circumference **a** 5π cm **b** 16π m

4 Calculate the circumference of the circles with these diameters, giving your answers correct to 1 decimal place. **a** 8 cm **b** 17 cm **c** 39.2 cm

5 Find the circumference of the circles with these radii, giving your answers correct to 1 decimal place. **a** 78 mm **b** 39 mm **c** 4.4 m

6 Find the diameter of the circle with circumference **a** 14 cm **b** 1.6 m

7 Find the radius of the circle with circumference **a** 8.4 m **b** 10 cm

Give all answers to the following questions correct to 3 significant figures.

8 Find the perimeter of a semicircle with diameter 12 cm.

9 Find the perimeter of a semicircle with radius 1.2 cm.

10 Find the perimeter of a quadrant of a circle with diameter 2.5 m.

11 Find the perimeter of a quadrant of a circle with radius 6 cm.

12 Calculate the perimeter of the shape shown.

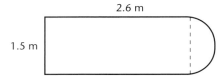

13 Find the area of the circles with these radii. **a** 17 cm **b** 4.3 cm

14 Find the area of the circles with these diameters.
a 18 cm **b** 3.64 m **c** 2120 mm

15 The area of a circle is 30 cm². Calculate its radius.

16 The area of a circle is 45 m². Calculate its diameter.

17 The area of a circle is 120 cm². Calculate its circumference.

18 The circumference of a circle is 19.48 m. Calculate its area.

19 Find the area of a semicircle with diameter 16 cm.

20 Find the area of a quadrant of a circle with radius 2.4 cm.

21 The area of a semicircle is 72.6 cm². Calculate the perimeter of the semicircle.

22 Calculate **a** the area **b** the perimeter.

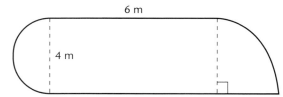

23 A flowerbed is a circle with radius 1.2 m. Mr Green wants to put compost on the flowerbed. The instructions tell him to use 2 litres of compost per square metre. What amount of compost should he use for the flowerbed?

24 A metal plate is a rectangle 8 cm by 12 cm. A circular hole with diameter 5 cm is cut out of the plate. Calculate the area of metal that remains.

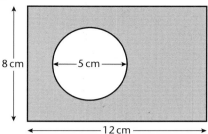

Exercise C

1 Find the arc length of each of these sectors.
Give your answers to 3 significant figures.

a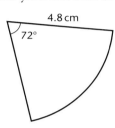

4.8 cm
72°

b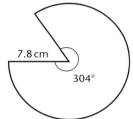

7.8 cm
304°

c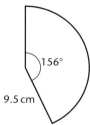

156°
9.5 cm

d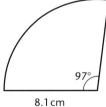

97°
8.1 cm

e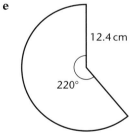

12.4 cm
220°

2 Calculate the area of each of the sectors in question **1**.
Give your answers to 3 significant figures.

3 Calculate the perimeter of each of these sectors.
Give your answers to 3 significant figures.

a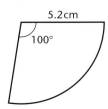

5.2 cm
100°

b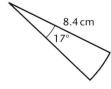

8.4 cm
17°

c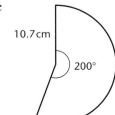

10.7 cm
200°

4 Calculate the sector angle in each of these sectors.
Give your answers to the nearest degree.

a

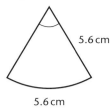

5.6 cm

5.6 cm

b

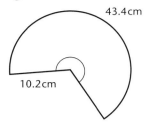

43.4 cm

10.2 cm

c

8.4 cm

6.5 cm

d

3.8 cm

Area = 8.2 cm²

e

Area = 50 cm²

7.3 cm

f

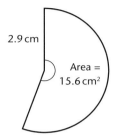

2.9 cm

Area = 15.6 cm²

5 Calculate the radius of each of these sectors.
Give your answers correct to 1 decimal place.

a

8.4 cm

45°

b

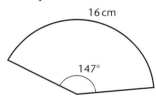

16 cm

147°

c

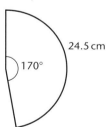

24.5 cm

170°

d

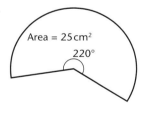

Area = 25 cm²

220°

6 O is the centre of the circle.
Angle AOB = 120°.

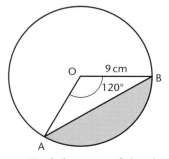

O   9 cm   B

120°

A

a Find the area of the shaded segment.
b Find the perimeter of the shaded segment.

7 O is the centre of the circle. Angle AOB = 74°

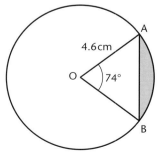

A

4.6 cm

O   74°

B

a Find the area of the shaded segment.
b Find the perimeter of the shaded segment.

8 O is the centre of the circle. Angle AOB = 116°

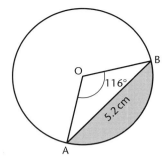

   a Find the area of the shaded segment.
   b Find the perimeter of the shaded segment.

Exercise D

1 Find the volume of a cube with edge 5 cm.

2 Calculate the volume of each cuboid. All measurements are in metres.

a 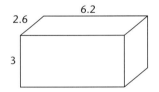   **b**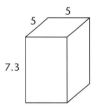

3 A cuboid is 1.2 m by 1.8 m by 2 cm. Calculate its volume
   a in cubic centimetres   **b** in cubic metres.

4 Find the height of a cuboid which is 15 cm long by 8 cm wide and has a volume of 480 cm^3.

5 A wooden block is 10 cm wide, 2 cm long and 4 cm high. How many of these blocks will fill a 1-metre cube?

6 A box is 10 cm high, 5 cm long and 3 cm wide. Calculate the surface area of the box.

7 A biscuit tin is 12 cm long, 5 cm wide and 6 cm deep and has a lid. Calculate the surface area of the tin.

8 The surface area of a cuboid is 180 cm^2. Its length is 8 cm and its breadth is 6 cm. Calculate the volume of the cuboid.

9 The volume of a cube is 512 m^3. Calculate the surface area of the cube.

10 Calculate the volume of the shape shown at the top of the next page.

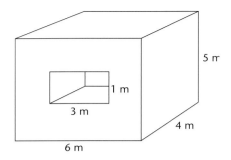

Find the exact volume of each of these prisms.

11

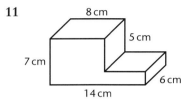

8 cm
5 cm
7 cm
6 cm
14 cm

12

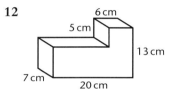

6 cm
5 cm
13 cm
7 cm
20 cm

13

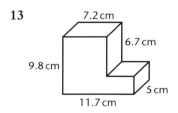

7.2 cm
6.7 cm
9.8 cm
5 cm
11.7 cm

14

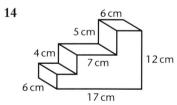

6 cm
5 cm
4 cm
12 cm
7 cm
6 cm
17 cm

15

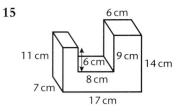

6 cm
11 cm
6 cm
9 cm
14 cm
7 cm
8 cm
17 cm

16

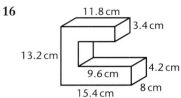

11.8 cm
3.4 cm
13.2 cm
9.6 cm
4.2 cm
15.4 cm
8 cm

17 Find the volume and surface area of the prism.

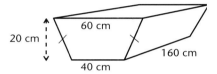

20 cm
60 cm
160 cm
40 cm

18 The volume of the prism is $0.675 \, \text{m}^2$.

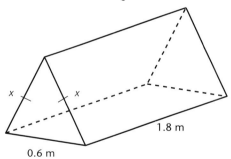

x
x
1.8 m
0.6 m

Calculate the length of the sides marked x.

Exercise E

1 Calculate the volume of this cylinder.

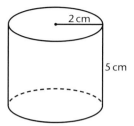

2 The volume of a cylinder is $1\,000\,\text{cm}^3$.
Its radius is $10\,\text{cm}$. Calculate its height.

3 The volume of a cylinder is $250\,\text{cm}^3$.
Its height is $12.3\,\text{cm}$. Calculate its radius.

4 Find the surface area of the closed cylinder.

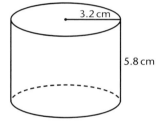

5 Find the surface area of the cylinder open at both ends.

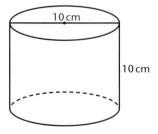

6 This cylinder holds 10 litres. The cross-sectional area is $200\,\text{cm}^2$. Find the height of the cylinder.

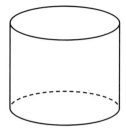

7 Calculate the surface area of the solid shown.

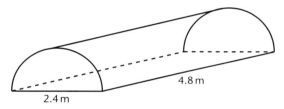

8 The volume of a cylinder is $209\,\text{cm}^3$. The height of the cylinder is $6.5\,\text{cm}$. Calculate the radius of the cylinder.

9 The curved surface area of an open cylinder is $72.1\,\text{cm}^2$. The radius is $1.82\,\text{cm}$. Calculate the height of the cylinder.

10 A circular pond of radius $1.3\,\text{m}$ is to be filled with water to a depth of $65\,\text{cm}$. Calculate the volume of water in litres.

Exercise F

1 For each cone calculate
  i the curved surface area   **ii** the total surface area   **iii**   the volume.

a

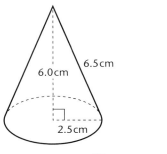

b

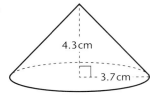

c

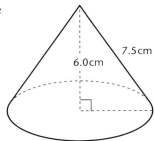

d

e

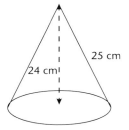

2 This conical glass holds 150 ml. Its top radius is 4 cm.

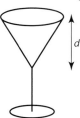

Find its depth, *d*.

3 A cone has slant height 15 cm and the area of its base is 254 cm^2.
Find its total surface area.

4 A cone has radius 10 cm and its volume is 2 513 cm^3.
Find its curved surface area.

5 The curved surface area of a cone is 754 cm^2 and its slant height is 20 cm.
Find its volume.

6 Work out the volume of each frustum.

a
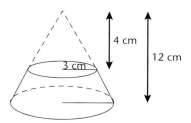

b
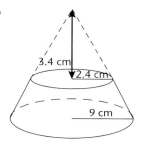

c
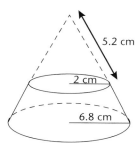

7 Work out the total surface area of each of the frustums in question **6**.

8 A sphere has radius 5 cm. Find
   a   its surface area   **b**   its volume.

9 A hemisphere has diameter 9 cm. Find
   a   its total surface area
   b   its volume.

10 A sphere has volume 30 cm³. Find its surface area.

11 A sphere has surface area 100 cm². Find its volume.

12 A football has outer diameter 310 mm and inner diameter 304 mm. Work out the volume of plastic required to manufacture one of these footballs. Give your answer in cm³.

13 A hemispherical glass bowl has an internal diameter of 22 cm and is 7.5 mm thick throughout. 5% of the original glass is removed when the pattern is cut into the bowl. Calculate the volume of glass remaining.

14 A cone and a cylinder have equal bases and equal heights.
   a   If the volume of the cylinder is 36 cm³, find the volume of the cone.
   b   If the volume of the cone is 19 cm³, find the volume of the cylinder.
   c   If the combined volume of both shapes is 64 cm³, find the volume of the cylinder.

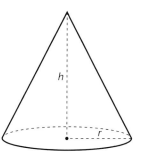
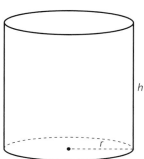

15 A metal sphere of radius 4 cm is melted down and then cast as a cube. What are the dimensions of the cube?

16 A cylindrical glass bowl of radius 15 cm has water in it with floating candles. 20 spherical glass marbles of radius 1.2 cm are placed in the bowl.

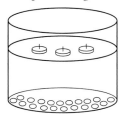

By how much does the water level in the bowl rise?

17 A cone has a base radius of 4.3 cm and perpendicular height of 8.4 cm.
It has the same volume as a sphere. Find the radius of the sphere.

18 The diagram shows a cone of radius 4.8 cm attached to a cylinder of height 5.8 cm which is attached to a hemisphere. The total height of the object is 17 cm.

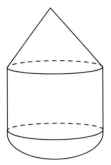

Calculate
a the total volume of the object
b the total surface area of the object.

19 A salt pot is in the form of a hollow cylinder of diameter 3 cm and height 3 cm with a hemispherical shell fixed on top.
Salt is poured into the pot to a depth of 2.5 cm.
The pot is inverted, with the hole covered, so the flat base of the cylinder is horizontal.

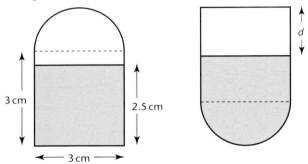

Find the distance, d, from the top of the salt to the flat base of the cylinder.

20 The volume of this frustum is 1 422 cm³.
a Find the volume of the original cone to 1 decimal place.
b Find the curved surface area of the frustum.

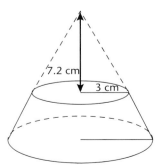

This chapter is about

- using a ruler and compasses to do constructions
- drawing loci.

1 a Make an accurate full-size drawing of this triangle.

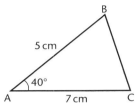

   b Measure the length of BC on your diagram.

2 a Make an accurate full-size drawing of this triangle.

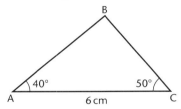

   b Measure the length of AB on your drawing.

3 a Make an accurate full-size drawing of this triangle.

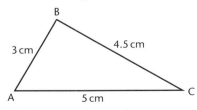

   b Measure the angles on your drawing.

4 a Construct the triangle ABC where AB = 7.4 cm, AC = 6.5 cm and angle BAC = 52°.

   b Measure BC.

5 a Construct the triangle DEF where DF = 7.7 cm, angle EDF = 48° and angle EFD = 55°.

   b Measure DE.

6 a Construct the triangle RST where RS = 5.8 cm, RT = 7.2 cm and ST = 6.3 cm.

   b Measure the angle SRT.

7 a Make a scale drawing of the triangle ABC using a scale of 1 cm to 5 m with AB = 25 m, angle CAB = 46°, angle CBA = 71°.

   b i Measure the length of BC on your drawing.

   ii Write down the actual length of BC.

8 a Draw this triangle accurately.

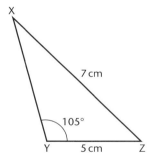

   b Measure angle YXZ.

9 AC is the shorter diagonal of a kite. AC is 4.6 cm. The sides AB and BC are each 5.8 cm and the sides AD and CD are each 8.3 cm.

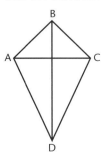

   a Draw the kite accurately.

   b Measure the length of diagonal BD.

   c Calculate the area of the kite.

10 Draw a horizontal line 7.2 cm long and label it AB. Construct the perpendicular bisector of the line AB.

11 Draw a horizontal line 10 cm long and label it XY. Mark the point P that is 4 cm to the right of X and 2 cm above XY. Construct a line perpendicular to XY that passes through P.

12 Draw an angle of 110°. Construct the bisector of the angle.

13 Construct a triangle with sides of length 9 cm, 8 cm and 6 cm. Construct the bisectors of each of the three angles. What do you notice?

14 Mark a point A. Draw the locus of points which are 3.6 cm from the point A.

15 Draw a line 8 cm long. Draw the locus of points which are 2 cm from the line.

16 Construct a triangle ABC with AB = 10 cm, BC = 8 cm and AC = 5 cm.
Show, by shading, the locus of points which are inside the triangle and closer to A than B.

17 The diagram shows a swimming pool 50 m long by 20 m wide.

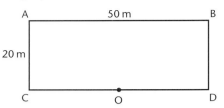

The shallow end extends to a distance 20 m from AC. Amanda's mother is at O, halfway along CD.
a Using a scale of 1 cm to 5 m, make a scale drawing of the pool.
b Amanda's mother says that Amanda can swim anywhere in the shallow end or within 15 m of her. Show, by shading, all the places that Amanda can swim.

18 Construct a triangle ABC with AB = 11 cm, AC = 7 cm and BC = 9 cm. Show, by construction, the locus of points inside the triangle which are equidistant from A and B and also nearer to AB than AC.

19 Construct a rectangle ABCD with AB = 10 cm and AD = 6 cm. Show, by construction, the locus of points inside the rectangle for which all these statements are true.
a The points are nearer to D than A.
b The points are nearer to DC than DA.
c The points are less than 7 cm from A.

CHAPTER 40

Compound measures

This chapter is about

- using compound units
- knowing and using the formulae to find speed, distance and time
- knowing and using the formulae to find density, mass and volume
- knowing and using the formulae to find pressure, force and area.

Exercise A

1 Find the average speed in mph for each journey.

   a 70 miles in 2 hours   **b** 31 miles in 30 minutes

   c 12.5 miles in 15 minutes   **d** 96 miles in $1\frac{1}{2}$ hours

   e 18 miles in 20 minutes   **f** 115 miles in 1 hour 40 minutes
   g 23 miles in 12 minutes   **h** 1.5 miles in 120 seconds

   i $\frac{1}{4}$ mile in 20 seconds   **j** 0.2 miles in $\frac{1}{2}$ second

2 Find the average speed in km/h for each journey.

   a 144 km in 3 hours   **b** 11 km in 10 minutes

   c 217 km in $3\frac{1}{2}$ hours   **d** 24 km in 40 minutes

   e 64 km in $\frac{3}{4}$ hour   **f** 9.4 km in 5 minutes

   g 190 km in 3 hours and 10 minutes   **h** 38 km in 24 minutes

   i 2 metres in 3 seconds   **j** 8 mm in $\frac{1}{3}$ second

3 Faith travels 28 miles in 35 minutes. Calculate her average speed in mph.

4 Andrew is on a high speed train. It travels 280 miles in $2\frac{1}{2}$ hours.
   Calculate the average speed of the train.

5 Find the time taken to travel 100 miles at an average speed of 60 mph.

6 Find the distance travelled on a journey that took $1\frac{1}{2}$ hours at an average speed of 70 km/h.

7 Barney travels 15 km in 18 minutes. Calculate his average speed in km/h.

8 Copy and complete the following table that shows the average speed, time taken and length of each journey.

	Average speed	Time taken	Length of journey
a	45 mph	30 minutes	
b	74 mph	$1\frac{1}{4}$ hours	
c	50 km/h		180 km
d	60 km/h		54 km
e		45 minutes	75 km
f		8 minutes	12 miles
g	36.4 km/h	6 minutes	
h	48 km/h		1 200 m

9 A train leaves the station at 09.10 and arrives at its destination at 10.40. The distance travelled is 84 miles. Find the train's average speed.

10 A car leaves Omagh at 4.25p.m. and travels 14 miles to Ballygawley. It arrives at Ballygawley at 4.45p.m. Find the average speed of the car for this journey.

11 A bus left Limavady and travelled 12 miles to Coleraine. It left Limavady at 07.45 and travelled at an average speed of 48 mph. What time did it arrive in Coleraine?

12 a Find the average speed of a car which travels 75 km in 1 hour 15 minutes.
   b A car travels 15 km in 14 minutes. Find its average speed in kilometres per hour. Give your answer correct to 1 decimal place.

13 A train is travelling at 22 m/s. How far does it travel in $1\frac{3}{4}$ hours? Give your answer in km.

14 A runner's average speed is 3.2 metres per second. How long does it take her to run 1.2 km? Give your answer in minutes.

15 Convert each of these speeds to m/s.
   a 36 km/h   **b** 66 km/h   **c** 90 cm/min **d** 8.28 km/h   **e** 0.9 km/h

16 Convert each of these speeds to km/h.
   a 20 m/s   **b** 48 m/s   **c** 0.2 m/s   **d** 12 cm/s   **e** $\frac{1}{3}$ m/s

Exercise B

1 Find the density in g/cm^3 of a substance which has mass 300 g and volume 150 cm^3.

2 Find the density of a substance with mass 1.2 kg and volume 400 cm^3
   a in kg/cm^3 **b** in g/cm^3.

3 A block of density 3.125 g/cm^3 has volume 400 cm^3. Find its mass in grams.

4 A block of density 5.25 g/cm^3 has volume 720 cm^3. Find its mass
   a in grams   **b** in kilograms.

5 A mass of 280 g has density 3.5 g/cm^3. Find its volume.

6 A mass of 3 kg has density 7.5 g/cm^3. Find its volume.

7 Find the density of each object in g/cm^3.

a

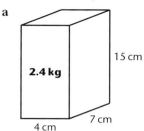

b

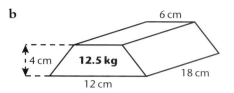

c

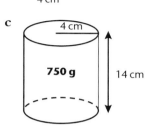

d

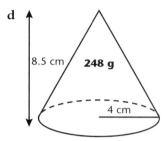

8 A cuboid measures 8 cm by 6 cm by 2 cm. Its mass is 75 g. Find its density in g/cm^3.

9 A cube has sides of length 5 cm. Its mass is 7 500 g. Find its density in kg/cm^3.

10 Find the mass of each object shown.

a

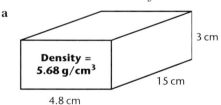

b

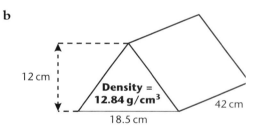

11 A cuboid's length is 10 cm and its breadth is 6 cm. The density of the cuboid is 2.5 g/cm^3. Given that the cuboid's mass is 900 g, find its height.

12 A solid cone has radius 6.8 cm and slant height 17.4 cm. Given that it has a mass of 1.2 kg, decide if it will float in water. The density of water is 1 g/cm^3.

13 A cylinder's height is 8 cm. The density of the cylinder is 1.6 g/cm^3. Given that the cylinder's mass is 540 g, find its diameter to 2 decimal places.

Exercise C

1 What is the pressure when a force of 80 N acts on an area of 25 m^2?

2 What is the pressure when a force of 100 N acts on a rectangle measuring 8 m by 5 m?

3 What is the pressure when a force of 45 N acts on a square of side 3 m?

4 What is the pressure when a force of 200 N acts on a circular area of diameter 10 m?

5 What is the pressure when a force of 330 N acts on a circular area of radius 60 cm?

6 Copy and complete the following table.

	Force	Area	Pressure
a	60 N	20 m^2	___ N/m^2
b	___ N	8 m^2	1.5 N/m^2
c	4 N	___ m^2	0.8 N/m^2
d	90 N	___ m^2	6 N/m^2
e	___ N	28 m^2	5 N/m^2
f	50 N	___ m^2	0.625 N/m^2
g	12 N	5 000 cm^2	___ N/m^2
h	___ N	25 000 cm^2	8 N/m^2

7 A force exerts a pressure of 8 N/m^2 on an area of 4.5 m^2. Calculate the size of the force.

8 A force exerts a pressure of 30 N/m^2 on a triangle with base 80 cm and perpendicular height 2 m. Work out the size of the force.

9 A force of 18 N exerts a pressure of 1.5 N/m^2 on a square. Work out the area of the square.

10 A force of 600 N exerts a pressure of 15 N/m^2 on a circle. Calculate the circumference of the circle.

11 A force of X Newtons is exerted on a rectangle. The rectangle is then enlarged by a scale factor of 2. If the force is unchanged Dermot thinks that the pressure will be halved. Is he correct? Explain your answer.

12 Jack exerts a force of 300 N on an area A. Sam exerts a force of 240 N on an area B. Area A is five times bigger that area B. How many times bigger is the pressure exerted by Sam than the pressure exerted by Jack?

CHAPTER 41

Transformations

This chapter is about

- knowing the properties of a reflection, rotation, translation and an enlargement
- drawing the image of a shape after each of these transformations
- finding the original shape given the image and the transformation
- describing fully the transformation given the original shape and its image
- finding the image after two transformations.

Exercise A

1 Copy the diagram.

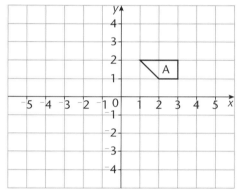

   a Reflect trapezium A in the x-axis. Label the image B.
   b Reflect trapezium A in the line $y = 1$. Label the image C.
   c Reflect trapezium A in the y-axis. Label the image D.
   d Reflect trapezium A in the line $x = 3$. Label the image E.

2 Draw a pair of axes and label them -3 to 3 for x and y.

   a Draw a triangle with vertices at $(-1, 1)$, $(-1, 3)$ and $(-2, 3)$. Label it A.

   b Reflect triangle A in the line $x = \dfrac{1}{2}$. Label the image B.

   c Reflect triangle A in the line $y = x$. Label the image C.
   d Reflect triangle A in the line $y = -x$. Label the image D.

3 Look at this diagram.

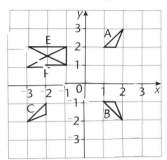

Describe fully the single transformation that maps
a   triangle A on to triangle B
b   triangle A on to triangle C
c   triangle E on to triangle F.

4 Look at this diagram.

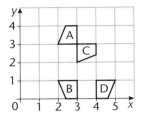

Describe fully 3 different reflections in the diagram.

5 Copy the diagram.

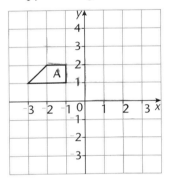

a   Rotate trapezium A through 180° about the origin. Label the image B.
b   Rotate trapezium A through 90° clockwise about the point (0, 1). Label the image C.
c   Rotate trapezium A through 90° anticlockwise about the point (−1, 1). Label the image D.

6 Copy the diagram.

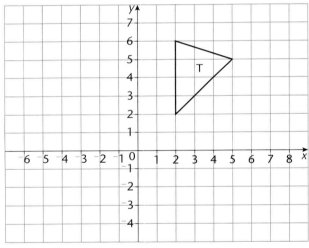

a   Rotate triangle T through 90° anticlockwise about the point (5, 5). Label the image A.
b   Describe fully the transformation that maps triangle A on to triangle T.
c   On the same diagram, rotate triangle T through 180° about the point (0, 3). Label the image B.
d   Describe fully the transformation that maps triangle B on to triangle T.

7 Look at this diagram.

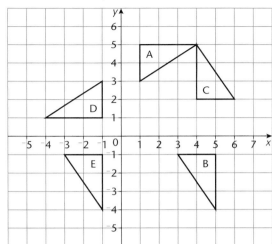

Describe fully the single transformation that maps

a A on to B   **b** A on to C
c A on to D   **d** D on to E.

8 Look at the diagram.

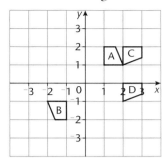

Describe fully the single transformation that maps
a trapezium A on to trapezium B
b trapezium A on to trapezium C
c trapezium A on to trapezium D.

9 Copy the diagram.

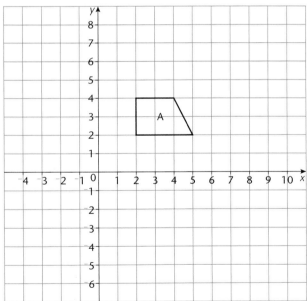

a Translate shape A by $\begin{pmatrix} 0 \\ -6 \end{pmatrix}$. Label the image B.

b Translate shape A by $\begin{pmatrix} 4 \\ 3 \end{pmatrix}$. Label the image C.

c Translate shape A by $\begin{pmatrix} -6 \\ 4 \end{pmatrix}$. Label the image D.

d Translate shape A by $\begin{pmatrix} -4 \\ -7 \end{pmatrix}$. Label the image E.

10 Draw a pair of axes and label them −3 to 5 for x and y. Draw a triangle with vertices at (2, 1), (2, 3) and (3, 1). Label it A.

a Translate triangle A by $\begin{pmatrix} 2 \\ 1 \end{pmatrix}$. Label the image B.

b Translate triangle A by $\begin{pmatrix} -5 \\ -3 \end{pmatrix}$. Label the image C.

c Translate triangle A by $\begin{pmatrix} 2 \\ -4 \end{pmatrix}$. Label the image D.

11 Look at this diagram.

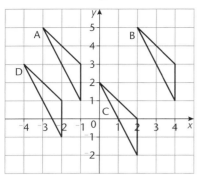

Describe fully the single transformation that maps
  a triangle A on to triangle B
  b triangle A on to triangle C
  c triangle A on to triangle D
  d triangle B on to triangle D.

Exercise B

1 Draw a pair of axes and label them 0 to 7 for both x and y.
   a Draw a trapezium with vertices at (1, 2), (1, 3), (2, 3) and (3, 2). Label it A.
   b Enlarge trapezium A by scale factor 3, with centre of enlargement (1, 2). Label the image B.
   c Describe fully the single transformation that maps trapezium B on to trapezium A.

2 Draw a pair of axes and label them 0 to 6 for both x and y.
   a Draw a triangle with vertices at (0, 6), (3, 6) and (3, 3). Label it A.
   b Enlarge triangle A by scale factor $\frac{1}{3}$, with the origin as the centre of enlargement. Label it B.
   c Describe fully the single transformation that maps triangle B on to triangle A.

3 Draw a pair of axes and label them 0 to 8 for both x and y.
   a Draw a triangle with vertices at (2, 1), (2, 3) and (3, 2). Label it A.
   b Enlarge triangle A by scale factor $2\frac{1}{2}$, with the origin as the centre of enlargement. Label it B.
   c Describe fully the single transformation that maps triangle B on to triangle A.

4 Look at this diagram.

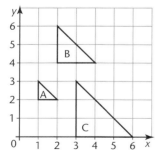

Describe fully the single transformation that maps
  a triangle A on to triangle B
  b triangle B on to triangle A
  c triangle A on to triangle C
  d triangle C on to triangle A.

5 Draw a set of axes. Label the x-axis and the y-axis from −7 to 6.
   Plot the points (1, 2), (1, 3) and (3, 3) and join them to form a triangle.
   Enlarge the triangle with scale factor of −2 and centre of enlargement (0, 0).

6 Draw a set of axes with both the x-axis and y-axis from −10 to 6.
   Plot the points A(2, 2), B(4, 2) and C(2, 4) and join them to form a triangle.
   Enlarge triangle ABC by a scale factor of −3 using the point (1, 1) as the centre of enlargement.

7 Look at this diagram.

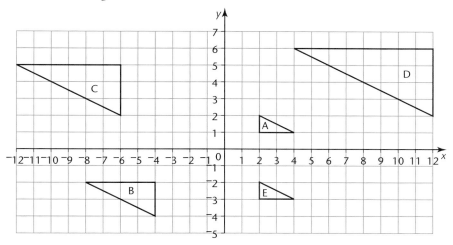

Describe fully the single transformation that maps
a triangle A on to triangle B
b triangle A on to triangle C
c triangle A on to triangle D
d triangle C on to triangle E
e triangle B on to triangle E
f triangle C on to triangle A.

Exercise C

1 Look at the diagram.

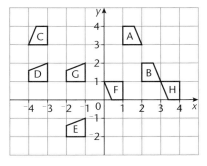

Describe fully the single transformation that maps
a shape A on to shape B
b shape A on to shape C
c shape A on to shape D
d shape D on to shape E
e shape A on to shape F
f shape E on to shape G
g shape B on to shape H
h shape H on to shape F.

2 Look at the diagram.

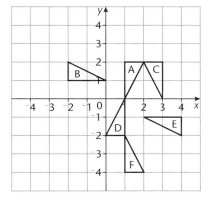

Describe fully the single transformation that maps
a triangle A on to triangle B
b triangle A on to triangle C
c triangle A on to triangle D
d triangle A on to triangle E
e triangle A on to triangle F.

3 Draw a set of axes. Label the x-axis from 0 to 10 and the y-axis from -3 to 5.
Plot the points (1, 2), (1, 4) and (4, 4) and join them to form a triangle. Label it A.
  a Reflect triangle A in the line $x = 5$. Label the image B.
  b On the same grid, reflect triangle B in the line $y = 1$. Label the image C.
  c Describe fully the single transformation that is equivalent to a reflection in the line $x = 5$ followed by a reflection in the line $y = 1$.

4 Draw a set of axes. Label the x-axis from 0 to 9 and the y-axis from -4 to 4.
Plot the points (1, 3), (3, 3) and (3, 2) and join them to form a triangle. Label it D.
  a Rotate triangle D through 90° clockwise about the point (0, 0). Label the image E.
  b On the same grid, rotate triangle E through 180° about the point (5, -2). Label the image F.
  c Describe fully the single transformation that is equivalent to a rotation through 90° clockwise about the point (0, 0) followed by a rotation through 180° about the point (5, -2).

5 Draw a set of axes. Label the x-axis from -5 to 5 and the y-axis from 0 to 6.
Plot the points (-3, 1), (0, 1) and (-2, 3) and join them to form a triangle. Label it A.
  a Translate triangle A by the vector $\begin{pmatrix} 4 \\ 0 \end{pmatrix}$. Label the image B.
  b Translate triangle B by the vector $\begin{pmatrix} -3 \\ 2 \end{pmatrix}$. Label the image C.
  c Describe fully the single transformation that is equivalent to a translation by the vector $\begin{pmatrix} 4 \\ 0 \end{pmatrix}$ followed by a translation by the vector $\begin{pmatrix} -3 \\ 2 \end{pmatrix}$.

6 Draw a set of axes. Label the x-axis and the y-axis from -6 to 6.
Plot the points (1, 1), (2, 1) and (1, 3) and join them to form a triangle. Label it D.
  a Enlarge triangle D with scale factor 2 and centre of enlargement (0, 0). Label the image E.
  b Rotate triangle E through 180° about centre of rotation (0, 0). Label the image F.
  c Describe fully the single transformation that is equivalent to an enlargement with scale factor 2 and centre (0, 0) followed by a rotation through 180° and centre (0, 0).

7 Draw a set of axes. Draw a triangle with vertices at (1, 1) (1, 2) and (4, 2). Label it A.
Draw the image of A after a reflection in the line $y = -x$ followed by a rotation through 90° anticlockwise about (0, -1). Label the image B.

8 Shape P is reflected in the line $x = 4$ and then translated $\begin{pmatrix} 5 \\ -2 \end{pmatrix}$ to give shape Q. Describe the two transformations that will map Q back to P.

9 What is the inverse of these transformations?
  a Rotation 90° clockwise about (1, -5) followed by translation $\begin{pmatrix} -4 \\ 7 \end{pmatrix}$.
  b Enlargement scale factor 3.5 about centre (-2, 3).
  c Rotation 90° anticlockwise about (-3, 0) followed by reflection in $y = x$.
  d Translation $\begin{pmatrix} 6 \\ 2 \end{pmatrix}$ followed by translation $\begin{pmatrix} -2 \\ 3 \end{pmatrix}$.
  e Reflection in $y = 3$ followed by translation $\begin{pmatrix} -7 \\ 4 \end{pmatrix}$.

CHAPTER 42   Questionnaires

This chapter is about

- understanding and knowing how to use the handling data cycle
- designing a recording sheet
- designing a questionnaire
- saying why questions are unsuitable
- suggesting alternative questions
- designing a questionnaire to test a hypothesis.

1   Design a recording sheet to investigate the main food eaten for breakfast by Year 11 students.

2   Marge is designing a questionnaire to find out opinions on a new supermarket being built on the edge of town. Here is one of her questions.

Do you not agree that a new supermarket on the edge of town will be harmful to our town centre?

           Yes ☐   No ☐

   a   Name one thing wrong with the question.
   b   Name one thing wrong with the response section.

3   Ned is designing a questionnaire to find out the eating habits of people in his community. One of his questions is:

How often do you eat fruit and vegetables?

     1–4 times ☐   4–8 times ☐   8–12 times ☐   More than 12 times ☐

   a   Name one thing wrong with his question.
   b   Name two things wrong with the response section.

4   Lucy wants to survey the teachers in her school for her Health and Social Care project. One of her questions is:

How old are you?

   a   Name one thing wrong with this question.
   b   Write a more suitable question and response section in order to gather the information that Lucy requires.

5   Rhiannon wants to know what activities people will want to do while on holiday at her family's caravan park. She asks this question:

What activities do you like to do?

     Tennis ☐   Swimming ☐   Crazy golf ☐   Football ☐

   a   Name one thing wrong with her question.
   b   Name two things wrong with her response section.
   c   Rewrite the question and add a more suitable response section.

6   You wish to investigate how regularly students at your school visit the cinema.
    You decide to write a questionnaire to find out.
    Write down two useful questions which would help you gather the necessary information.

7   A local radio station is about to start broadcasting.
    They decide to send out a questionnaire to find out the sort of programmes that are preferred.
    Comment on each of these questions and where necessary write a more suitable one with a response section.
    a   What type of programme do you like most?
    b   Do you listen to the radio in the afternoon?
    c   Do you like competitions and phone-ins?
    d   How much do you earn?
    Are there any other questions that you think will be useful for the radio station?

8   Design a questionnaire to test each hypothesis. Include two suitable questions and their response sections.
    a   Children who live in the countryside go to the play park less than children who live in the town.
    b   Students who text more on their mobile phones do not perform as well in exams compared to those who text less.
    c   People who eat carrots have better eyesight than people who do not eat carrots.
    d   People who are musical are also good at Maths.

CHAPTER 43 Statistical diagrams

This chapter is about

- understanding what is meant by frequency
- distinguishing between continuous and discrete data
- constructing a frequency distribution table
- drawing and interpreting frequency diagrams
- drawing and interpreting frequency polygons
- drawing and interpreting stem and leaf diagrams
- drawing and using scatter graphs
- knowing and recognising the three types of correlation
- using flow diagrams.

Exercise A

1 These are the marks of 30 students in a test.

32	42	23	37	28
12	37	6	5	37
17	18	31	29	27
11	21	37	28	37
22	31	23	47	23
12	24	34	41	43

   a Draw a grouped frequency table for the data. Use groups 0–9, 10–19, etc.

   b Draw a frequency diagram.

2 As part of a survey, Eve measured the heights, in centimetres, of the 50 teachers in her school. Here are her results.

168	194	156	167	177
180	188	172	170	169
174	178	186	174	166
165	159	173	185	162
163	174	180	184	173
182	161	176	170	169
178	175	172	179	173
162	177	176	184	191
181	165	163	185	178
175	182	164	179	168

   a Draw a grouped frequency table for the data. Use groups $155 \leqslant h < 160$, $160 \leqslant h < 165$, etc.

   b Draw a frequency diagram.

3 This frequency diagram shows the times taken by a group of girls to run a race.

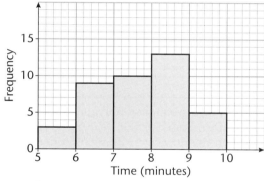

   a What is the modal class interval?

   b How many girls took part in the race?

   c What percentage of the girls took less than 7 minutes?

   d Work out an estimate for the mean time.

4 This table shows the heights of 40 plants.

Height (h cm)	Frequency
$3 \leqslant h < 4$	1
$4 \leqslant h < 5$	7
$5 \leqslant h < 6$	10
$6 \leqslant h < 7$	12
$7 \leqslant h < 8$	8
$8 \leqslant h < 9$	2

Draw a frequency polygon to show this data.

5 This table shows the times taken for a group of children to get from home to school.

Time (minutes)	Frequency
$0 < t \leqslant 5$	3
$5 < t \leqslant 10$	15
$10 < t \leqslant 15$	27
$15 < t \leqslant 20$	34
$20 < t \leqslant 25$	19
$25 < t \leqslant 30$	2

Draw a frequency polygon to show this data.

6 The table below shows the frequency distribution for the vital lung capacities of students in Year 12.

Vital lung capacity (litres)	Frequency girls	Frequency boys
$3.4 \leqslant x < 3.6$	3	2
$3.6 \leqslant x < 3.8$	7	5
$3.8 \leqslant x < 4.0$	26	24
$4.0 \leqslant x < 4.2$	17	25
$4.2 \leqslant x < 4.4$	4	7
$4.4 \leqslant x < 4.6$	1	3

a Draw a frequency polygon to show the information for the Year 12 boys.
b On the same axes draw a frequency polygon to show the information for the Year 12 girls.
c How many students are in Year 12?
d Using the frequency polygons, write down one comparison between the boys' and girls' vital lung capacities.

7 These are the test results for 30 students who sat a Geography test.

29 39 46 17 21 30 43 47 39 31
18 32 45 50 41 32 37 39 22 14
28 39 40 47 38 31 25 29 35 42

a Draw a stem and leaf diagram to show these results.
b Use your diagram to find
  i the median
  ii the mode
  iii the range.

8 The ages of the first 25 men and the first 25 women to enter a supermarket on a Wednesday were recorded.

Men

52 62 73 37 69 62 77 76 65 47
47 48 61 39 27 51 21 47 58 57
61 74 82 81 64

Women

62 61 63 37 23 52 44 64 51 53
52 42 33 77 49 41 84 82 71 64
40 67 61 52 54

a Draw a back-to-back stem and leaf diagram to show the ages of men and women.
b What is the difference between the median age for the men and the median age for the women?

9 Use the stem and leaf diagram below to answer the questions that follow.

3	4 5 8 9
4	0 2 3 6 8 8
5	1 3 3 5 6 7 9
6	4 6 8
7	0 2

Key 3 | 8 = 3.8 cm

a Find the range.   **b** Find the median.
c Find the mean.   **d** How many values are less than 2 inches?

10 The stem and leaf diagram shows the results for the boys and girls from 11T in a recent Maths test.

Boys		Girls	
5	**0**	1 6	
9 9 3 3	**1**	2 2 7 8	
8 7 6 6 4	**2**	0 5 6 6 7	
9 9 9 6 5	**3**	1 2 4 5	
2 1	**4**		

Key 3 | 1 = 31%

From the stem and leaf diagram find
a the median score for the boys
b the range for the girls
c the number of students in class 11T
d the range for the class
e the mode for the boys
f the mode for the class.

Exercise B

1 For each of the graphs below, describe the correlation.

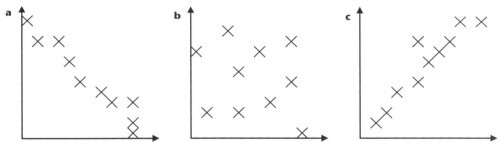

2 This table gives the average maximum temperature and the average rainfall in June for 10 different cities.

Temperature (°C)	18	19	20	21	22	23	24	27	28	29
Rainfall (mm)	47	64	60	73	84	37	49	29	63	47

a Draw a scatter diagram to show this information.
b Comment on the relationship, if any, between the temperature and the rainfall.
c Another city has an average maximum temperature of 23 °C in June.
What can you deduce from the scatter diagram about the average rainfall in this city?

3 Bill grows tomatoes. As an experiment he divided his land into eight plots.
He used a different amount of fertiliser on each plot.
The table shows the weight of tomatoes he got from each of the plots.

Amount of fertiliser (g/m²)	10	20	30	40	50	60	70	80
Weight of tomatoes (kg)	36	41	58	60	70	76	75	92

a Draw a scatter diagram to show this information.
b Describe the correlation shown in the scatter diagram.
c Draw a line of best fit on your scatter diagram.
d What weight of tomatoes should Bill expect to get if he uses 75 g/m² of fertiliser?

4 The table shows the prices and mileages of seven second-hand cars of the same model.

Price (£)	6 000	3 500	1 000	8 500	5 500	3 500	6 000
Mileage	27 000	69 000	92 000	17 000	53 000	82 000	43 000

a Draw a scatter diagram to show this information.
b Describe the correlation shown in the scatter diagram.
c Draw a line of best fit on your scatter diagram.
d Use your line of best fit to estimate
   i the price of a car of this model which has covered 18 000 miles
   ii the mileage of a car of this model which costs £4 000
   iii the price of a car of this model which has driven 2 000 miles.
e Did you use interpolation or extrapolation to answer each of the parts in **d**? Comment on their reliability.

Exercise C

Look at the flow diagrams and follow the instructions until you get the answer.

1 Start with $x = 4$ and $y = 8$.

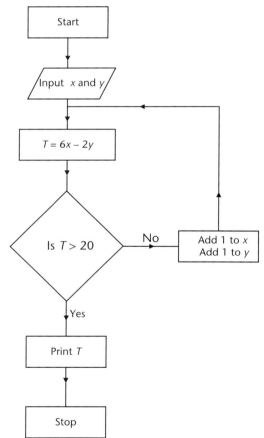

2 Start with $A = 3$ and $B = 2$.

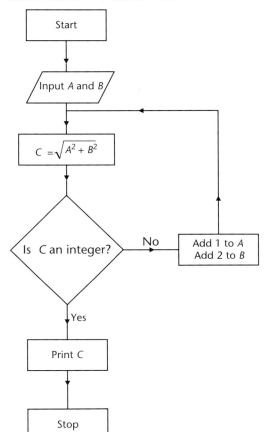

3 Start with $m = 0$ and $n = 3$.

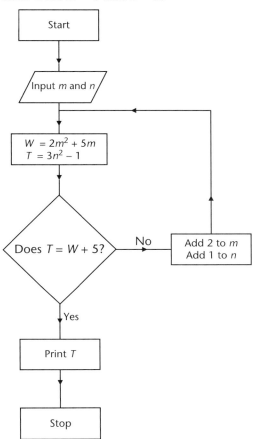

CHAPTER 44 Statistical averages and spread

This chapter is about

- defining and distinguishing between mode, median and mean
- understanding what is meant by the range
- calculating the range, mode, median and mean of a frequency distribution
- finding the limits of the median and modal class of a grouped frequency distribution
- calculating an estimate for the mean of a grouped frequency distribution
- using means in problems
- comparing distributions
- choosing the most appropriate statistical average.

Exercise A

In questions **1** to **4**
   a find the mode   **b** find the range   **c** calculate the mean   **d** find the median.

1

Number	Frequency
1	3
2	2
3	4
4	5
5	7
6	2
7	0
8	1
9	0
10	1

2

Number of drawing pins in a box	Number of boxes
98	5
99	14
100	36
101	28
102	17
103	13
104	7

3

Number of snacks per day	Frequency
0	23
1	68
2	39
3	21
4	10
5	3
6	1

4

Number of letters received on Monday	Frequency
0	19
1	37
2	18
3	24
4	12
5	5
6	2
7	3

5 The table shows the numbers of trains arriving late at a station in the month of May.

Number of trains arriving late each day	Frequency
0–4	18
5–9	9
10–14	3
15–19	0
20–24	1

  a Write down the modal class.
  b In which class does the median lie?
  c Calculate an estimate of the mean.

6 The table shows the numbers of days the students in Year 11 were absent last year.

Number of days' absence	Frequency
0–3	13
4–7	18
8–11	9
12–15	4
16–19	0
20–23	1
24–27	3

  a Write down the modal class.
  b In which class does the median lie?
  c Calculate an estimate of the mean.

7 The table shows the heights of sunflowers in centimetres.

Height of sunflower in centimetres (x)	Frequency
$100 \leqslant x < 110$	6
$110 \leqslant x < 120$	13
$120 \leqslant x < 130$	35
$130 \leqslant x < 140$	29
$140 \leqslant x < 150$	16
$150 \leqslant x < 160$	11

  a Write down the modal class.
  b In which class does the median lie?
  c Calculate an estimate of the mean.

8 The table shows the times taken to complete a race in minutes.

Time to complete race in minutes (x)	Frequency
$54 \leqslant x < 56$	1
$56 \leqslant x < 58$	4
$58 \leqslant x < 60$	11
$60 \leqslant x < 62$	6
$62 \leqslant x < 64$	2
$64 \leqslant x < 66$	1

  a Write down the modal group.
  b In which group does the median lie?
  c Calculate an estimate of the mean.

9 The table shows the heights of shrubs in metres.

Height of shrub in metres (x)	Frequency
$0.3 \leqslant x < 0.6$	57
$0.6 \leqslant x < 0.9$	41
$0.9 \leqslant x < 1.2$	36
$1.2 \leqslant x < 1.5$	24
$1.5 \leqslant x < 1.8$	15

  a Write down the modal group.
  b In which group does the median lie?
  c Calculate an estimate of the mean.
  d Given that measurements are taken to the nearest 10 cm, what is the height of the shortest possible shrub?

Exercise B

1 The frequency table shows the number of pets owned by students in Year 9.

Number of pets	Frequency
1	9
2	5
3	2
4	x
5	1

Given that the mean number of pets was 2.1, calculate the number of students who own four pets.

2 David rolled a dice a number of times and calculated the mean to be 3.2.

Number on dice	1	2	3	4	5	6
Frequency	10	5	10	4	4	

The number of times he rolled a six is missing from the table. Work out how many times he rolled a six.

3 The frequency table shows the number of siblings that students in Year 11 have.

Number of siblings	Frequency
0	5
1	10
2	x
3	4
4	2

Given that the mean number of siblings is 1.75, calculate the number of students who have two siblings.

4 An estimate for the mean height is 161 cm. Find n.

Height of student (cm)	Frequency
$140 < h \leqslant 150$	8
$150 < h \leqslant 160$	17
$160 < h \leqslant 170$	15
$170 < h \leqslant 180$	n
$180 < h \leqslant 190$	3

5 The mean height of 5 students is 149 cm. A sixth student is measured and has a height of 161 cm. Calculate the mean height of all six students.

6 Mr Black recorded the mean test score of the 25 students in his class as 72%. He then realised that the lowest score was 50%, not 60%. What is the actual mean test score for his class?

7 In a class of 25 students the mean for a Maths test is 64%. In a second class of 20 students the mean for the same test is 55%. Find the mean for both classes.

8 The mean contents of a sample of 50 packets of crisps is 35.4 grams. The mean contents of a sample of 30 different packets of crisps is 34.6 grams. Find the mean of all the packets sampled.

9 The mean pocket money for the 14 boys in a class is £1.74. The mean pocket money for the 10 girls in the same class is £1.88. Find the mean pocket money for the whole class.

10 Matt has x marbles. Daniel has 3 more than Matt and Beth has 6 less than Matt. The mean number of marbles is 6. Form an equation and solve it to find the number of marbles that Matt has.

11 Tim and Zoe grow potatoes. They each dig up 10 potatoes and weigh them.
Here are their results, in grams.

Tim	Zoe
190	230
170	210
180	270
120	90
160	210
100	180
320	260
210	270
220	110
270	250

Use suitable calculations to compare their results. Give reasons for your decisions.

12 Two friends, Aoife and Clare, in Year 11 record the number of homeworks they receive **each week** from their subject teachers for a whole year.

Homeworks set (Aoife)	Frequency	Homeworks set (Clare)	Frequency
0–1	2	0–1	0
2–3	3	2–3	0
4–5	5	4–5	3
6–7	8	6–7	8
8–9	9	8–9	13
10–11	9	10–11	11
12–13	2	12–13	4
14–15	1	14–15	0

Use suitable calculations to compare their results.

13 The lengths of 200 leaves from a variety of trees were measured at each of two sites, one exposed and one sheltered.
The results are shown in the table.

Length of leaf (l cm)	Frequency	
	Exposed site	Sheltered site
$7 < l \leqslant 8$	8	0
$8 < l \leqslant 9$	25	2
$9 < l \leqslant 10$	62	14
$10 < l \leqslant 11$	53	31
$11 < l \leqslant 12$	32	88
$12 < l \leqslant 13$	17	42
$13 < l \leqslant 14$	2	18
$14 < l \leqslant 15$	1	5

Make three comparisons between the leaves taken from the two sites.

14 The following amounts are the salaries of a sample of 10 employees at a large factory.

£17 300   £15 690   £21 650   £12 480   £47 340   £17 000   £72 600   £14 550   £18 890   £28 750

  a Calculate the mean salary for this sample.
  b Calculate the median salary for the sample.
  c Which average is more appropriate?

15 State the most appropriate statistical average to use in each of these. Give a reason for your answer.
  a A shop owner ordering in men's trousers for the autumn and winter season.
  b A teacher wants to find the average height of students in Year 10.
  c A scientist wants to find the number of bacteria in a sample of 12 colonies.
  d A writer wants to know the average number of words per page.
  e A farmer wants to know the average amount of fuel used each month.
  f An electronics shop wants to order in different makes of mobile phone.
  g A shopkeeper wants to work out the average takings for a month.
  h A café owner wants to order in fillings for sandwiches.

CHAPTER 45

Cumulative frequency curves and box plots

This chapter is about

- defining and using limits and boundaries
- knowing what is meant by quartiles
- finding quartiles and interquartile range
- drawing and interpreting cumulative frequency curves
- drawing and interpreting box plots.

Exercise A

1 The cumulative frequency diagram represents the heights of trees in a park.

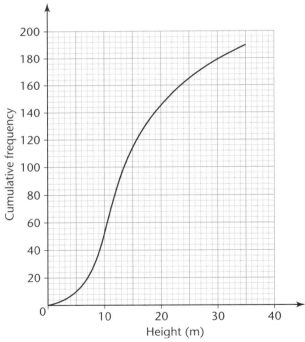

a How many trees are there in the park?
b How many trees have a height of less than 12 m?
c What percentage of the trees have a height of more than 26 m?

2 The frequency table shows information about the masses of 100 letters.

Mass(m grams)	Frequency
$30 < m \leqslant 40$	7
$40 < m \leqslant 50$	12
$50 < m \leqslant 60$	24
$60 < m \leqslant 70$	32
$70 < m \leqslant 80$	18
$80 < m \leqslant 90$	5
$90 < m \leqslant 100$	2

a Copy and complete the cumulative frequency table.

Mass(m grams)	Cumulative frequency
$m \leqslant 40$	7
$m \leqslant 50$	19
$m \leqslant 60$	
$m \leqslant 70$	
$m \leqslant 80$	
$m \leqslant 90$	
$m \leqslant 100$	

b Draw a cumulative frequency diagram.
c Use your diagram to find the median and the interquartile range of these masses.

3 The heights of 150 plants were measured. The results are shown in the table.

Height (h cm)	Frequency
$0 < h \leqslant 10$	15
$10 < h \leqslant 20$	23
$20 < h \leqslant 30$	36
$30 < h \leqslant 40$	42
$40 < h \leqslant 50$	24
$50 < h \leqslant 60$	10

a Draw a cumulative frequency table and diagram for these heights.
b Find the median and the interquartile range of the heights.
c Use your cumulative frequency diagram to estimate the number of plants over 37 cm in height.

Exercise B

1 The graph below shows the cumulative frequency of marks obtained in a theory test.

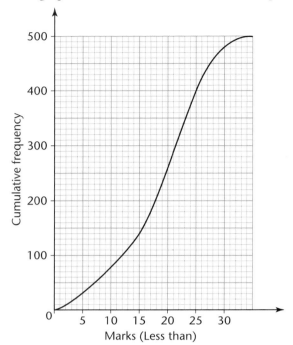

Use the graph to estimate
a the upper quartile
b the median
c the interquartile range
d how many obtained more than 10 marks
e how many obtained between 20 and 30 marks.

2 The graph below shows the cumulative frequency of times taken by students to complete an assignment.

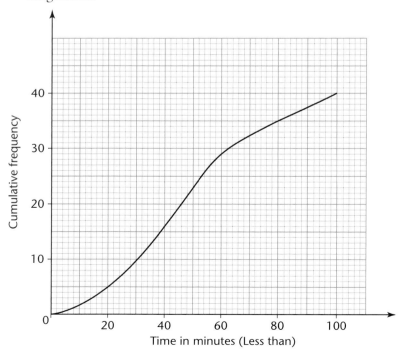

Use the graph to estimate
a the median
b the interquartile range
c how many took less than 20 minutes
d how many took between 30 and 60 minutes
e how many took between 80 and 100 minutes.

Exercise C

1   Draw a box plot to show this data.
Minimum = 2
Maximum = 14
Median = 9
Lower quartile = 5
Upper quartile = 11

2   Draw a box plot to show this data.
Minimum = 5
Median = 10
Lower quartile = 8
Upper quartile = 11
Range = 12

3   Draw a box plot to show this data.
Minimum = 21
Maximum = 32
Median = 26
Upper quartile = 30
Interquartile range = 6

4   Draw a box plot to show this data.
Maximum = 44
Median = 35
Lower quartile = 33
Range = 14
Interquartile range = 9

5   Draw a box plot to show this data.
2, 3, 7, 7, 10, 11, 15, 15, 16

6   Draw a box plot to show this data.
17, 6, 9, 10, 11, 16, 4, 4, 7, 15

Exercise D

1 The box plots show the distribution of the weekly wages of men and women who work at a factory.

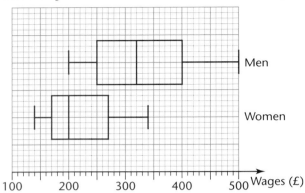

   a For the men, write down
      i   the median
      ii the range
      iii the interquartile range.
   b For the women, write down
      i   the median
      ii the range
      iii the interquartile range.
   c Make two comparisons between the weekly wages of men and women at this factory.

2 The box plots represent the times, in minutes, spent making mobile phone calls by a sample of 50 boys and 50 girls in one week.

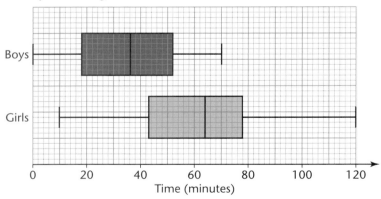

   Make two comparisons of the times spent making mobile phone calls by the boys and the girls, stating which statistics you use.

3 These marks were obtained by two students in 12 tests.

Alice	6	7	9	9	10	13	18	20	21	21	22	24
Ronnie	3	11	12	13	14	14	16	16	17	17	18	28

   a Draw a box plot for Alice's marks.
   b Draw a box plot for Ronnie's marks.
   c Make two comparisons of the two sets of marks.

4 The cumulative frequency graph shows the heights of 200 girls and 200 boys.

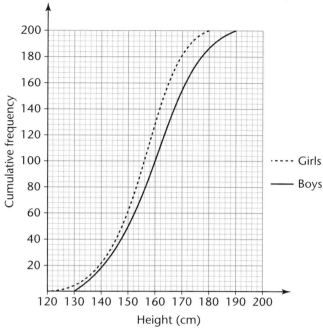

a Use the graph to find for the girls:
  i the median
  ii the interquartile range
  iii the range.
b Use the graph to find for the boys:
  i the median
  ii the interquartile range
  iii the range.
c Make two comparisons of the heights of the girls and the heights of the boys.
d **i** Draw a box plot for the girls.
  ii Draw a box plot for the boys.

5 The table shows the mean and interquartile range of Kevin and Dermot in their last twenty 400 m races.

	Mean	Interquartile range
Kevin	45.2 seconds	4.7 seconds
Dermot	46.1 seconds	2.1 seconds

Which runner would you select for the team? Give your reasons.

6 The table shows the mean and the interquartile range of the lifetime of two makes of battery.

	Mean	Interquartile range
Make A	23.4 hours	10.6 hours
Make B	19.3 hours	1.7 hours

Which battery would you purchase in each of these cases?
a You need a single battery. Give your reasons.
b You are a buyer for a company which uses 50 batteries a week. Give your reasons.

CHAPTER 46

Histograms and sampling

This chapter is about

- drawing histograms
- interpreting histograms
- understanding population and sample
- understanding bias
- knowing the different sampling methods
- using stratified random sampling.

Exercise A

1 The table summarises the distribution of the heights of 120 children.

Height(h cm)	Number of children
$75 < h \leqslant 100$	15
$100 < h \leqslant 120$	20
$120 < h \leqslant 140$	32
$140 < h \leqslant 160$	44
$160 < h \leqslant 180$	9

Draw a histogram to show this information.

2 The table summarises the distribution of the money raised for charity by runners in a sponsored race.

Amount raised ($£x$)	Frequency
$0 < x \leqslant 50$	6
$50 < x \leqslant 100$	22
$100 < x \leqslant 200$	31
$200 < x \leqslant 500$	42
$500 < x \leqslant 1000$	15

Draw a histogram to show this information.

3 The table summarises how much the workers at a factory earn each week.

Amount earned ($£$)	Frequency
$0 < x \leqslant 250$	25
$250 < x \leqslant 400$	30
$400 < x \leqslant 450$	55
$450 < x \leqslant 500$	30
$500 < x \leqslant 1000$	80

Draw a histogram to show this information.

4 This histogram shows the distribution of time spent watching TV in a week by a group of people.

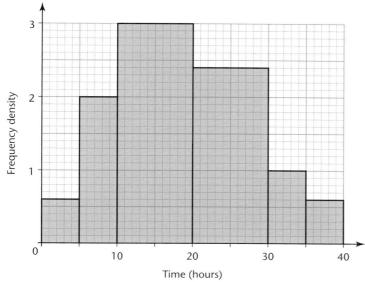

a How many people watched TV for 5 to 10 hours?
b How many people watched TV for more than 15 hours?
c How many people watched TV for less than 35 hours?

5 The histogram shows the distribution of money raised in a sponsored race.

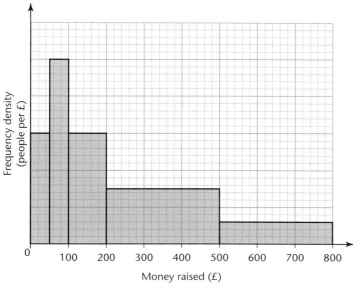

Fifteen people raised £50 or less.

a Copy and complete the frequency table for this data.

Money raised (£)	Frequency
$0 < m \leqslant 50$	15
$50 < m \leqslant 100$	

b Calculate an estimate of the mean amount of money raised.

6 This histogram represents a distribution of waiting times in an outpatients department one day.

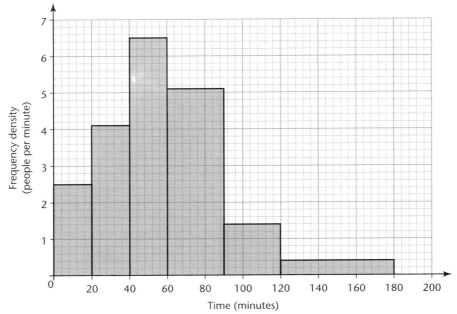

a Work out how many people waited more than 2 hours.
b Work out how many people waited less than 1 hour.
c Calculate an estimate of the mean waiting time.

7 The two histograms show the ages of the passengers on two buses one morning.

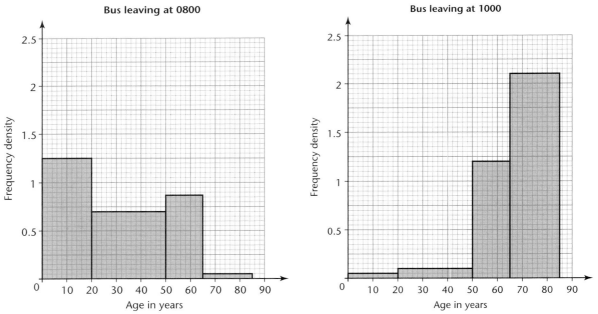

Write down two comparisons of the histograms.

Exercise B

1   You need to obtain a representative sample of 1 000 people for an investigation into how often people eat out at restaurants.
    Comment on the following methods for obtaining the sample.
    a   By choosing 1 000 names from the telephone directory.
    b   By stopping 1 000 people at random outside the railway station.
    c   By asking 100 restaurants to supply 10 names each.

2   Roisin wants to conduct an investigation to find out how much time Year 8 students spend watching TV.
    At her school there are five classes in Year 8, each of 30 students.
    a   How should she obtain a stratified sample of 10% of Year 8?
        In each class, 40% are boys and 60% are girls.
    b   How should this affect Roisin's sample?

3   In a large company there are four departments.
    These are the numbers of employees in each of the departments.

Department	Number of employees
A	175
B	50
C	250
D	125

    a   Explain why the directors should take a stratified sample.
    b   How many workers from each department should be selected for a sample of 60?
    c   Describe how the directors should select their sample.

4   Comment on the method of sampling in each of these cases.
    a   To find out how much support there is for a local football club, an interviewer stops the first ten people leaving a football match.
    b   To find out how long cars stay in a pay-and-display car park, a researcher reads the time on the ticket of every tenth car in the car park.
    c   A quality control technician takes samples from the production line at the beginning and end of each shift.

5   A researcher decides to take a sample of 200 people to find out what proportion of the population owns a car. Comment on these possible methods for choosing the sample.
    a   Ask people returning to a car park in the evening.
    b   Select names at random from the electoral register.
    c   Ask people at random at a bus station one morning.

6   The table shows the number of employees working at two different locations in a large factory.

	Factory floor	Office
Number of staff	240	60

    The company director wants to take a stratified sample of 30 staff. Work out the number of employees he should survey from
    a   the factory floor
    b   the office.

7 The two-way table shows the numbers of people who live in a small village.

	Adults	Children
Male	80	33
Female	110	27

A sample of 50 residents is to be questioned about the proposed development of a new recreational area in the village.
  a Given that the sample is stratified by gender, work out the numbers to be selected.
  b Given that the sample is stratified by age, work out the numbers to be selected.

8 A stratified sample was taken of Year 8 students according to their method of travelling to school.
Eight in the sample travel by bus, five in the sample travel by car and three in the sample walk to school. Altogether 40 students travel by bus.

How many students in Year 8   **a** travel by car   **b** walk   **c** are there?

9 A stratified sample was taken of the men, women, boys and girls at a shopping centre. In the sample there were 9 men, 12 women, 3 boys and 6 girls.
At the shopping centre there were 80 women altogether.
  a How many men were there at the shopping centre altogether?
  b How many children were there at the shopping centre?

10 There were 128 builders, 40 engineers and 8 foremen at a large building site. A stratified sample was taken of the workers. 32 builders were in the sample.
  a How many foremen were in the sample?
  b How many engineers were in the sample?

11 At a conference, 75 physiotherapists, 200 doctors, 325 nurses and 100 radiographers attended. A stratified sample was taken of those attending the conference. 39 nurses were in the sample. How many people were in the sample altogether?

12 To monitor the number of birds of a particular species, 100 are trapped and tagged. The next month a sample of 60 birds of the same species are caught. 24 of them are found to be tagged. Calculate an estimate of the size of the population of this species of bird in the selected area.

13 To estimate the population of perch in a lake, 30 of the fish were caught, marked and released back into the lake. After 3 months a second sample of 20 perch was caught. Three of them were found to be marked. Calculate an estimate of the size of the population of perch in this lake.

14 A sample of 50 rabbits was taken from woodland in County Tyrone. Each rabbit was marked and released back into the woodland. A year later a sample of 30 rabbits was taken from the same area of woodland. Three of the rabbits in this sample were marked.
  a Calculate an estimate of the population of rabbits in this woodland area.
  b State one fault with the experiment and suggest how it could be improved.

CHAPTER 47 Probability

This chapter is about

- understanding and using relative frequency as an estimate of probability
- knowing and using the fact that the sum of all the probabilities is 1
- knowing that the probability of something happening is 1 minus the probability of it not happening
- knowing what is meant by mutually exclusive events
- knowing what is meant by independent events
- knowing what is meant by dependent events
- knowing and using the addition law of probability
- knowing and using the multiplication law of probability
- finding the probability of combined events
- using tree diagrams for independent events
- using tree diagrams for dependent events
- using systematic listing strategies
- using the product rule for counting.

Exercise A

1 Pete rolls a dice 200 times and records the number of times each score appears.

Score	1	2	3	4	5	6
Frequency	29	34	35	32	34	36

    a Work out the relative frequency of each of the scores correct to 2 decimal places.
    b Do you think that Pete's dice is fair? Give a reason for your answer.

2 In a survey, 600 people were asked which flavour of crisps they preferred. The results are shown in the table.

Flavour	Frequency
Plain	166
Salt and vinegar	130
Cheese and onion	228
Other	76

    a Work out the relative frequency for each flavour correct to 2 decimal places.
    b Explain why it is reasonable to use these figures to estimate the probability of the flavour of crisps that the next person to be asked will prefer.

3 The owner of a petrol station notices that in one day 287 out of 340 people filling their car with petrol spent over £20.
    Use these figures to estimate the probability that the next customer will spend
    a over £20   **b** £20 or less.

4 Jasmine made a spinner numbered 1, 2, 3, 4 and 5.
She tested the spinner to see if it was fair. The results are shown below.

Score	1	2	3	4	5
Frequency	46	108	203	197	96

   a Work out the relative frequency of each of the scores correct to 2 decimal places.
   b Do you think that the spinner is fair?
      Give a reason for your answer.

5 The probability that Eric will have meat for lunch is 0.85.
What is the probability that he will not have meat for lunch?

6 The probability that I will have a drink of grapefruit juice with my breakfast is $\frac{5}{7}$.

What is the probability that I will not have a drink of grapefruit juice with my breakfast?

7 Nina, Christine and Jean are the only entrants for the Mathematics prize.
The probability that Nina wins is 0.3 and the probability that Christine wins is 0.45.
What is the probability that Jean wins?

8 Brian has a black, a green and a blue tie.
He decides to wear a tie.
The probability that he selects the black tie is 0.28 and the probability that he selects the green tie is 0.51.
What is the probability that he selects the blue tie?

9 In an experiment, four mutually exclusive outcomes are possible: A, B, C and D.
$P(A) = 0.12$, $P(B) = 0.34$ and $P(C) = 0.27$. Find $P(D)$.

10 A biased five-sided spinner is numbered 1 to 5.
The table shows the probability of obtaining some of the scores when it is spun.

Score	1	2	3	4	5
Probability	0.37	0.1	0.14		0.22

The spinner is spin once. Work out the probability of getting
   a 4   **b** 1 or 2   **c** 2 or 5   **d** an even number   **e** a prime number.

11 When Mrs Smith goes to town, the probability that she goes by bus is 0.5, in a taxi is 0.35 and on foot is 0.15. Find the probability that she goes to town by
   a bus or taxi   **b** bus or on foot.

12 In any batch of televisions made by a company, the probabilities for the number of faults per television are as follows.

Number of faults	0	1	2	3	4	more than 4
Probability	0.82	0.09	0.06	0.02	0.007	p

   a Find the value of p.
   b What is the probability that any particular television will have
      i one or two faults?   **ii** more than three faults?
      iii fewer than two faults?   **iv** at least one fault?

13 Katie has three black pens, five blue pens and two red pens in her bag.
She selects a pen at random. What is the probability that the pen is
   a red or blue?   **b** black or red?   **c** not red?

14 The probabilities of rolling each number on a biased dice are given in the table.

Number	1	2	3	4	5	6
Probability	0.1		0.09			0.27

Getting a 6 is three times as likely as getting a 4. Getting a 5 is twice as likely as getting a 2.
  a Work out the probability of getting
    i 4   **ii** 5   **iii** 2

  b The dice is rolled once. What is the probability of getting
    i 1 or 3? **ii** an odd number? **iii** a factor of 6? **iv** a prime factor of 10?

Exercise B

1 An ordinary dice is thrown and a coin is tossed.
What is the probability of getting a 6 and a tail?

2 The numbers on the menu show the probabilities that Richard chooses these dishes.

Menu		
First course	**Second course**	
Soup (0.6)	Spaghetti Bolognese	(0.1)
Melon (0.4)	Lamb Biryani	(0.7)
	Chicken & Mushroom Pie	(0.2)

What is the probability that Richard chooses soup and lamb biryani?

3 I take an ordinary dice and roll it twice.
What is the probability that I get an even number both times?

4 From experience, the probability of Sam winning a game of Solitaire is 0.7.
Sam plays two games.
  a What is the probability that Sam wins the first game but loses the second?
  b What is the probability that he loses both games?

5 On her way to work, Rami must go through a set of traffic lights and over a level crossing.
The probability that she has to stop at the traffic lights is 0.4. The probability that she has to
stop at the level crossing is 0.3. These probabilities are independent.
  a What is the probability that Rami does not have to stop at the traffic lights or the level
crossing?
  b Rami thinks that the probability she has to stop just once is 0.28.
Show why she is wrong.

6 On the way home I pass through three sets of traffic lights.
The probability that the first set is on green is 0.5.
The probability that the second set is on green is 0.6.
The probability that the third set is on green is 0.75.
Calculate the probability that
   a I do not have to stop at any of the lights
   b I have to stop at at least one set of traffic lights
   c I have to stop at exactly two of the sets of lights.

7 A box contains seven red pens and three blue pens.
Jane takes a pen from the box and keeps it. Susan then takes a pen from the box.
Both girls take their pens without looking. Find the probability that Jane and Susan
   a both take red pens   **b** both take blue pens
   c take pens of the same colour   **d** take pens of different colours.

8 A bag contains two red counters, three white counters and four blue counters.
Three counters are drawn from the bag without replacement. Find the probability that
   a they are all red   **b** they are all blue
   c there is one of each colour   **d** there are at least two of the same colour.

9 A box contains four red, five blue and six green counters.
Two counters are selected at random, without replacement. Calculate the probability that
   a they are both the same colour   **b** exactly one of the counters is blue.

10 The letters of the word PREPOSSESSING are placed in a box.
A letter is selected and then replaced in the box and a second letter is then selected.
Find the probability that
   a the letter S is chosen twice   **b** the letter G is chosen twice   **c** a P and an E are chosen.

Exercise C

1 For breakfast, Lydia has either tea or coffee, and then either muesli, toast or grapefruit.
   a Draw a tree diagram to show all the possibilities.
   b Given that she is equally likely to choose any combination, find the probability that
      i she has coffee and grapefruit
      ii she does not have tea or grapefruit.

2 There are fifteen balls in a bag. Eleven of them are red and
the rest are white. Elias takes a ball at random, notes its
colour and replaces it. He then repeats the operation.
   a Copy and complete the tree diagram to show the choices.
   b Find the probability that Elias chooses
      i two white balls
      ii one of each colour.

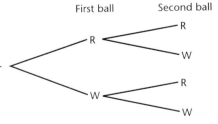

3 When Dani goes to work, she either goes in the car, cycles or catches the bus. At lunchtime she either goes to the canteen, the gym or shopping. The probability that she goes by car is 0.5 and the probability that she cycles is 0.2. The probability that she goes to the canteen is 0.1 and the probability that she goes to the gym is 0.4.

  a Draw a tree diagram to show her possible choices.
  b Find the probability that
    i she goes on the bus and goes shopping
    ii she cycles and goes to the gym or the canteen.

4 On his way to work, Mr Green always buys a newspaper and a drink. He buys the *Daily Gazette* or the *Daily News* and for his drink he buys either mineral water or cola or orange.

The probability that he buys the *Daily Gazette* is $\frac{1}{3}$.

The probability that he buys a mineral water is $\frac{1}{5}$.

The probability that he buys a cola is $\frac{2}{5}$.

  a Draw a tree diagram to show all the possibilities for the two purchases.
  b Find the probability that
    i he buys the *Daily Gazette* and mineral water
    ii he buys the *Daily News* and not cola.

5 A class has 8 boys and 12 girls. Two students are selected at random from the class.
  a Draw a tree diagram to represent the two choices.
  b Calculate the probability that
    i they are both girls
    ii one is a boy and one is a girl.

6 A drawer contains six red socks and seven black socks. Two socks are taken from the drawer at random.
  a Draw a tree diagram to represent the two choices.
  b Calculate the probability that
    i both socks are black
    ii at least one sock is red.

7 The probability that it will rain on Saturday is 0.3.
  If it rains on Saturday, then the probability that it will rain on Sunday is 0.5.
  If it doesn't rain on Saturday, then the probability that it will rain on Sunday is 0.3.
  a Draw a tree diagram to show the probability of rain during the weekend.
  b Find the probability that it will rain on at least one day at the weekend.

8 The probability that the plum tree in my garden will produce more than 50 kg of plums in a given year is 0.6.
  If the plum tree produces more than 50 kg of plums one year, then the probability that it will produce more than 50 kg of plums in the following year is 0.8; if it does not, then the probability is 0.4.
  a Draw a tree diagram to represent the probabilities of whether or not the plum tree will produce more than 50 kg of plums in two consecutive years.
  b Calculate the probability that the tree will
    i produce more than 50 kg of plums in both years
    ii produce more than 50 kg of plums in just one of the two years.

9 On his way to work, Owen has to drive through two sets of traffic lights.
The probability that the first set is green when he gets there is 0.6.
If the first set is green, then the probability that the second is also green when he gets there is 0.9, otherwise the probability is 0.2.
Find the probability that Owen will have to stop at just one set of lights.

10 It is estimated that 3 out of every 10 cars over 15 years old will fail the M.O.T. test because of a problem with their lights.
Of those that pass on lights, it is estimated that 4 out of 10 will then fail on brakes.
Of those that pass the test so far, 25% will then fail because of a problem with steering.
Find the probability that a car over 15 years old will pass these three checks of the M.O.T. test.

11 The probability that it will rain on a given day is 0.4.
If it rains one day, the probability that it will rain the next day is 0.7. If it does not rain that day, the probability that it will rain the next day is 0.2.
Find the probability that
a it rains for three consecutive days
b on three consecutive days it rains at least once.

Exercise D

1 In a school there are two entrance doors and four exit doors. How many different ways could a student enter and exit this school?

2 On a menu there are five choices of starter and six choices of main course. How many different choices for a starter and a main are there?

3 James has three new shirts and four new ties. How many different choices has he got if he plans to wear a new shirt and a new tie on Monday?

4 David wants to use each of the digits 4, 5, 6 and 7 once each to make a four-figure password for his tablet. How many possible passwords can he create?

In Defence
of Bolshevism

Max Shachtman

**"The only true prophets are those who carve out
the future they announce" — James Connolly**

Edited and introduced by Sean Matgamna

In Defence of Bolshevism

Max Shachtman

Edited and introduced by Sean Matgamna

ISBN 978-1-909639-43-0

Published September 2018 by Phoenix Press
20E Tower Workshops
Riley Road
London SE1 3DG

Printed by Imprint Digital, Exeter EX5 5H

Introduction : the British labour movement and Bolshevism

Part 1: Under the Banner of Marxism

Part 2 : The party that led the revolution

Part 3: Lenin, Bolshevism, and Rosa Luxemburg

Part 4: Appendices

Other than the introduction and as indicated, all the texts here are by Max Shachtman. Sources for items in parts 2 and 4 are given at the end of the book, after the Index.

Illustrations: between pages 179 and 186

Poster for May Day 1920
The Spider and The Fly
Retreating before the Red Army
The Tsar, the Priest, and the Kulak
1st May
Comrade Muslims
To The Deceived Brothers
Year of Proletarian Dictatorship
The Entente under the Mask of Peace
October 1917 — October 1920
Have you enrolled as a volunteer?
Capital and Co.

A note on the text of *Under the Banner of Marxism*

The main text in this book, *Under the Banner of Marxism*, was a reply by Max Shachtman in 1949 to a document by a hitherto prominent comrade, Ernest Erber, announcing his withdrawal from organised revolutionary Marxist politics. I have thinned out the reference to Ernest Erber and his political history in the first section. I have cut a part early on in Shachtman's polemic which details the still half-wrecked state of world capitalism at the beginning of 1949, when *Under the Banner of Marxism* was published. Nothing argued in the body of *Under the Banner* is built on that. It is of interest only as a historical record of the thinking of the Workers Party (to which Shachtman and Erber belonged) on that aspect of things at that point. I have cut an entire final section of the text which dealt with what Third Camp politics was and should be in relation to the Third World War, which a wide range of opinion then thought likely or certain to come soon. That again is of interest only in relation to the particular issue and the thinking of the Workers Party at that point.

Acknowledgements

Gemma Short helped produce this book and designed the cover. Cathy Nugent did a lot of work. So did Martin Thomas, who also made valuable criticisms and suggestions. Dave Ball helped with proof-reading.

Introduction: the British labour movement and Bolshevism

"World history would indeed be very easy to make, if the struggle were taken up only on condition of infallibly favourable chances" — Marx, letter to Dr Kugelmann, 18 April 1871

"The tradition of all dead generations weighs like a nightmare on the brains of the living... They anxiously conjure up the spirits of the past to their service, borrowing from them names, battle slogans, and costumes in order to present this new scene in world history in time-honoured disguise and borrowed language..." — Marx, The 18th Brumaire

"It is necessary to find the particular link in the chain which must be grasped with all one's strength in order to keep the whole chain in place and prepare to move on resolutely to the next link" — Vladimir Lenin, 1918

In the Russian Revolution of October 1917 the Bolshevik Party led the workers to take and hold state power. We mark the hundredth anniversary of that Revolution at a time when the British labour movement is striving, for the second time in living memory, to renew and reconstruct itself in politics.

The defeats which the labour movement suffered in the 1980s and 90s were inflicted, I will argue, because, though the labour movement had great weight in society and in politics, it had not been able to emulate the Bolsheviks and to take state power. The Thatcher Tories launched their offensive in 1979 and we lost much of what we had gained over many decades, including the right to free trade-unionism. This introduction will explore why that was, and the part played in it by the neo-Bolshevik organisations.

The Labour Party is still what the Blairite changes in the 1990s and after made it. The structural alterations under Blair radically changed the old Labour Party from a comparatively open and liberal membership organisation to a leadership-heavy, careerist-infested, and seriously depoliticised entity. Political discussion, and any politics other than those of the leadership, played little part. Three years after the Corbyn victory of 2015, none of the Blairite structural changes has yet been reversed. Not one.

The "new" Labour Party is a hybrid mix of old and new, inexperience and veteranship. A lot of the new people are politically near-virginal. A large proportion of the leading activists among the new leftist youth seem to be people making, or hoping to make, careers in the Labour Party or in the surrounding galaxy of NGOs ["non-governmental organisations"] and think-tanks. They behave accordingly.[1] There is at the same time a surprising growth of the political influence of the old Stalinist paper, the *Morning Star*.

In general the new Labour Party members, unlike, for instance, those of the "Benn surge" in the 1980s, who mostly had some background as trade-unionists and community activists, are short of prior direct personal involvement. The Corbyn surge is something new in politics, a phenomenon of the social media age, and its possibilities are still being elaborated.

The sounding of the Corbyn trumpet has reawakened older long-ago and long-dormant leftists, including one-time Trotskyists. They have returned to politics older, but not politically wiser, and not always all that left-wing either. The new left tends more to flabbiness than to militancy. Age dims the energies and empathies of youth, but not, it seems, old animosities. Some returnees bring with them the prejudices of the old Trotskyist groups. For example, they are an important component among the party's absolute anti-Zionist advocates of the destruction of Israel.

The idea that this left is a left is by and large only a matter of the contrast with the Labour Party in the age of Blair and Brown. So far, the Corbyn left is a politically inchoate preliminary ingathering. It is more an embryo than a mature and viable birth.

It is enormously under-educated about politics. Within it there are anti-book and anti-intellectual currents impatient with knowledge and theory beyond internet scuttlebutt. Many do not understand that socialism is inseparable from the working class, and thus they have a very under-elaborated idea of socialism as comprising class struggle, radical democracy and working-class power — indeed no conception of socialism at all as the negation of capitalism.

Jeremy Corbyn himself is deeply mired in the politics of the *Morning Star* left. He is surrounded by what someone aptly called "posh Stalinists", Seamus Milne and Andrew Murray for instance. What will happen to this politically half-emerged "Corbyn left" if the Labour Party forms a government? It will be rudely politicised in the unfolding of political events; political differentiation will begin its work of sifting and realigning. If Labour loses the next General Election something like that will come also, from a different direction.

This introduction will try to answer the following questions: what can the Bolsheviks and the revolution they led offer in the way of positive advice and precept to the work of self-renewal which the labour movement is now engaged in? What can the experience of the Bennite attempt to renovate the political labour movement in the 1980s tell us not to do again? What might the neo-Bolshevik organisations in Britain contribute to that work? What did they do last time round? How do things stand now after three years of the Corbyn surge?

---

1. The "third sector" employed 891,000 people in 2016, almost double its numbers in 2004, over five times as many as the whole automotive industry and four times as many as the whole rail sector. Only a minority of those 891,000 jobs are left-NGO lobbying-and-campaigning jobs. But there are enough of those, and the political temperature is low enough, that many young radicals who would otherwise have become left-wing workplace trade-union activists now choose an NGO office "career" instead. To a large extent they behave accordingly.

The Russian revolution

The memory of Vladimir Lenin, who led the Russian workers' revolution, is still buried in one historical sarcophagus with that of Josef Stalin, who led the bureaucratic counter-revolution against the Bolshevik Russian workers.

The Russian Revolution of October 1917 is widely seen as the inauguration of one of the worst tyrannies, or the worst, in history. An image of the Stalinist totalitarian state, ruling by terror and frequent mass murder, conjures itself up out of a near-century of truth about Stalinism and lies — lies "friendly" and hostile — about the Bolsheviks. Stalin's counter-revolution is mechanically read backwards on to the Bolshevik revolution. The calendars of the revolution and of the counter-revolution are broken up, shuffled like a pack of cards and jumbled together as if they are one set, not two. They are radical opposites. In one the working class gained power; in the other they lost it. In one the Bolsheviks led the workers to political and social power, in the other the Bolsheviks were massacred along with vast numbers of workers and farmers. We have to cut through jungles of myths, malice, lies, and fostered not-knowing to reach the truth about the Bolsheviks and Vladimir Lenin.

In 1917 a great network of workers' councils, "soviets", covered Russia. To a large extent even before the October Revolution it ruled the country, contesting for authority with the government. Delegates were elected to the councils by a variety of local bodies, and those bodies had the right to recall their delegates and elect new ones when they disagreed with what they were doing, or for any reason, whenever they pleased.

Led by the Bolsheviks, the workers and farmers defeated the old ruling classes and destroyed their power of ownership and of coercion. It was the people, the workers, the working-class Red Guards in St Petersburg, Moscow and other towns, who made the October Revolution. What they thought they were doing was liberating themselves from exploiting-class rule and beginning to liberate humanity from all future such rule.

Measured by history, the world around it, what it did, and its aspirations, it was the most democratic movement in all the history of class society. In other states, among them the most advanced capitalist states, women did not yet have the vote. In Britain a very large number of men who did not fit the property qualifications did not have the vote either. In France women got the right to vote as late as 1947; in Switzerland, strange as that is, only in 1971.

The workers determined what happened. For example, after taking power, the Bolsheviks did not want to nationalise industry, yet. They thought that was premature because Russia was so immensely backward. They wanted to go on using the services of the capitalists. For the time being they wanted only workers' control in the factories. Yet the Bolshevik government did nationalise industry, in 1918. The workers in the factories drove out the bosses and their servitors, as the peasants had driven out the landlords, and demanded that the government nationalise industry. The government complied.

It is difficult today even for socialists, and usually impossible for academic historians, to imagine, grasp, understand the visionary enthusiasm which possessed those who made and led the Russian revolution. They dared act on aspirations and hopes which to hostile contemporaries, as to historical outsiders, seemed utopian and millenarian. They were neither utopian nor millenarian. They had a perfectly rational idea of ends and means, of what they wanted and how it could be achieved.

Though they knew and said that Russia alone was far too backward and underdeveloped for the building of a socialist society there, they thought, and proved, that the working class could take power there. The Bolsheviks saw the Russian revolution as the first stage of an international socialist revolution. It would be followed by the spreading of the workers' revolution to more advanced countries. They acted on the belief that the revolution would spread, that their example would ignite Europe.

They were right, as well as ultimately mistaken, in their hopes. They did ignite Europe. There was a revolution in Germany in 1918 that removed the Kaiser and led to the ending of the Great War. The workers still followed the majority Social Democrat leaders who had supported the war and became ministers of the Kaiser in the last days before he abdicated. The majority within the German workers' councils (soviets) was held by those right-wing Social Democrats. They saved capitalism in Germany; inadvertently, they prepared the way for Hitler and the horrors that engulfed Europe and the rest of the world in the middle of the 20th century.

The Russian Soviets inspired the imagination of people throughout the world. In Britain in July 1917 a great representative labour movement convention assembled in Leeds and declared for soviets in Britain. Soviet republics were declared in Hungary and in Bavaria in 1919, though they were quickly stifled by the armed counter-revolution. In London, dockers refused to load ships of ammunition for the Poles who were fighting the Bolsheviks.

A more surprising example: Ireland. Though Ireland, both Orange and Green, Nationalist and Unionist, had highly militant working classes in Dublin and Belfast, for the most part outside north-east Ulster it was then a very backward country, a place where bishops, priests, nuns, and friars were a great social power. Yet, even in Ireland, in the south, in the countryside, strike committees in small dairy processing factories ran up the red flag during strikes and declared themselves soviets. Limerick is the third most important city in the 26 counties. In Limerick in 1919 the Workers' Council (equivalent of the trades council in Britain) declared itself to be a soviet and vied for control of the city with the British Army. It issued money, permits to travel and did other things governments do. It too was defeated.[2]

Blow upon blow, the workers were defeated in Germany, and defeated everywhere. Bolshevik Russia was left isolated.

The Bolshevik government was doomed because of its isolation, the underdevelopment of the economy and the society, and the additional ruination of the country by the civil wars and no fewer than 14 foreign anti-Bolshevik invading armies.[3] No socialism could be reared up in those conditions. The isolated revolution was, inevitably, overthrown. But not in the way the Bolsheviks

2. See *The Story of the Limerick Soviet*, by Rayner Lysaght

3. UK, Canada, Australia, Japan, Czechoslovakia, Greece, USA, France, Serbia, Romania, Italy, China, Germany, Austro-Hungary

had believed it could, and in isolation would be defeated.

Civil war and the soviets

Karl Marx wrote: "Men make their own history, but they do not make it as they please; they do not make it under self-selected circumstances, but under circumstances existing already, given and transmitted from the past".[4] The Bolsheviks too, of course, lived and worked in conditions, which were not of their own making or, much of the time, of their shaping — conditions which were inimical to what they wanted to do in Russia.

The Bolsheviks were forced to learn that they had limited choices if they were not to let the Revolution be overwhelmed in a bloody tsunami of White Guard assault and foreign intervention. In September 1918, after the Left SR rising against the soviet government of July 1918 and Fanya Kaplan's attempt to kill Lenin, the Bolsheviks launched the Red Terror. As are all such things, it was ugly, bloody and, inevitably, sometimes unjust, and all in all, a very crude and costly instrument of political struggle.[5] The Red Terror, and the way it was organised, regulated and conducted, added to the brutalising and bloodying of Russian society that the Great War, then the civil war and the wars of intervention by hostile powers against the Bolshevik workers' revolution — of which the Red Terror was an interlocking part — had engendered.

It is impossible not to regret much that the Bolsheviks did in the Red Terror; and all too easy for people living in the comparatively peaceful and safe foothills of the early 21st century to second-guess the Bolsheviks, who lived on the slopes of an erupting volcano. They had experienced the imperialist butchery in the World War. The world they lived in, inside Russia and in face of the intervening armies, was brutal, terrible, merciless. The Bolsheviks themselves suffered great casualties. They had to steer, act and fight to prevail in the conditions in which they, isolated, found themselves, and which constrained them and dictated terms of existence to them.

They lived in the memory of the slaughter of the workers after the defeat of the Paris Commune in 1871. In early 1918 they saw the White Terror in Finland, which according to Victor Serge killed a quarter of the Finnish working class (then, of course, a minority of the population). From January 1919 they had the example of what the counter-revolution did in Germany, where military — proto-fascist — gangs under the aegis of the Social Democratic government killed the revolutionary workers and their leaders. Rosa Luxemburg was beaten to death. Karl Liebknecht was shot. So were thousands of other workers. From August 1919, the White Terror raged in Hungary. The Bolsheviks themselves faced the atrocities of the Whites. Large numbers of workers and Bolsheviks had been killed. The greatest slaughter of Jews before Hitler's Holocaust had been inflicted by the Whites during the civil war in Ukraine.

The Bolsheviks had disbanded the disintegrating Tsarist army in February

---

4. *The Eighteenth Brumaire*

5. It was run by a political police, the Cheka, which came to act more or less autonomously, under the leadership of a Polish Bolshevik, Felix Dzherzhinsky. He had a long history of imprisonment for revolutionary activity and of selfless devotion to the working-class movement; his comrades believed him to be incorruptible.

1918. To fight the Civil War, they, and directly Trotsky, created the Red Army. After terrible years they won the wars, but Russia was ruined. Production in the factories was much less than it had been in 1913, and it did not reach the 1913 level again until about 1927.

In the Civil War, the Red Army fought under the political control of the workers, and directly in each section under the control of Bolshevik commissars. Nevertheless, military imperatives forced them to create a hierarchy and a command structure, and a heavy bureaucracy to seize food from hoarding peasants to feed the cities and the army.

Throughout its existence, up to 1917, the Bolshevik party had advocated a bourgeois democratic republic, a revolution like the 1789-94 revolution in France, believing that it would best be realised by way of a temporary Jacobin-type dictatorship. Until early 1917, the Bolsheviks thought that only such a radical bourgeois revolution was possible in Russian conditions. It would be the work of a revolutionary coalition government of peasant and of working-class organisations. Lenin expected that this government, like its Jacobin predecessor in France, would be overthrown after it had done its path-breaking, ground-clearing work in Russian society.

In 1917, after the first, February, revolution that overthrew the Tsar, Lenin and others saw the possibility and the necessity for the workers to take power in Russia, and won the Bolshevik party to that proposition at its conference in April 1917. The workers' revolution in October 1917 was profoundly democratic, but democracy was slowly battered into dormancy by the events of the civil war. The Bolshevik soviet government had to outlaw the parties — the SRs and the majority of the Mensheviks — that fought against it in the civil war. The vitality of the soviets was further sapped by workers going into the Red Army, in whose wars many died. In the course of the civil war, too, with industry languishing for lack of supplies, many starving workers left the towns and returned to their villages. Stage by stage, the soviets atrophied.

As the main battles of the civil war were won, in March 1921, at their 10th party congress, the Bolsheviks decided on a "New Economic Policy" under which small-scale capitalism was revived under licence from the workers' state. In that dangerous transition, and remembering the precedent of the French counter-revolution against the Jacobins in 1794, at the same congress the Bolsheviks decided to ban "factions" in the party (though explicitly still allowing "platforms"). They thought it would be only a temporary measure.

Dictatorship of the proletariat

The Bolsheviks declared that their new order was a dictatorship of the proletariat. That is a term a Marxist socialist would hesitate to use today. It conjures up only images and memories of totalitarian states ruling over a society stifled and crushed — of a great tank bearing down bloodily on unarmed or badly armed people. The Bolsheviks, and the workers who used the term — what did they mean by it?

Dictatorship was an institution resorted to by the ancient Roman Republic. A dictator was elected in an emergency. What Marx meant when he talked about the dictatorship of the proletariat — what the Bolsheviks meant, what the workers who made the revolution meant — was a dictatorship of the working class which they understood as a "democratic dictatorship", a majority dictatorship, against the old ruling classes and their defenders.

In 1905 and until early in 1917, Vladimir Lenin and the Bolsheviks had advocated a "democratic dictatorship of the proletariat and peasantry" in Russia.[6] Not because they didn't recognise an oxymoron — a contradiction in terms — when they made one, but because by "dictatorship" they understood the activity of a mass popular, working class, plebeian, democratic regime that would exercise dictatorship over the old ruling classes and their institutions. It would override the laws, customs, interests, persons and institutions of aristocratic rule. The society which quelled the aristocrats and landlords and their defenders would function as a radical, self-governing democracy.

The "dictatorship of the proletariat" in 1917 and after meant — in terms of what those who set it up wanted to set up — a democratic class dictatorship exercised by the elected workers' councils, the soviets, whose leaders took power out of the hands of the unelected Kerensky government on 25 October (new style, 7 November) 1917.

Stalinism gave the formula "dictatorship of the proletariat" only connotations of state tyranny against the people by a new autocracy. In 1917 the Bolshevik organisation was no malign dictatorship-in-waiting, ready to extend itself as a brute physical power over society. Its central organising apparatus consisted of only half-a-dozen people around the secretary, Yakov Sverdlov, and its files, so Trotsky and others attest, of scraps of paper in Sverdlov's pockets.

Of course, no class is clinically sealed off from other classes in society. Traditions, ideas, beliefs acquired from diverse influences affect what people do in a time of flaring class conflict. Menshevik railworkers opposed the Bolshevik revolution. Religion tied some workers and farmers to the ruling class. Here the idea of a working-class democratic dictatorship is more complicated and problem-bound than in the simple formula.

In judging the Bolsheviks and their revolution, their ideas are an important part of the history, but the will and intentions of the Bolsheviks were not the sole factor determining what happened.

The Russian counter-revolution

"The worst thing that can befall a leader of an extreme party is to be compelled to take over a government in an epoch when the movement is not yet ripe for the domination of the class which he represents and for the realisation of the measures which that domination would imply. What he can do depends not upon his will but upon the sharpness of the clash of interests between the various classes, and upon the degree of development of the material means of existence, the relations of production and means of communication upon which the clash of interests of the classes is based every time. What he ought to do, what his party demands of him, again depends not upon him, or upon the degree of development of the class struggle and its conditions...." — Frederick Engels, The Peasant War in Germany.

---

6. See for example Lenin's *Two Tactics of Social-Democracy in the Democratic Revolution*, bit.ly/2-tactics

The Bolsheviks had thought that in isolation they would be overthrown by invasion, and by the Russian bourgeoisie and the rich peasants, in alliance with the invaders. They defeated the invading armies and the internal White armies of counter-revolution. The counter-revolution that came was not a bourgeois counter-revolution. It was anti-bourgeois as well as anti working class. Elements of the Bolshevik party itself, from within the state officialdom, led by Josef Stalin, seized control. They raised up a brutal and increasingly savage state bureaucracy as a new exploitative class ruling over the working people. They destroyed the Russian labour movement. They erected grim Stalinist edifices of tyranny and exploitation.

At the end of the 1920s the Stalinists, continuing to call themselves "communists" and "working class" and "Bolshevik", expanded nationalised industry and forcibly collectivised the farms. They did those things with the utmost brutality and reckless inhumanity, killing millions of people.

That Stalinist "dictatorship of the bureaucratic autocracy" — brutal, exploiting class power pretending to be socialism — would last until 1991.

The Russian Revolution led to ruinous civil war, and ultimately, after the revolution had won its battles with the old-style reactionaries, to its destruction by a new form of counter-revolution hatched out from a segment of the Bolshevik Party and the old state bureaucracy. Isn't the unavoidable conclusion, then, that it should never have taken place?

That judgement is plausible only from outside of the history inside which the Bolsheviks had to act, and act urgently. They believed that the only alternative to the workers taking power then would be some form of military dictatorship. General Kornilov had tried to impose a military dictatorship in August 1917, two months before the October revolution. If the Bolsheviks had not seized the initiative in October 1917, then the alternative would have been very bloody counter-revolution. Rosa Luxemburg, critical as she was of some things the Bolsheviks did under pressure, agreed with that judgement: "The Bolsheviks' ... October uprising was not only the actual salvation of the Russian Revolution; it was also the salvation of the honour of international socialism".

Revolution, by "the law of its nature demands a quick decision: either the locomotive drives forward full steam ahead to the most extreme point of the historical ascent, or it rolls back of its own weight again to the starting point at the bottom; and those who would keep it with their weak powers half way up the hill, it drags down with it irredeemably into the abyss".[7]

Vladimir Lenin

Vladimir Lenin died of natural causes in January 1924. In the 18 months during which he lay sick and dying, Stalin's counter-revolution was consolidating itself. Lenin's comrade and companion Nadezhda Krupskaya stated in 1926 that in her opinion if Lenin had lived and been able to function a couple of years longer, then he would be in a Stalinist jail (or have ended his life in a

---

7. After her release from prison in November 1918, Rosa Luxemburg did not choose to publish her criticisms of the Bolsheviks written in jail earlier that year (from which these passages are quoted). They were first published in 1922, under the title *The Russian Revolution*, by her lawyer Paul Levi, who had fallen out with the German Communist Party and the Communist International.

Stalinist homicide chamber, as Robespierre, the leader of the Jacobins in the French Revolution, had ended his in the guillotine).

With Lenin there have been two streams of misrepresentation, that of the bourgeoisie and that of the Stalinists. Since the collapse of the USSR they have, more or less, flowed into one. The triumphant bourgeoisie took over not only the territory of the USSR and its empire of satellite states in Eastern Europe, but also part of its spiritual and intellectual empire, and the central Great Lie of the bureaucratic Stalinist ruling class — the claim that the Stalinist system was the legitimate and necessary continuation of the 1917 Bolshevik Russian working class revolution and was a form of socialism.

Socialism? That was Stalinism. Stalinism? That was socialism.

Vladimir Lenin posthumously came to be the captive — hypocritically worshipped — of people who were in deed, word, politics, class allegiance, social identity, and explicit anti-egalitarianism, in their very being, his social and political opposites and mortal enemies — the Stalinist autocracy that overthrew the Bolsheviks. Most of those who wanted to learn from Lenin and 1917 could thereafter see and approach the revolution and the Bolsheviks only through Stalinism, its putative successor. Bolshevism was buried under the foundation stones of Stalinism; it is still half buried under its ruins.

As Stalin and the bureaucrats he led subverted working-class rule in Russia, and crushed the Bolsheviks, the anti-Bolshevik counter-revolutionaries turned the memory of the revolution into quasi-religious mummery and the politics of the international communist working class movement, which they controlled, into a great and prolonged historical masquerade. They labelled their Bolshevik enemies Trotskyists, White Guards, counter-revolutionaries and fascists. Bolshevism being prepared for the Stalinist guillotine had a hood forced over its head daubed with the word "Trotskyism". Stalin killed most of the leading Bolsheviks of the October revolution and the civil war period. He killed more Bolsheviks, socialists and communists than all his contemporary big-scale butchers put together — Hitler, Mussolini, Franco — killed.

With Lenin dead, his memory, his deeds, his likeness and the iconography of the revolution he led were annexed and used as theirs by the Russian equivalents of those in France who had guillotined Maximilien Robespierre.

The ban on Bolshevik factions was intended to be temporary. At the same congress Lenin, defining Russia as a workers' state "with bureaucratic deformations", successfully defended the right of the workers to have independent trade unions, and the right to strike. The ban was a disastrous mistake. It came to be in the interests of the all-ensnaring bureaucracy — whose real rule Trotsky later dated from 1922 — to enforce the ban, and to make a party ideal of it. The ban came to be "exported" to Communist Parties outside Russia.

This was part of a broader shift in the Bolsheviks' ideas about democracy. Democracy was the banner under which the German Social Democrats and others repressed the revolutionary workers. The Social Democrats' criticism of the Bolsheviks' dispersal of the Constituent Assembly in January 1918 (in favour of soviet rule, which was more democratic) counterposed parliaments, as representing a fuller and broader democracy of all classes, to soviets as representing only workers and other plebeians.

The result of these many pressures was a shift by the Bolsheviks and the other parties of the Communist International into disparaging all general talk of democracy. That took on a logic of its own and merged with the need of the Russian bureaucracy to glorify the arrangements that let it rule in the USSR, and without which it could not have continued to rule.

Who were the Bolsheviks?

The people who led and consolidated the Russian revolution, the Bolsheviks, who were they? What were they? They were a working-class party, a class-struggle party, and a Marxist party. They believed, with the Communist Manifesto, that the communists have no interest apart from the interests of the working class. They were guided by the logic and needs of the working-class struggle.

By their work over 15 and more years, the Bolsheviks had trained themselves to be men and women of exceptional courage, energy, dedication, discipline, and spirit. Without that spirit, the workers' rule would have collapsed within a few weeks or months of October 1917.

What they did was rooted in the proposition that the class struggle which the workers have to fight — if they are not to let themselves be beaten down and robbed more brutally by the bourgeoisie — takes place not on one but on three main fronts: the economic, the political, and the battle of ideas. The unique and irreplaceable role of the Marxists is in the realm of ideas, for Marxism and against bourgeois ideas. They worked to integrate the three fronts into a coherent strategy of class war and, ultimately, of a battle for working-class power in society.

They were a party of fully-committed activists, whose members practised majority rule and majority discipline in action, but a party of politically educated members who could think for themselves and did so. They expected themselves and their comrades to think for themselves and had the contempt of thinking people for those who did not. It was a democratic party that could and regularly did judge and criticise the work of the party leaders, their deeds and proposals, and if so minded could remove them from office. The Bolsheviks were honest rational people, self-respecting militants guided and self-guided by reason and by Marxism.

Being a Marxist party meant that they used Marxism to analyse the world around them; they took the results of that analysis seriously in what they did and said, and knew how to learn from new working-class experience. They gave themselves and others honest accounts of what they were doing. They told the truth about it afterwards. They saw what they did and tried to do in a historical perspective and measured and judged themselves in that perspective. They worked to keep themselves clean of demagogy, that is, of what Lenin denounced as "phrasemongering".

They strived for clarity, and for honesty. Read Lenin on the national question for instance. In these and other writings the spirit and mind of Bolshevism flares bright and clean. Lenin is honest. He does not bluff. He tells the truth. When he is retreating, or changing, he says so. When he makes a mistake and knows it, he says so. His attitude to the party and to its members is profoundly and consistently democratic. He sets up an honest and responsible dialogue with the party members and with the working class. The proportion of his writings given over to persuasion and polemic with other comrades is irrefutable evidence of that. He abhors the substitution of demagogy and emotive "phrasemongering" for thought about the realities. The truth is always concrete, he insisted.

Lenin was certainly the central leader of the Bolsheviks, but he had to win his leadership again and again with every big problem that the party encountered. He was often in the minority, and it was taken for granted by Lenin himself and by the other Bolsheviks that he could and would be. He was repeatedly questioned, challenged, outvoted, had rival ideas counterposed to his own and preferred to them by Bolshevik party members no less loyal to the party and its purposes than he was. There was free discussion. Even during the civil war it was taken for granted that people starting from one point and wanting to get to the same place can disagree, and that reason and argument and subsequent experience are the only rational, democratic ways to resolve such differences. The party was able to lead a great democratic revolution in 1917 because it was itself profoundly democratic. It would not have been able to do that had it not itself been democratic.

The Bolsheviks had a conception of what they were and how they related to each other that Lenin summed up in 1906 thus: there would be full democracy in the party in reaching decisions. Once a decision was reached and an action was agreed upon, discipline would take over in the action. Even if people in the party disagreed with the decision they would carry out the action. But there was no question of forever curtailing discussion and having just one thinking element in the party, the leadership. Members were not required to lie in public about what they themselves thought. Discipline was a quality in party action, not the requirement to cease to think, or pretend to think differently from how you really thought. All that was Stalinism, not Bolshevism.

"Democratic centralism", for the Bolsheviks as for other Marxists of their time, was as Lenin described it in 1906: keen, open debate, linked with prompt collective effort in action. "Universal and full freedom to criticise, so long as this does not disturb the unity of a definite action; it rules out all criticism which disrupts or makes difficult the unity of an action decided on by the Party".[8]

The idea that the Bolshevik party was a machine of obedience to the central leadership, organised around an omniscient and infallible leader, with few if any democratic rights for its members, is a Stalinist invention (a version of which was taken over by some post-Trotsky Trotskyists). The Bolshevik party was above all else, a democratic, a tremendously democratic party. The Stal-

---

8. *Freedom to Criticise and Unity of Action*, May 1906, bit.ly/fc-ua. The term "democratic centralism" had in fact been coined by the Mensheviks in 1905, as they adapted to Russia more vigorous organisational ideas agreed by the German Social Democrats at their Jena congress in September 1905. When the Social Democrats abjured "democratic centralism", after World War One, what it mostly meant was that their parliamentarians and officials were shielded from democratic control, and that most of the membership was organised as a passive, only quarter-engaged mass, easily manipulated by demagogy.

inised myths of Bolshevism has little to do with the party that led the October Revolution.

The Bolsheviks believed that it was their duty to learn from all past and all current working class experience and bring the lessons of the past and honest and conscientious analysis of the present to the working class now. The revolutionary party had to be the true memory of the working class. Only in strict day-to-day truthfulness could it be that true memory. They recognised that as well as teaching history and Marxism and offering analyses of contemporary society, they had to learn from the working class, and from new working-class struggles and developments. The way they did that was one of the great strengths of the Bolsheviks.

Soviets — workers' councils — first appeared in 1905, in the revolution that was eventually crushed. The Bolsheviks initially were suspicious of them. Weren't they too loose, too politically amorphous? But they learned quickly. They concluded that Soviets were the historically specific form of working class rule. They supported the soviets in 1905, and they were able in 1917 to take the lead in a mass movement of workers organised in the soviets who at first did not have a clear goal other then that they wanted to be rid of their rulers and, some of them, to have some sort of socialism. The Bolsheviks were able to lead the workers — that is, propose clear ideas about what needed to be done, win the workers to accept those ideas or versions of them amended in dialogue with the broader working class, and give practical purpose and guidance to the whole movement.

Otherwise the revolution would probably have been defeated.

Victor Serge was a Bolshevik ex-anarchist who lived to write his memoirs. In those memoirs he wrote: "Together with a few others, this man [Lenin] had endowed an immense movement of faltering masses with a political consciousness that was supremely clear and resolute... Without it, the minds of those who marched would have been several degrees dimmer, the chances of chaos, and of defeat amid chaos, immeasurably greater".

The Bolsheviks and Marxism

Most of the traits alleged from the mid-20s to be special to the Bolshevik party — that it was "a party of a new type", one of unquestioning obedience to Lenin, one of mechanical military discipline — were characteristic not of Bolshevism at all but of the "Bolshevik parties" reshaped and remade in the image of the developing Russian counter-revolution.

In one way, however, Lenin's Bolshevik party was one of a kind — in its relationship to Marxism. It had much in common with the German Social Democratic party, on which the Bolsheviks modelled themselves. What was unique was the quality of the Marxism which the Bolshevik party hammered out in ideological conflict with every sort of opponent, from the agrarian-socialist populists of Narodnaya Volya and the Social Revolutionary party to the other Russian Marxists, the Mensheviks.

Its Marxism was concerned first with social analysis, of how the working class related to other classes and what the class conscious workers needed to understand and do; and what it needed to understand about some of the non-

Stalinists later doctored this picture of Lenin speaking to remove Trotsky

Bolshevik Russian and international Marxist currents. Bolshevism was unique in that it was able to unite sharp and continuously-honed Marxism with the tremendous energy, combativity, creativity, and audacity of the Russian working class.

The Third International and the Fourth

"... The year 1919... The entire structure of European imperialism tottered under the blows of the greatest mass struggles of the proletariat in history and when we daily expected the news of the proclamation of the soviet Republic in Germany, France, England, in Italy. The word 'soviets' became terrifically popular. Everywhere these soviets were being organised. The bourgeoisie was at its wits' end. The year 1919 was the most critical year in the history of the European bourgeoisie... What were the premises for the proletarian revolution? The productive forces were fully mature, so were the class relations; the objective social role of the proletariat rendered the latter fully possible of conquering power and providing the necessary leadership. What was lacking? Lacking was the political premise; i.e. cognisance of the situation by the proletariat. Lacking was an organisation at the head of the proletariat, capable of utilising the situation for nothing else but the direct organisational and technical preparation of an uprising, of the overturn, the seizure of power and so forth — this is what was lacking".[9]

As the French Revolution haunted the imagination and the political lives of the Bolsheviks, so too the Bolsheviks have haunted the memory and the imagination of consistent socialists after 1917. They do that still today, a hundred years after. Socialists, were, naturally, inspired by the Russian Revolution. They looked to its example for guidance on what they themselves should do, and not do. Before 1917 there had been a workers' revolution in Paris, in 1871.

9. Leon Trotsky, *The First Five Years of the Communist International*, Vol 2, p.193

The Paris Commune held power for nine weeks and was then crushed, with tens of thousands of Communards massacred and thousands more deported to New Caledonia, in the Pacific. The Bolsheviks took power and held it. Trotsky reports a conversation with Lenin in late 1917, when Lenin expressed astonishment that the Bolshevik government had lasted longer than the Paris Commune.

Russia was the revolution where the workers succeeded in clarifying their ideas, in acting effectively, in winning control. It could be the model for future revolutions. The Bolsheviks tried after the October revolution to reshape the international labour movement, to transplant Bolshevism, to create all across the world a communist movement that would make working-class revolutions, and in conditions more favourable than Russia. Socialists all over the world set out with a just humility to go to school with the Bolsheviks.

The prime "Lesson of October" was said to be that it proved above all else the need for an adequate revolutionary party, a party like the Bolshevik Party. But what the Bolshevik Party was, before and during the revolution, was soon, being radically falsified by the Russian Stalinists and their anointed leaders of the Communist International. A bureaucratic and then totalitarian palimpsest was superimposed on Bolshevism and what it did in and after October.

By 1924 Trotsky was noting how little "success" the real history of the real revolution had had. He wrote:

"We met with success in the October Revolution, but the October Revolution has met with little success in our press. Up to the present time we lack a single work which gives a comprehensive picture of the October upheaval and puts the proper stress upon its most important political and organisational aspects... Numerous documents and considerable material have been issued bearing on the pre-October history of the revolution and the pre-October history of the party; we have also issued much material and many documents relating to the post October period. But October itself has received far less attention. Having achieved the revolution, we seem to have concluded that we should never have to repeat it. It is as if we thought that no immediate and direct benefit for the unpostponable tasks of future constructive work could be derived from the study of October; the actual conditions of the direct preparation for it; the actual accomplishment of it; and the work of consolidating it during the first few weeks" (*The Lessons of October*).

The Third, Communist, International was founded in March 1919 to regroup socialists throughout the world who wanted to learn from the Bolsheviks. The tragedy of it was that they started to create Bolshevik parties in situations where revolutionary crises were already breaking over their heads. They were disastrously unready, and Stalinism would soon overwhelm them.

When the Third International was founded, European revolution seemed imminent. The First Congress was a badly attended, almost a scratch gathering, held in Moscow, deep in the heart of a Russia under siege. Under the banner of the Russian Revolution, they founded the Third International, proclaimed its principles, and presented a bill of indictment of the theory and practice of the preceding socialist movement, the Second International, which had collapsed in 1914. With few exceptions, its parties had backed their national governments and beat the war drums that marched workers out to kill other workers, under the different national banners. The Third International drew sharp political lines between themselves and Social Democracy.

From the beginning a defining question was what exactly the new International's relationship with Moscow, with the successful state-power wielding workers' revolution, should be. The German Spartacusbund, the organisation of Liebknecht and Luxemburg, which had recently (December 1918) become the Communist Party of Germany, had taken the advice of the now-dead Rosa Luxemburg and opposed the immediate declaration of a new International. In existing conditions it would be too much dominated by one section, the Russian. Their delegate at the Moscow Conference, Hugo Eberlein, was persuaded there to vote for proclaiming the Third International there and then.

The International was headed by Gregory Zinoviev, one of Lenin's closest co-workers before the revolution. It was centralised to counter the tendencies towards national self centredness that had wrecked the old international when the countries went to war in 1914. But Zinoviev and his co-workers gradually made the International an over-centralised and bureaucratised organisation, increasingly micro-managed from Moscow.[10]

Rosa Luxemburg, who knew the Bolsheviks, and knew Lenin, with whom she had friendly personal relations, wrote that the Bolsheviks would be the last to think that their exigent actions should be copied everywhere. Lenin wrote in *Left Wing Communism* — which dealt with work in bourgeois democratic countries — that Russian Bolshevik schemas could not be followed mechanically there. In 1922, Lenin commented on a resolution on organisation adopted by the Third Congress of the Communist International, in 1921: "The resolution is too Russian, it reflects Russian experience. That is why it is quite unintelligible to foreigners, and they cannot be content with hanging it in a corner like an icon and praying to it. Nothing will be achieved that way." Terrible things could be, and were, "achieved".

The Second Congress, in Moscow in July-August 1920 had a greatly improved attendance by delegates of political parties, and was, in terms of organisation, the real foundation Congress. There and at the Third and Fourth Congresses the International tackled many questions. The National and Colonial Questions, the nature of Communist Parties, relations with the reformist labour movements, and many others. They confronted the ultra-left, who in recoil from the 1914 collapse and what they thought had led to it, rejected work in bourgeois parliaments, and believed that communists should not work in reformist trade unions. The Fourth Congress in late 1922 was the last attended by Vladimir Lenin, already struck down by illness and soon to be incapacitated by it.

There were two distinct but connected phases in the early Comintern that culminated in its political destruction, that of the Zinovievists and then the Stalinists. The Fifth World Congress in June-July 1924 — Lenin was dead — declared the need to "Bolshevise" the parties of the International. The parties

---

10. The *Theses on the role of the Communist Party* adopted by the Second Congress of the Communist International in 1920, written by Gregory Zinoviev, were heavily biased towards "organisational-machine", "apparatus politics" ideas.

were mostly ramshackle, improvised, untrained. In many respects they needed re-modelling. They would be remodelled until they were like, not the real Bolshevik party, but a mutation of it with the internal life dominated by demagogy and a perpetual hunt for "deviations". Stalinism would grow out of the "Zinovievism" of the early Third International.

Trotsky commented on what happened in the US Communist Party. His verdict would hold for many others, including post-Trotsky Trotskyist organisations: "A young party representing a political organism in a completely embryonic stage, without any real contact with the masses, without the experience of a revolutionary leadership, and without theoretical schooling, has already been armed from head to foot with all the attributes of a 'revolutionary order', fitted with which it resembles a six-year-old boy wearing his father's accoutrement".

"Official Bolshevism" was soon modelled not on the Bolshevism of 1917 but on the greatly transformed and bureaucratised post-Lenin ruling party of 1924 and after. Over time this brought destruction on the Communist Parties and the International as a revolutionary force. National section leaders could be and were appointed from Moscow, irrespective of their support in their own party. A bureaucratic ideal-model "Bolshevik" party was detached from the real Bolshevik Party and, progressively, from what a party had to be and do if it was to lead the working class to emancipate itself from wage slavery in a socialist revolution. This "Bolshevism" became the ideal of parties bred and trained up to serve as local pawns and make-weights for Moscow's foreign policies.[11]

Progressively, the highest communist virtues came to be discipline, obedience, complete surrender to Moscow of political mind and conscience, of the right to have or express an independent opinion. That didn't happen overnight or in one instalment, but it happened. A political and organisational culture was created that would soon have an autonomous life of its own, even among some anti-Stalinists. Bureaucracy, intimidation, and the cumulative destruction of honest dealing, truth and independent mindedness in politics by demagogy, bit-by-bit crowded out free, that is real, discussion about the real issues. Over time "official Bolshevism" introduced the position of a designated supreme leader, in every party, rewriting the history of Bolshevism and of Lenin, who had been primus inter pares, not a cult-armoured party dictator.

Trotsky wrote later of the mid-20s that then "the bureaucracy was still only feeling its way" in exercising the power it had acquired by 1922.[12] In 1924 a tremendous outcry against "Trotskyism", to which was counterposed a caricatural, falsified "Leninism"and Bolshevism, destroyed rational, honest discussion. Soon the baying, hysterical heresy-hunting against anyone questioning the sacred official line would target Bolshevism in the Communist International — its political programme of 1917, its perspectives, its old lead-

---

11. By the late 1930s the Comintern (Russian-appointed) political representatives ran the Communist Parties. See Trotsky's *The Comintern and the GPU* (bit.ly / c-gpu), and Francis King and George Matthews (eds., with an introduction by Monty Johnstone), *About Turn: The British Communist Party and the Second World War: The Verbatim Record of the Central Committee Meetings of 25 September and 2-3 October 1939*.

12. *Stalin*, p.398

ers, its old methods.

In 1925-6 Zinoviev and Kamenev split with Stalin, who controlled the party machine. With the Trotskyists of the 1923 Opposition, they formed a United Opposition. From November 1926 Bukharin, then Stalin's close associate, replaced Zinoviev at the head of the Communist International. Sections and elements flaked off from the Communist Parties or were expelled. By 1929 Stalin had broken with Bukharin and his supporters, and a mere tool of Stalin, Dmitry Manuilsky, had taken control of the Communist International. Central control became absolute. Parties and party members had imposed on them — and they imposed on themselves: that was an essential part of it — ideas about the world that were, according to all tests of fact and reason, false to reality, often nonsensical. In the "Third Period", for instance, immediate revolution was at hand everywhere. In Germany the Social Democrats, not the Nazis, were the main enemy. Forced obedience and denial bred hysteria. That too, adroitly managed by the demagogues, could be utilised in controlling the members, and it was. It was an essential ingredient of the whole Stalinist phenomenon. A political movement emerged that had some of the central traits of a messianic religion, its members some of the key traits of religious devotees.

As someone said about other people in another cause, they would do anything, "make any alliance, pay any price, bear any burden, meet any hardship, support any friend, oppose any foe", to assure the survival and the success of the socialist state. They would follow any line decreed by Moscow, say and do anything thought useful, whether truth or lies, wear any political masks that would help them bamboozle and manipulate useful dupes, say the opposite today to what they were saying yesterday, and might say again tomorrow. They would do everything and anything Russia's rulers thought would be to their advantage, even to allying with the Nazis in Germany before 1933, and in Europe and the world in 1939-41. It was always a case of ask not what your adopted political country can do for you, but what you can do for your adopted country, the lustred "Soviet Union".

Various leaders of Communist Parties, in Europe, for example, initially sided with Trotsky, if only to defend his right to a respectful hearing in the conflict that broke out inside the Russian party late in 1923. Nevertheless, Stalinism eventually triumphed in the Communist Parties. They became what the Stalinists called "Parties of a New Type". The forces of authentic Bolshevism were reduced to very small numbers: the Trotskyists and, episodically, a few others.

At one end of the Bolshevik period, the anti-war Zimmerwald Marxist left of 1915-6, which included the Bolsheviks, went from being, in terms of organisation, very little, to victory in the October 1917 Revolution, and then the creation of mass Communist Parties in a number of countries. At the other end of the period, the forces of revolutionary Marxism went from being a large mass of people all over the world able to shape and aspire to shape events, to being only atolls in a hostile sea, and all of it within a ten or twelve year timespan.

Rearguard: the Bolsheviks after Bolshevism

We need to describe and account for the neo-Bolsheviks who played a big part in the British left in the watershed period between 1979 and 1985. If there was a painful lack of anything like a Bolshevik party, there was no shortage of self-avowed Bolsheviks.

"Trotskyism" started as an opposition to the rising bureaucracy, with Lenin's attempt from his deathbed to remove Stalin from power. The Left Opposition of 1923 was sparked into being in protest against a decision to use the police to regulate the internal affairs of the ruling party. It had sympathisers outside Russia. It had specific policies — restoration of democracy in the ruling party, unshackling of the labour movement from the state, a stronger drive to industrialise. It had objections to specific policies of those who ran the Communist International, on the might-have-been German revolution of 1923, on China in 1926-7, on the 1926 British general strike, and it objected to the bureaucratic regime in the Communist International and in its parties.

In August 1927, Stalin told the Opposition: "these cadres can be removed only by civil war", and he told the truth. After December 1927 the Opposition was expelled from the Party. Its leaders were deported to remote areas of the USSR; then, in January 1929, Trotsky to Turkey.

Until about 1932, in the chaos of forced-march industry-building and the murder-forced concentration of peasants in collective farms, the Opposition expected that the Stalin regime would be shattered by a bourgeois-rich-peasant uprising. It would fall apart, and the scattered forces of the real Bolshevik party would coalesce, become a party again, and take power. That perspective is misleadingly called the Opposition's "reform the regime" period. It was to be "reform" after other forces, and Stalinist brute ineptitude, had shattered the regime. From 1933, and in explicit naming words from 1936, the Opposition stood for a new working-class revolution. It called it a "political" revolution, but it would be a revolution that shattered the bureaucratic state and the ruling oligarchy and put the working class in power again. This "workers' state" needing a new workers' revolution was, so Trotsky believed, a freak historical formation, radically unstable.

In 1937 Trotsky argued that — in a world of capitalist convulsions, decline and regression — the nationalised economy would be progressive whether or not Russia was a degenerated workers' state. At the end, in 1939-40, Trotsky abandoned the idea that the nationalised property in and of itself was progressive even compared to capitalism in hopeless decay (as he thought it was). Now, he declared, it would be progressive only if and when the working class overthrew the autocracy and won state power from them.[13]

When Russia, together with Hitler's Germany, invaded and partitioned Poland in September 1939, and then invaded Finland (which fought back with some success) and took territory from Romania, Trotsky said that he refused to use the same term, "imperialism", for "the expansionist policy of finance capital" and for Russia, but Russia was imperialist "in the widest sense of the

---

13. "The conquests of the October revolution will serve the people only if they prove themselves capable of dealing with the Stalinist bureaucracy, as in their day they dealt with the Tsarist bureaucracy and the bourgeoisie". *Letter to the Workers of the USSR*, May 1940.

word". Though denouncing Stalin's policy in Finland, he sided with Russia in the Russo-Finnish war of November 1939 to March 1940, thinking it would or might merge into the world war that had started in September 1939.

Trotsky had seen the nationalised property in Russia as historically progressive, and then 'only progressive if...', in the dynamic and perspective of the 1917 workers' revolution. Large-scale nationalised property was then unique to Russia; it had been possible only because the working class had destroyed Tsarism, the bourgeoisie, and their state; and the Russian workers, overthrowing the bureaucracy, would preserve it and subordinate it to their own economic plan. The world war now underway would decide in "just a few years or even a few months"[14] between a new workers' revolution and bourgeois restoration. Stalinism, to Trotsky, was a historically unviable freak socio-economic formation.

In fact Russia came out of the war, still under the Stalinist autocracy, as the greatest military power in Europe and the second greatest power in the world, with a new empire stretching into Europe a hundred miles west of Berlin and, in Asia, into Korea and Manchuria; and more-or-less comprehensively nationalised economies became numerous, instituted without workers' revolution by Stalinists or by radical nationalists.

The Trotskyists had split in 1939-40 over Finland and "the regime" in the American Trotskyist organisation. After Trotsky's death on 21 August 1940, those who had opposed him on Finland evolved the view that Russia was not in any sense or degree a workers' state, but a historically new form of exploitative class society. Those were the Heterodox. The ones who had supported Trotsky on Finland and maintained their own "regime" in the SWP (USA) called themselves the Orthodox. They evolved, inside the old name of "degenerated workers' state", a new theory of the Russian Stalinist state.

Trotsky's theory was: Russia was a product of the October Revolution; the bureaucrats were social and political usurpers who had no necessary economic role to play in the collectivised economy, and whose autocratic rule was shortterm, unstable, and unviable. The aberrant socio-economic formation could not survive invasion and could not engage in large-scale expansion. Now the same name ("degenerated workers' state") was wrapped around characteristics which were, in terms of the war and the post-war world, the very opposite of Trotsky's criteria.

Now the bureaucracy plainly had an organic role in its own form of nationalised economy. It was capable of spreading that form of bureaucratically-planned economy to other countries, and did that in Eastern Europe, and indirectly, in other countries. This mutant type of bureaucratic nationalised economy was rooted in the Stalinist counter-revolution against Bolshevism, not in the October revolution of the workers.

If the Orthodox compared the facts of the second half of the 1940s with what they, after Trotsky, had been saying in 1939-40, they were in logic and self-loyalty compelled either to change the criteria or to abandon the idea that Russia was a degenerated workers' state. They wobbled a bit in the mid-1940s, but held on to the formula, while changing its content and criteria radically.

---

14. *The USSR in War*, September 1939.

After 1948, nationalised economy became for them the criterion for a special ("deformed") type of workers' state, such as Stalin's satellites in Eastern Europe and later China and others — without the historical framework and the perspectives which made Trotsky think the Russian nationalised economy (uniquely, then, the by-product of the workers' October revolution) worth defending.

But why, outside of the context of a workers' revolution, as in Trotsky's account of Stalinist Russia, is nationalised economy a criterion? Not to speak of being a sufficient criterion on its own? Why, linked to the October Revolution only by those who had subverted and overthrown it, could a nationalised economy be more important to socialist purposes, be historically nearer to socialism, than the liberties and opportunities for the working class to organise of the advanced bourgeois-democratic capitalist states? Why is a bureaucratic nationalised economy in an underdeveloped or comparatively underdeveloped country closer to the ultimate socialist social economy than the gigantic economic entities of advanced capitalist society?

For Trotsky the answer was in the past, in the 1917 Revolution. For the post-Trotsky Trotskyists it was in the socialist future to which they thought the Stalinist states were "in transition". (For them the completion of that transition would require at some point a "political revolution" in some, though not all, Stalinist states).

They redefined the Fourth International as a one-tendency "Orthodox" organisation. From about 1947 most of the small organisations of both Orthodox and Heterodox Trotskyists, haemorrhaging members, shrivelled. From about 1949, and unrestrainedly from the outbreak of the Korean war that started in mid-1950, the Orthodox Trotskyists were drawn into supporting the expansion of Stalinism. This was a continuation of their politics on Russian expansion in the war and immediately afterwards.[15]

Advancing Stalinism was both bad and good, and the conclusion was to side with it for the good: it wasn't capitalism, it eliminated the bourgeoisie, and it created a nationalised economy that was by analogy with Trotsky's — now anachronistic — assessment of Russia, progressive. In an overview of Stalinism and the advanced capitalism confronting it, that made no Marxist or socialist sense. The totalitarian and semi-totalitarian regimes bore down with special politically cauterising severity on the working class, blocking free thought, discussion, organisation. That was one of the factors that prepared the triumph of market capitalism and its ideologies in Russia and Eastern Europe in 1989-91.

All in all, with some lapses, the Orthodox Trotskyists maintained a international socialist opposition to Stalinism in Russia and in the satellite and other states. But they were pulled, as by gravity, into backing the Russian state outside Russia — that is, into backing the Russian imperialism whose existence even "in the widest sense" they, unlike Trotsky in 1939, still denied even though now Russia had a vast empire.

The Heterodox at first saw Stalinism as the freakish socio-economic formation it was for Trotsky. Stalinism's survival in World War Two, its wartime ex-

---

15. See *The Two Trotskyisms Confront Stalinism.*

pansion and postwar consolidation, and its further advances in China, Vietnam, Cuba, etc., convinced them that Stalinism was advancing barbarism. Their dilemma was that they saw advanced capitalist conditions as better for the working class and for socialist possibilities, and logically, in a world where the Third Camp of working-class socialism was minuscule, that put them, to a greater or a lesser extent, and however reluctantly, on the side of advanced capitalism.

Unlike the Orthodox, who took both comfort and sorrow from the advance of Stalinism, but mainly comfort, the Heterodox could take no comfort from it, but only horror, perplexity, and despair at the advances of Stalinism. Some of them, including Max Shachtman from about 1960, resigned themselves to accepting that capitalism was for now the alternative, and a progressive alternative, to Stalinist barbarism.

The Orthodox Trotskyists with Stalinism were like the woman in the old song who "live not where I love; and love not where I live". In 1953 they split. One side, led by James P Cannon, recoiled from where the politics they had established and voted for were leading the Fourth International. Logically they were questioning the whole evolution of the Orthodox Trotskyists — themselves — since the 1939-40 split and the death of Trotsky. That was an operation of pulling themselves up by their vital roots that they were not willing, or by that point able, to undertake.[16]

The Cannonite groups of 1953, including, importantly for what concerns us, the British one led by Gerry Healy, were or quickly became, to one degree or another, regime-heavy "Zinovievite" organisations in which the rule of "Little Great Men" also played the part of suppressing ideological contradictions and potential conflicts.

Zinovievo-Trotskyism

Russia proved the need for a "Bolshevik-type" party positively; Germany and other countries proved it negatively. The case for a revolutionary party and for making the building of a revolutionary party central is indeed a powerful one. Compare all the failed revolutions after 1917 with October, and you immediately see a clear distinction: the presence or absence of a Bolshevik Party. Or the ineptitude of those trying to use the Bolsheviks as a model. In a number of history-shaping situations the workers were willing to fight, or fought, to make a socialist revolution. They lacked a worthy leadership, or had

---

16. The Fourth International after World War Two was shaped by its conception of "deformed workers' states" — Yugoslavia, China, Vietnam, Cuba, etc. At its congress in April 1948 it declared that the East European Stalinist states, Russia's satellites, were capitalist police states. Only weeks later, when Yugoslavia and Russia fell out, they switched, within days, to writing letters to the "comrades" leading the Yugoslav CP, praising it as "a revolutionary workers' party". In the next year or so they switched to defining all the capitalist police states", with no change other than Stalinist tightening-up, as workers' states, deformed workers' states. By that switch, and its response to the Maoist overturn in China, and other Stalinist anti-bourgeois revolutions, it ruled out the idea that a revolution might be regressive and reactionary. It committed itself to a view in which you never knew where a new deformed workers' state might emerge, where it might be heading, and what initial designation it would have had. It took up, so to speak, the stance of open-mouthed millenarian votaries towards "the current historical process" and of after-the-event rationalisers of what had happened.

a treacherous one, and were defeated. The Trotskyist movement in Trotsky's time had to analyse a number of situations in these terms, beginning with 1918-19 and then 1923 in Germany.

But in the eight decades since Trotsky died, those proclaiming themselves to be in the Bolshevik and Orthodox Trotskyist tradition have laid down a terrible tradition of their own on "the Party". They have built a culture, a force of cultural inertia, and a heritage of sectism in different organisations and traditions, that owe more to Zinoviev's Comintern than to Lenin's Bolshevik organisation or the Fourth International in Trotsky's time.

Since the early 1940s at least, among Orthodox Trotskyists, organisations have been built in which the leader, or leader and chosen coterie, typically plays the role of archbishop, pope, authoritative interpreter of doctrine. Unity in action with other currents may generate dialogue about differences, so generally it is shunned. A low level of Marxist and socialist culture is both cause and product and again product and cause of such regimes[17].

"Building the Party" became the central focus, the universal recipe, the all-explaining explanation. The question is thereby begged: what exactly is a revolutionary party? In fact the whole notion of what a Bolshevik-type party is was filtered for those who came after through the Zinoviev phase of the Communist International.

The delusion — derived from Stalinism — grew that the Bolshevik party was able to do what it did because it was a large, tightly-organised "machine" in

---

17. One British organisation today, not the smallest and not at all the youngest, is reported to operate a system where discussion on the leading committee rarely leads to a vote. There is discussion. Differing views may be voiced. And then the organisation's Little Great Man (who is second to none in sagacity, as in his admiration of the SP's Peter Taaffe) sums up the discussion and announces his conclusion. And that is the decision.

Even if he were clever and knowledgeable — I don't say especially clever or especially knowledgeable — that would be a preposterous way for averagely rational beings to behave, let alone people proclaiming themselves to be Marxists, Bolsheviks and Trotskyists. It is preposterous, but it is mainly sad and pitiable — and destructive of the capacities both of the Little Great Man and of the others.

A startling example of that emerged in an exchange between Taaffe and the present writer. Taaffe complained about AWL: "They have now adopted Shachtman's position, characterising the Soviet Union in the past as 'state capitalist'" (bit.ly/taaffe-2011). In fact Shachtman always emphatically rejected the idea that the USSR was state-capitalist, arguing after 1940 that it was "bureaucratic collectivist", a form of society distinct both from a workers' state and from capitalism.

The theoretical understanding of the Stalinist states was for many decades central to working-class politics, and its own idea of expanding Stalinism as "proletarian Bonapartism" certainly shaped the Militant/Socialist Party. The implications of thinking Russia and similar states were some sort of workers' states (degenerated or deformed), or capitalist or state-capitalist, or an entirely new sort of socio-economic formation (bureaucratic-collectivist), were politically enormous.

More than half a century after Taaffe, exaggerating a little, started calling himself a Marxist, and almost as long a time that he had been a full-time organiser, this Little Great Man, this teacher and Dear Leader of his organisation, was profoundly and astonishingly ignorant. There was a political reason for his confusion. The picture the Grant tendency drew of Stalinism was that of a new form of society spreading across the world. It was a variant of "bureaucratic collectivism" onto which they arbitrarily pasted a "deformed workers' state" label. No wonder Taaffe is confused!

the hands of a Great Man. It could be replicated by mimicking and parroting what was taken to be Bolshevism. "Bolshevism" came to be reshaped and corrupted, creating a strain of Zinovievist-Trotskyism.

Each organisation lives in a closed-in world, has shibboleths which are above question, has its own in-house Marxism, its own Little Great Man, and is run as more or less a one-faction or — commonly — one-person or few-persons tyranny.

A typical Little Great Man is a Bonaparte in politics as well as in organisational matters. The manners of the tyrant can vary from those of the greedy, preternaturally hungry Galway peasant Gerry Healy, living at the end like a king or a Hollywood mogul, to that of the sabra intellectual Tony Cliff, who turned the SWP into a tight personal ayatollahship but whose personal and family life were modest and frugal. Every minority has to submit to being entirely suppressed and forbidden to express its own views in public. The idea of discipline in action is turned into a requirement always to have or pretend to have "unity" in ideas and responses, and into a public monopoly for the leading faction or the Little Great Man.

Where, despite the attempts to stifle them, differences burst out (almost always in the top layer), the minority either submits to being completely silenced, or it splits away. It may be persecuted and discriminated against, so that it has little choice but to split. It may be expelled. Short of that, since the meagreness of internal discussion works to stop like-minded and like-opinioned people from identifying each other, developing their ideas, and forming a tendency or a faction, many individuals just drop out.

Those regimes have created an archipelago of big-ramparted entities, behind high ideological tariff walls, with no dialogue between groups about their differences, and rarely unity in action. Worse, the internal and often the external life of the organisations is dominated by bullying and demagogy, and that makes discussion difficult even when the formal procedures license it or are said to license it. This system is an engine for splits and the multiplication of organisations, and for the progressive destruction of the collective political intelligence of the organisation.

It can be traced back not only to Zinovievism in general, of which it is a dialect, but specifically to the politically-murderous "Zinovievist" character of the discussion in the SWP-USA in 1939-40 and the practice of treating the organisation as the sole property of the majority, depriving minorities of full party citizenship and, for most of the time, of public utterance. In 1939-40 Zinovievist-Trotskyism "made its bones".[18]

The version of "the party" current in many Orthodox Trotskyist organisations is based on a series of historical misunderstandings. The model of the Party they have is not even remotely like the Bolshevik Party as it really was.

The lack of a revolutionary party can only be singled out as the determining central fact of a situation when all or most of the other conditions for revolutionary working-class development are present — a militant, aroused, combative working class, already educated in some degree of general socialist

---

18. Then came the substitutionist caricature parties. James P Cannon was the living link here, the Buonarotti of Zinovievism.

aspirations. Vladimir Lenin explained what a revolutionary situation is: for socialist revolution, the lower classes must not want to live in the old way, the upper classes must be unable to live in the old way, and there must be an alternative available. Much of that set of conditions can develop along with a rapid development of a revolutionary party, where it already exists on some scale. But the idea of the lack of a party being the yes-or-no question has to presuppose those conditions already developed to a large degree.

To many the cry for "the party" becomes a mystified call, a yearning, for wholeness, in conditions where a great deal more is absent — all the other aspects of Russia that made it possible for the Bolshevik Party to be what it was and lead the revolution it did lead. The typical Zinovievist-Trotskyist organisation could only fall apart or be a hindrance in a real revolutionary situation.[19]

In order to learn from the Bolsheviks you have to be able to learn that workers' democracy is irreplaceable. You have to anticipate the workers' democracy that we advocate. No individual, no three or five (or twenty-five) person committee, has all knowledge, nor can properly lay claim to all the power to think and all the initiative for discussion and action. The implication otherwise is that they cannot make mistakes, they will never be wrong, because there is no procedure for recognising in discussion that they are wrong, no provision for what to do if they are wrong. There is no normal mechanism for rectifying mistakes, short of split — no democratic mechanism for establishing that the Little Great Man is mistaken other than his own will. All initiative is in the hands of a central figure, or a small committee. Trotsky wrote: "Since sectarians, as in general every kind of blunderer and miracle-man, are toppled by reality at each step, they live in a state of perpetual exasperation, complaining about the 'regime' and the 'methods' and ceaselessly wallowing in small intrigues. In their own circles they customarily carry on a regime of despotism". When the Catholic Pope claims to be infallible as he pronounces on "faith and morals", the assumption is that God guides him and speaks through him. The kitsch-Bolshevik Little Great Man has to be his own god and guardian. This "Leninist democratic centralism" has inevitably, over the decades, led to organisational fragmentation, political and intellectual corruption, and ideological decay.

More: the prosperity of the group, "the Revolutionary Party", becomes the single guiding principle of life. The revolutionary party, here and now, is the embodiment of the Socialist Revolution. There are no interests higher than the revolution and here and now, its present embodiment, the revolutionary party. Tony Cliff of the SWP used to put it, with typical crassness: "tactics contradict principles".[20] In other words, there is no principle but the principle that principles are subordinate to tactics, which may be in contradiction with them. The

---

19. For the Healy organisation, for which see below, its annual summer camp was central to its life-cycle. Healy would have the biggest tent in the small camp city. Part of the camp agenda, as always, would be the ritual humiliation of scapegoats, usually the most devoted people. In July 1972 the crisis over five jailed dockers led to immediate strikes by a quarter of a million workers, and the scheduling of a one-day general strike by the TUC, broke while the camp was in session. They decided to continue with the camp, despite what was happening. It was only a geographical version of the organisation's isolation from the working class and its profound indifference to it.

20. See *The Left in Disarray*, by Sean Matgamna, pp. 336ff.

only principle is that there is no principle.

This post-Trotsky Trotskyist tradition is in fact a strange hybrid of the time-serving, bureaucratised mainstream of the pre-World-War-One Second International for which the party's own well-being, its vote-count, its fate, growth, decline or suppression, was the measure of more or less everything, of the ripening of conditions for social transformation — and of the Stalinist International with its rigid bureaucratic structures and procedures, its infallible leadership.

Trotsky in 1938 compared those bourgeoisies in Europe, the German and the Italian, who had made the fascists masters of the state, to a man going helter-skelter down hill on a toboggan, with his eyes closed, not knowing what will happen next. Something like that is true of all the bureaucratic neo-Trotskyist parties.

The Zinovievo-Trotskyist model has hindered progress even where it has been negated. In France, the Mandelite organisation (LCR, then NPA) was always more open, more alive with debate, than the British groups. Despite that, its trend since 1991 has been to invert the Zinovievo-Trotskyist tradition rather than to rediscover pre-1917 Bolshevik norms — in fact, to drift towards social-democratic norms: a party of "adherents" rather than activists, one where the members who hold official union positions are outside party supervision, one where the party's publications are desultorily produced and little circulated. That inversion can bring no progress. And the inversion can itself be inverted, turned back, as in Britain the SWP was after 1968.

July 1972: Vic Turner (front) and Bernie Steers are carried on shoulders after being released from Pentonville Jail, where with three other dockers they had been held under the Tory anti-union laws.

The British and Russian labour movements

The pre-Blair British and the pre-1917 Russian labour movements were different models of labour movement. The British movement grew up unplanned, groping its pioneering way; the Russian movement was shaped at the start by Marxists, their analysis of capitalist society and their perspectives on history.

Comparing the Catholic trade unions in Italy and the socialist trade unions in Germany, Vladimir Lenin famously concluded that in one case working-class trade-union needs and drives had interacted with "the consciousness of priests", in the other with the consciousness of Marxists.

"To say that ideologists (politically conscious leaders) cannot divert the movement from the path determined by the interaction of environment and elements is to ignore the simple truth that the conscious element participates in this interaction and in the determination of the path. Catholic... labour unions in Europe are also an inevitable result of the interaction of environment and elements, but it was the consciousness of priests... and not that of socialists that participated in this interaction".[21]

The story of the Bolshevik faction and party was of a determined effort by Marxists in varying conditions to shape the labour movement on Marxist lines, and then retain that shape. They offered the early labour movement Marxist ideas about capitalism, and Marxist criticism of other sorts of socialists and of the "moderate" wing of the Marxist socialist movement. They sought, and most of the time created, social and political clarity of vision and perspective. Marxists had done the same in Germany earlier, and the Bolsheviks modelled themselves on the German Social Democratic Party and its experience.

The British labour movement was much older. It existed before Marxism did. To an extent Marxism was formed from out of the experience of the British working-class movement, in a number of its phases, including that of the first mass labour movement, the Chartists, from 1838. Frederick Engels became a communist in Manchester from what he saw of the proletariat and its potentials and possibilities there.

With the defeat of the Chartists in 1848 and after, the labour movement settled into a long existence as craft trade unionism and Liberal politics.

Labour movements

In Britain the early Marxists were faced not with a new and politically still-forming working class, as were the Bolsheviks. They existed where a long history had produced distinct types of workers' organisation, with deep roots: the trade unions, in alliance with the Liberal Party; and then their own political near-equivalent for reformist bargaining within the bourgeois political system, the pre-Blair Labour Party.

Bolshevism was built in the near-fresh territory of Russia; it was built consciously by revolutionaries who drew on the immense experience of the West European proletariat, including negative aspects of that experience. As has been said, Bolshevism's guiding idea was that the class struggle takes place

21. bit.ly/twde1901

on at least three levels: the economic front, the political and the ideological (the battle of ideas). The Marxist party co-ordinates the different aspects of what is essentially a single class struggle. On the ideological front the Bolsheviks saw themselves as the warriors of the proletariat, their work that of integrating Marxism with the struggles and the organisations of the working class and thus of building a revolutionary party.

The working class as a class aspiring to remake society suffers from the enormous disadvantage that it remains the basic exploited class until it emancipates itself by the political act of taking power. The difficulty is to get to that point. By contrast, the bourgeoisie, under feudalism, developed its own areas of the economy, and bases in the cities, initially islands in the feudal oceans. Even before it became the main ruling class, the bourgeoisie acquired wealth, culture, and the knowledge it needed of itself, society, and history. The working class can be nothing comparable within capitalist society.

The syndicalists before World War One — the De Leonists[22], Tom Mann, James Connolly, who were revolutionary politicians, not, a-political syndicalists — set out in the early 20th century to reorganise and regenerate trade unionism into industrial unions, unions of all the workers of all the crafts in a given industry. These new types of unions would be the working-class bastions and infrastructure in industry — the equivalents for the working class of the strong positions the bourgeoisie had built up in the cities of pre-capitalist society. Eventually, as with the unveiling of a sculpture, they would slough off the capitalist owners, and the structures of the Workers' Republic would be there — the industrial unions.

But industrial unions too could fall under the control of time-serving bureaucracies. The great revolutionary-syndicalist union federation in France, the CGT, collapsed into support for its own government in World War One as abjectly as the parliamentary socialists. In and after the Russian workers' revolution, Soviets, workers' councils, seemed to have gone much of the way to answering the conundrum of an aspirant ruling class that must remain the main exploited class up to the point of becoming the ruling class in society. But soviets could not be built at will outside revolutionary conditions.

Attempts to reshape the British workers' movement have been many: that of the first British Marxists, the SDF, under varying names, from 1883; of the socialists of the ILP, from 1893; of the non-Marxist middle-class Fabians (who, as Frederick Engels lampooned them, thought socialism was too good to be left to the working class), from 1884.

The early Communist Party — built initially from the remnants of the SDF, the BSP, and others — set out to reshape British labour movement politics, trying to learn from the Bolshevik experience. The Labour Party leaders, socialists of a looser definition, beat them to it. The Labour Party had at first no direct individual members, only members of the affiliated Trade Unions and socialist organisations, the Independent Labour Party, the Fabian Society and others.[23]

---

22. Daniel De Leon, fl. 1890-1914, was a US Marxist with influence in other countries including Britain, who argued for a socialist effort synchronising industrial unionism and revolutionary electoralism and was an early and just critic of the parties of the Second International.

23. The British Socialist Party, the main forerunner of the Communist Party, was an affiliate of the

In 1918 the Labour Party set up Constituency Labour Parties, which individuals could join (and to which the affiliated organisations would elect delegates). For the first time the party, as distinct from the affiliated socialist organisations, proclaimed socialism to be its objective, in a reform-for-the-workers rather than self-emancipatory shape: but the new Clause Four was chiefly a device to placate the rank-and-file, rather than an imperative guiding principle for most of the parliamentary leaders.

After the Stalinists took over the Communist Party, the Trotskyists — those of them who knew what they were about — set their goal as that of reshaping the labour movement and the Labour Party into organisations that would better represent the interests of the working class and of socialism. In the mid-90s the Blairites reorganised the party; to a serious degree they reorganised it out of existence.

The Benn "surge" in the early 1980s was an attempt to reshape the labour movement. The people of the Corbyn "surge" are now trying, once again, to reshape the Labour Party.

The post-1945 settlement and the working class before Thatcher

In 1945 the British workers — and a lot of others too — voted for a socialist revolution. The Tory war leader Winston Churchill, who was genuinely popular and admired, was voted out, and Labour, led by Clement Attlee, Deputy Prime Minister since 1940, voted in.

They didn't understand the revolution they voted for as either a Jacobin or a Bolshevik revolution. The British plebeians and workers had a long tradition of their own Parliamentarianism — petitions, pressure, demands for the vote. From way back in history people had petitioned and lobbied Parliament about their grievances, their desired laws and legal protections. The Chartists, the first mass labour movement in history, focused on petitioning Parliament to expand the suffrage (which they and their antagonists understood as giving them the power to reshape society to their own needs, as those who had the vote shaped it to theirs: the petition was the "Charter".)

A shift of social and political weight and of latent power occurred in Britain during and after World War Two. Whatever it was for the ruling class and for such as Churchill, for the working class it really was a war against fascism, a war to avoid being conquered by the Nazis and to free those who had been conquered. In the army, lecture halls turned into discussion clubs, where politics was freely debated. When the press set up a chauvinist howl against striking engineers, accusing them of stabbing the soldiers in the back, the "Workers' Parliament" in Cairo — a soldiers' discussion club — famously passed a reso-

---

Labour Party from 1916. When the BSP merged with smaller groups in 1920-1 to form the CP, the Labour-affiliated BSP could simply have informed the Labour Party that it had changed its name, and put the onus on the Labour Party to exclude it. But the dominant mood of revolutionary enthusiasm militated against the CP doing that, although many CP members, individually, were active in local Labour Parties. Lenin and others had to persuade the CP to apply for affiliation. The Labour Party leadership weathered the post-World-War-One upsurge of working-class militancy sufficiently to reject the application and set about trying to root out CPers from local Labour Parties, at first none too vigorously. The definitive rejection came only at the Labour Party conference of 1925.

lution in their defence. The soldiers asserted that the right to strike was one of the things they were fighting for. The workers were determined not to go back to the system they had endured before the war. To a serious extent, the ruling class had lost control. In 1945, when the ruling class's preferred parties lost the general election to a Labour landslide, most of the soldiers voted Labour.

The workers who looked to the Labour Party and Parliament didn't get the displacement of bourgeois rule and capitalism, but they got a number of big, basic reforms, of which the NHS is one of the most important survivors. From the 1945 election on, the labour movement attained a tremendous power in Britain. Even after the Labour party lost its Parliamentary majority in 1951, the labour movement remained tremendously strong, and continued to grow in strength. In the conditions of economic boom and full employment the working class gained a great power of de facto control in industry — mainly negative control, the ability to veto by resisting what the employers and their government wanted. In the factories and in whole industries workers could and did, in practice, challenge the right of the capitalist class to rule as it liked, in its own interests and by its chosen methods.

With the welfare state as safety net; full employment and a prolonged seller's market in labour-power; industrial organisation in the trade unions, whose membership was steadily growing; and political organisation in the Labour Party, the working class was an immense power in British capitalist society.

The shop stewards' movement had expanded during the war and continued to expand up to the late 70s. Large numbers of strikes, over seemingly small things like tea breaks, were about the workers maintaining or attaining control over their conditions of work.

This was a syndicalist movement, though not avowedly that. As we've seen, the syndicalists before World War One — led by James Connolly, Tom Mann, Jim Larkin, and other committed socialists — aspired consciously to take control of society and put an end to capitalism. They worked to build the preliminary structure of the future workers' republic within capitalism, industrial trade unions. Eventually they expected to throw out the capitalists. The syndicalism of 1940 to 1980 had no such big idea of what it was, and was doing. It was in effect politically headless: it had no understood political purpose commensurate with the workers' struggles, drives, and half-formed desires which gave it strength.

In elections it voted Labour. A disproportionate number of the militant shop stewards were Communist Party in their politics. Apart from foreign policy considerations, that meant little. The Communist Party was far to the right of the Labour right wing in 1945, advocating that the wartime coalition with "progressive Tories" like Churchill continue. Formally from 1951, but in substance for a decade and a half before that, the CP said that the way to socialism was via Labour MPs in Parliament, and pressure on them from other "progressive" (CP) MPs. Until Labour lost its third election in a row in October 1959, the politically-conscious workers looked to the return of a Labour government that would pick up where the 1945-51 government had left off and take more industries into public ownership.

After the 1959 defeat the hopes and expectations that Labour would come back and continue from the 1945-51 government diminished. Emphasis shifted even more to industrial guerrilla warfare. The labour movement fought for what has been called do-it-yourself-reforms — higher wages and better working conditions, amidst the unprecedented, long-lasting capitalist prosperity. Market bargaining and the economic power of self-betterment it could bring to individuals and groups of workers partially replaced the old looking to Parliament for general social reforms. But Labour remained the fall-back politics of the militant workers. When industrial direct action destroyed the Heath Tory government in February 1974, it could only be replaced by a Labour Party government.

When the Tories won the 1951 election (though they got fewer votes than the Labour Party), the Marxists expected an immediate Tory drive to reverse the welfare state — a Tory social counter-revolution. It did not come. The Tory government denationalised steel, but embraced the welfare state and even expanded it. The big push of "counter-revolution" against Labour's constitutional revolution of 1945-8 came fully thirty years later — in Thatcher's offensive against the labour movement after 1979.

In the years of the long world capitalist prosperity there was in British capitalist society a stalemate between the ruling class and a labour movement which had grown so very powerful within it, and yet was unable — despite a cloudy mass socialistic sentiment in the trade unions, expressed in union resolutions for the nationalisation of their own and other industries — to overthrow capitalism and replace it. The situation lasted a long time, but it could not last indefinitely. Either the working class and the labour movement would push further, follow through the logic of what the labour movement was doing, and take political control of society. Or the bourgeoisie would eventually regain full control of affairs in "their own" bourgeois society.

After 1970 Heath failed in a first attempt at some of what Thatcher would later be able to do

The first Thatcher

The labour movement defeated the Labour government's attempt in 1969 to introduce anti-union laws ("In Place of Strife"), by using internal Labour and trade-union procedures before the proposals could become law. The Heath government, too, elected in mid-1970, tried to legislate to shackle the unions, as the Thatcher government would later succeed in doing.

There was a national miners' strike for 50 days, between 9 January 1972 and 28 February. "Flying pickets" of miners spread through the country, picketing coke depots, power stations, and other places using coal. The great battle symbolic of the early 1970s took place outside Saltley Coke Depot in Birmingham. Engineering workers struck to back the miners, and a mass of local trade unionists and flying-picket miners forced the coke depot to close its gates.

The dilemma of the ruling class was stated plainly in Edward Heath's reply to criticism that he should have used the army in Birmingham: "But should the soldiers have been issued with live ammunition?" (The British Army had shot dead 14 unarmed people on a nationalist demonstration in Derry just eleven days earlier).

Heath saw that as unthinkable. Would Thatcher have taken the same attitude during the miners' strike? Heath was still one of the Tories shaped in the World War and the post-World-War-Two social-political settlement.

Heath put laws on the statute book that led to a court order against dock-workers picketing container-packing depots. When dockers were jailed for picketing, a quarter of a million workers across the country immediately walked out on strike. Masses of people surrounded and laid siege to Pentonville Jail for the five days the dockers were inside. The TUC set the date for a one-day protest strike on 31 July. The government buckled and released the jailed dockers. Thereafter Heath's anti-union laws were largely unworkable.

Heath misjudged things during another miners' strike, early in 1974, and called a general election. He put the question to the electorate: "Who rules? Government or unions?"

A minority Labour government came out of the February 1974 general election. The trade union leaders' opposition to Labour's In Place of Strife in 1969 had regained for them a measure of militant rank-and-file respect. The explosion of rank and file militancy against the Heath Government in the early 1970s had raised the trade union bureaucracy to a position of unprecedented power under the Labour Government. An opinion poll in January 1977 showed that many people believed Jack Jones, leader of the TGWU (then Britain's biggest union: it merged into Unite in 2007), to be more powerful than the Labour Prime Minister. Jones probably thought so too.

They were right in the sense that the Labour government which came in on a wave of industrial militancy in 1974 could not have subsisted without the active collaboration and support of the trade union leaders. Yet the Labour government of 1974-9 was not the exercise of working-class power, but an exercise in dissipating our power, strength and effectiveness.

The Labour government repealed Heath's anti-union laws, and made such concessions to workers badly hit by inflation as controlling the price of milk.

Labour won a majority in a second 1974 general election, in October.

In September 1973 there had been a military coup against a left-wing government in Chile, one of the longest-lasting parliamentary systems in the world. Fear of such a coup was a factor in the behaviour of the Labour left around Michael Foot and of the left trade-union leaders in 1974 and after. In 1982 Michael Foot would rant in the *Observer* against the left in words that reflected that fear. "Those self-styled revolutionaries who speak today too readily of the resort to illegal methods or to street battles... should at least train to become soldiers or policemen — to face the storm troopers".[24] His response to such perceived threats was surrender in advance. In 1974, according to later testimony by then Chief of Staff Michael Carver, "fairly senior officers [had been] ill-advised enough to make suggestions that perhaps, if things got terribly bad, the army would have to do something about it"[25].

From mid-1975 the Labour government lurched into imposing pay curbs and making cuts at the diktat of the IMF.

Margaret Thatcher, entering Downing St in May 1979, quoted St Francis of Assisi: "Where there is discord, may we bring harmony. Where there is error, may we bring truth. Where there is doubt, may we bring faith. And where there is despair, may we bring hope". Unctuous hypocrisy to camouflage a naked class war against the working class and the labour movement.

---

24. *Socialist Organiser* replied: then organise to defeat the storm troopers, who might move against a left-wing government coming from a general election as much as against working-class direct action on the streets. See Michael Foot and Sean Matgamna, *Democracy, direct action and socialism*, p.57.

25. *Guardian*, 5 March 1980.

The left and the EU in the 1970s

Opposition to the European Union and to British membership (Britain joined in January 1973) was an article of faith on the left in the 1970s. It would continue to be so for a very long time after. Echoes of it still resonate on the Labour left. The issue played a central political role in determining the fate of the left in the 1970s.

European unity (the "United States of Europe", or the "Socialist United States of Europe") had been a left-wing cause since the 19th century. The European Union, or the European Economic Community as it was still called then, was a belated bourgeois attempt, in the wake of two world wars in 30 years, to unite capitalist Europe.

There were two possible left and socialist responses. To accept the progress made in its own way and with its own limitations by the bourgeoisie and to build working-class unity across Europe inside the EU, aiming to fully democratise and transform the bureaucracy-clogged EU. Or to advocate "Brexit", on the plea of those bourgeois limitations but for varying, not always respect-worthy, "real" reasons.

In the labour movement, the Stalinists, their sympathisers, and those they influenced, were against the EU because Russia did not want to face a united bourgeois Europe. Trade union officials secure — so they thought — in the political-bureaucratic structures of the national state opposed the EU.

Sections of the Labour right opposed the EU, counterposing to it the ties of the ex-Empire (British Commonwealth). Labour Party leader Hugh Gaitskell "unified" the Labour Party and co-opted much of the left by opposing Britain's first attempt to join the EU in 1961.

The Tory right opposed EU entry, and so did the *Daily Express*, then the dominant "middle-market" daily newspaper. Tony Benn was in the Cabinet in 1967 and supported British entry. In opposition Benn adapted to the "left consensus". Opposition to the EU became a central defining element of the "broad" left.

After Heath's February 1974 defeat by industrial militancy and the electorate, the left had something like a quasi-hegemony in the Labour Party. That left, led by Tony Benn, now made exit from Europe their central concern and goal. Wilson outplayed them by calling a referendum on it.

In the referendum the left united with the Tory right and others to press for exit. The clear defeat which exit suffered in the June 1975 referendum struck a paralysing blow at the left which had let the issue define it. The left had shot its bolt. Benn was demoted from Minister of Industry to Minister of Energy.

Demagogic anti-capitalist arguments against the "capitalist EU" were central to the campaign. In fact, of course, the alternative was a no less capitalist Britain. Benn used the argument for exit that movement to a socialist Britain would be blocked by EU rules. Soon there would come a time when Thatcher's tooth-and-claw version of capitalism ruled in Britain, and the EU rules, or, later, the EU Social Chapter, were a lot more reform-socialist than Britain's Thatcherite neo-liberal regime.

Yet the neo-Bolshevik groups all fell into line with the Stalinist-influenced trade union and Labour left. At first, in the early 1960s, they did not oppose EU entry. They openly scorned the attitude of the CPers and their friends, and advocated European working-class unity. Then they shifted, one after the other, in the years between a second British attempt to join in 1967, and its joining in 1973. The first to do so was the proto-WRP in 1967. The proto-SWP was the last, in mid-late 1971. All moved in response to the pressure of the CP-influenced Labour left.[26]

Labour: Party Against Government

The Labour Party at conference and in the Constituency Labour Parties had been highly critical of the Labour government since — to go back no further — 1976, when that government accepted an International Monetary Fund diktat to impose cuts. Unemployment, which had been very low for the previous 35 years, rose to over 1.5 million by the end of 1977, and remained over 1.4 million until the Labour government fell. The Labour Party, expressing itself at conference, and rank-and-file trade unionists, were at odds with the government on many issues, though the government could and did ignore Labour Party conference decisions. Where the Labour government was concerned, the Parliamentary Labour Party (heavily controlled by the Cabinet) was the only Labour Party that mattered. The PLP alone elected the Party leader.

The Tories, under the leadership of Margaret Thatcher from February 1975, were however unpopular, or very unpopular, for most of the time from 1975 through to early 1982, with only a brief and fateful poll lead in 1979... Thatcher achieved the ascendancy she eventually gained only because of inept, weak, and faltering Labour and trade union leadership, and because of the British "greatness" she parodied in the Falklands war of 1982.

In the 1970s a big part of the British working class tried, raggedly, to remake society. In the 1970s some 200 groups of workers occupied their workplaces during disputes. The working class was frustrated in that endeavour. Two of the three preconditions for revolution which Lenin once listed existed then: the rulers could not go on as before, and the people did not want to go on as before. The third condition, the availability of a viable alternative, was painfully absent. Everything that happened came from that.

The working class came to be unable to go forward because of the state of the labour movement, including the Labour Party. From 1976 demoralisation set in. Disgust with the Labour government; anti-climax to the stirring events of 1974; dismay at the Tory election victory in June 1979; the slump and ensuing mass unemployment; and the Tories' use of the state to legally hamstring the labour movement — all eventually combined to push industrial militancy into a prolonged decline.

---

26. The forerunner of AWL was the main group of the revolutionary socialist left to stick to the common view of the early 1960s and not oppose British entry into the EU.

Thatcher's counter-revolution

The Heath government of 1970 to 1974 that ended in seeming working class victory had been only the dress rehearsal. In 1979 when the Tories won the general election, the bourgeois counter-revolution against the labour movement began in earnest.

The first part of the 1980s was the reckoning time for a labour movement that had paralysed the ruling class and yet failed to destroy it and raise itself to the position of ruling class.

Though the possibility of it intimidated the Labour and trade union leaders[27] a military coup was only a faint threat in the 70s, one that never took off. Thatcher's "coup" after 1979 was brutally real. It was a "very British coup", as someone said, but a coup is what it was.

Thatcher's Tories came to power on a program of breaking the labour movement. In 1984 she would denounce the striking miners as "the enemy within". That was how she saw the whole labour movement. And the state she ran acted accordingly.

They used the state power to reshape society, and to quell and legally shackle the working class movement. The effective core of trade-unionism is working-class solidarity, in a workplace, in an industry, between workplaces, between different industries, between the workers of different countries. They outlawed it. As in a civil war — and in some limited respects that is what it was — during the miners' strike her government acted outside the law when necessary. They took control of pit villages like an occupying foreign army. Illegally, they controlled the movement of miners in and around the pit villages. Thatcher told the police, publicly, to do what they needed to break the miners. If necessary, she would legislate to make what they did legal. That is a statement that deserves to be bracketed with Tory leader Andrew Bonar Law's statement during the Home Rule crisis before World War One that "there are things stronger than parliamentary majorities".

Thatcher was able to launch the Tory offensive against a labour movement suddenly stricken by a tremendous economic slump that brought mass unemployment for the first time in 40 years.

The government used soaring unemployment — which rose from 5.3% in mid-1979 to over 12% in 1984 (3.3 million) — as a weapon of war. That was not only a rational measure from the Tories' point of view, but also an act of venomous class spite against the movement that had thwarted and embittered the ruling class. Thatcher's personal narrow-minded small-shopkeeper bigotry and spite against working-class people embodied the spirit of the new bourgeois triumphalism.

The Thatcher government:
 • Imposed anti-union laws in six instalments between 1980 and 1990. They outlawed the trade union rights which had existed in law, with some intermissions, since the 1870s and in fact for decades before that, and had helped make

---

27. And sent the Healyite Orthodox Trotskyists into a paranoid panic, from which they never would recover.

the labour movement the great force that it was.

• Drastically cut down the powers of local government, that is, of local democracy.

• Deliberately used the slump, and made its effects worse, to destroy the power of workplace organisation.

• Destroyed important parts of the militant working class by precipitately closing down industries. Trade union membership declined from 13.2 million in 1979 to 9.8 million in 1990 and 7.8 million in 1997.

• The NHS was undermined by the introduction of the internal market.

• Social benefits — on which the millions of Thatcher's newly unemployed had to live — were cut.

• In some parts of the country a generation of young people went from school to years of life on the dole, and with no training for future work.

• Council housing was sold off cheap, as a bribe to make sitting tenants Tory-friendly. The social housing stock was not and has not been replaced. About half of the properties sold to tenants under "right-to-buy" are now being let for rent by private landlords.

• A lot of people, many of them young people, were forced to sleep on the streets, something that had been unknown in Britain for decades.

• A poll tax was introduced, hitting the poorest people and favouring the well off.

The Tories had planned what they would do in the class struggle well ahead. Their "Ridley plan" for the miners' strike which they provoked in 1984-5 had been written in 1977. They used the slump strategically. The labour movement was faced with a fight for its existence in its old form.

The working class was in part destroyed as a class. Major industries were all but wiped out. Working-class youths, with no jobs and no training for future employment, were deprived of a big and formative part of their lives.

How things stood with the labour movement

A Marxist paper of the broader left at that time, *Socialist Organiser*, summed up how things stood for the labour movement confronted with the ruling class offensive in the worst world capitalist economic dislocation since World War Two.

"If we do not, in the relatively short period ahead, succeed in reorganising and politically transforming the existing labour movement, which is the only mass labour movement, and which holds the allegiance of millions of the most advanced workers, and if we fail to win it for revolutionary socialist politics, methods, and perspectives, then the working class will face a historic defeat".

The alternatives? "Either: the continued deterioration of the working class itself as capitalism rots around us.

"Or: a socialist society in which the economy is owned collectively by the producers, in a self-controlling and self-administering socialist democracy".

"Only the working class can create such a system... For this to become pos-sible, the existing labour movement — Labour Party and trade unions — must transform themselves organisationally, by a process of democratisation and by breaking the undemocratic power of cliques, bureaucrats, and uncontrollable

Parliamentarians within the organisations of the labour movement...

"Simultaneously, the labour movement must re-arm itself politically with the ideas and the immediate goal of a revolutionary socialist transformation of society".

There was "great urgency about the work of organising a non-sectarian and anti-sectarian left wing in the labour movement... Because such a left wing must unite the revolutionary left around a perspective of winning the existing labour movement to revolutionary socialism, it must also be a left wing which fights (by reason and argument) the sectarians who counterpose.... their own organisations to those of the mass labour movement in a way which is destructive of the work that needs to be done".[28]

Why did labour movement militancy collapse after 1979? Why did not the workers respond to the slump with a new surge of militancy? There is no mechanical relationship between working-class militancy and boom or slump.[29] How the working class reacts to something like the slump and the Tory onslaught it faced at the beginning of the 1980s is determined in large part by what the working class has experienced before and by the state of its organisations.

The slump at the end of the 20s pushed the British working class down quite badly, because of its previous defeat in the General Strike of 1926 and after. But in the USA the same slump — after the initial shock had worn off — helped to spark the working class for the first time into a magnificent and successful drive to create mass industrial trade unions.

Disappointment with the victories it won in the 70s shaped the way the British working class responded in the early 80s. So did the fact that in the 60s and 70s no adequate organisations of a rank-and-file militant trade-union type had been built, and neither had a halfway serious revolutionary political organisation with substantial roots in the working class.

If those organisations had been built — and they could have been built — in the 50s, 60s, and 70s, then the Thatcherites would most likely have been faced with a tremendous working-class revolt when they used the slump to drive workers out of the factories. The anti-union legislation would have provoked working-class revolt — whatever the trade union officials and Labour Parliamentarians said — and not helped beat it down.

What the labour movement should have done after 1979

The working class and the labour movement did not have to suffer defeat in 1979 and after. The Thatcher government might have been broken, as Heath was broken in 1974. At the beginning of the Thatcher era the movement had the strength to fight back and win. Even four bad years later, at a number of points during the 1984-5 miners' strike, other workers briefly joined the miners, or came near to joining them — dockworkers, the pit overseers who could have stopped all coal production, local government workers... Labour councils, in conflict with the government, might have coordinated a fightback with the miners. A coherent labour movement fightback was possible and might have

28. *Labour Democracy and the fight for a Workers' Government*, Socialist Organiser pamphlet, 1980.

29. Trotsky wrote on this topic: see the Workers' Liberty pamphlet at bit.ly/why-down.

defeated the Tories (see below, on Liverpool, for example).

What should the labour movement have done when Thatcher won the election and started to use the state to rip apart what the movement had constructed over many decades of effort and sacrifice?

Faced with murderous class war, an epoch-changing assault on the working class, the labour movement should have put itself on a footing for war. It should have done what Labour Party leader Michael Foot had promised but never tried to do: raised a storm of protest and resistance. Systematically countered the propaganda barrage from the Tories and their press, the savage baiting of the labour movement and its leaders, and of the Left. It needed occupations when closures were announced. It needed to organise and mobilise the youth of the working class[30]; to break trade-union and Labour Party collaboration with the government. Tony Benn called for that at one point: it could have made Britain ungovernable for the Tories. The labour movement needed to set itself the goal of creating a Workers' Government — that is, a government that would be for the working class what Thatcher was for the capitalist class.

It did not happen. Early in the Thatcher period, the TUC organised big demonstrations and adopted formal policies to defy Thatcher's anti-union laws. At the beginning of the Thatcher offensive the labour movement was still strong enough to mount a general strike. The Wales TUC advocated one. Against Tory determination all that shattered and shrivelled and came to nothing.

The contrast between Thatcher's energy and conviction and the Labour leaders' feebleness and bewilderment was discussed very often in the labour movement in the early Thatcher period. "Swap Kings with us and we'll fight you again", an Irish Jacobite officer is supposed to have said after the Battle at Aughrim between William of Orange and the wretched King James II, in 1691. Indeed.

There were alternatives to the catastrophe that happened. One of those possible alternatives was that the labour movement would use its expanding base in local government to resist the Tory national government — if the local government left would fight. Instead, the local government left slunk into their political graves.

---

30. In the spring and summer of 1981 there were large riots by young people in dozens of cities and districts , focused against the police. The Labour Party youth organisation, controlled by one of the pseudo-Bolshevik organisations, Militant, played no part in any of that. It behaved like a politically dead sect.

The fire last time: the Benn movement

Thatcher was one consequence of the failures of the 1974-9 Labour government. The so-named Benn movement of 1979-1983 was another. The Labour left upsurge began to gather force while Tony Benn was still in the Cabinet of James Callaghan's Labour government.

The Labour Party then was a living political party with a far more open and democratic structure than it has now. The Annual Conference had a fair measure of democracy, discussed many motions from constituency parties, and took political positions. The conference and the affiliated bodies, including the trade unions with their block votes, elected the National Executive, which had real power rather than being subordinate to the "Leader's Office" and the party machine.

When Labour lost the May 1979 general election, immediately the Labour Party erupted. The older Labour left, around the once-Bevanite paper *Tribune*, had shrunk and faded after its defeat in the referendum on European Union in 1975. A newer Labour left had begun to organise at a conference in July 1978 that set up the "Socialist Campaign for a Labour Victory". It included such central labour movement figures as Ken Livingstone and Ted Knight, then Labour borough councillors in London. Knight became leader of Lambeth council in May 1978, and Livingstone of the Greater London council in May 1981. The SCLV began to produce a monthly paper, *Socialist Organiser*, from October 1978: it later became fortnightly, and then weekly.

In 1980, the SCLV initiated the Rank and File Mobilising Committee, which — notionally — united the whole Labour left in an active campaign which made important gains. Mandatory reselection of MPs was won and opened the possibility of Labour Party rank and file control of the Parliamentary Labour Party. The election of Party leader by the whole labour and trade-union movement, not as before by the Parliamentary Labour Party alone, opened the possibility, in theory, of the labour movement electing and dismissing the prime minister.[31]

Thatcher and the local government left

In the first two Thatcher years, Labour, and specifically the Labour left, made electoral gains in local government. Local government was then much more powerful than it is now: the Thatcher government would cut down local-government autonomy, drastically.

The left-wingers who now took control of some local Labour councils had been prominent in the attempts to organise the Labour left. Would left-Labour

31. Before the September 1979 Labour Party conference *Workers' Action* wrote: "If the proposals for Labour Party democracy get through it will be the beginning of a major left/right struggle. The outcome will probably determine the character of the Labour Party for decades. Either: the left will go on from a victory on democracy at Brighton to consolidate the Labour Party as a genuinely socialist party seeking to overthrow capitalism on the basis of the class struggle of the working class, and build up a mass membership around such policies. Or: the left will be purged and the Labour Party transformed from its present ramshackle self into a tight and intolerant party modelled on the West German, Swedish and other Social Democracies". (Sean Matgamna, *Workers' Action* 153, 22 September 1979).

local government implement Tory cuts? In the 1978 Socialist Campaign for a Labour Victory conference those left-wingers had committed themselves, by way of a resolution passed without opposition, not to evade confrontation with the government by raising local taxes to compensate for Government cuts. A local tax on properties was then the councils' main source of revenue other than via central government.

Outright defiance of the class-war-mongering Thatcher government on cuts by Labour-left-controlled local government could have mobilised mass opposition to it. There were precedents of local government defying central government — Poplar council in London, in 1921, and more recently Clay Cross in Derbyshire, in 1972-4. Resistance could have spread across the country.

Leaders like Ted Knight (who headed Lambeth council from May 1978) and Ken Livingstone (then a councillor in Camden, and leader of the Greater London Council from May 1981) had talked a great fight. But now they buckled. Shortly after the general election of 1979, Knight's Lambeth council announced cuts. Forced to withdraw the cuts by a local Labour Party revolt, it raised rates by 49% in its April 1980 budget. Knight would, years later, after the miners' defeat, when the possibility of a general local government fight was gone, blunder into being disqualified from council position (for delay in setting a rate in 1985). But he led the way to surrender to the Tories in 1979-80.

The Labour left split. Those bent on surrender to Thatcher rationalised that local government was too weak to challenge the government. We had to wait for the "big battalions" to move, the trade unions in heavy industry. In the meantime the labour movement should not risk, in a clash with central government, losing the local "power" it had gained. A bird in the hand is worth two in the bush, as one half-wise vicarious "practical politician" put it.

But the labour movement had beaten governments in the 1970s — in battles not always initiated by "big battalions" — and it was not absurd to think it could do it again. In fact there was no argument the Labour-left local government compromisers used for avoiding confrontation with Thatcher that could not be, and had been, used equally by a Labour government faced with confronting the IMF. Local government became a school not of resistance in the tradition of Poplar and Clay Cross, but of right-wing surrender and the philosophy, as Ken Livingstone would self-approvingly put it, of "the cynical soft-sell" (*Tribune*, summer 1985).

It was now that the WRP, the most "Bolshevik" of the pseudo-Bolshevik groups, became a factor.

By 1979 the WRP was reduced to a membership of a few hundred. But in the Labour Party there were many ex-members, Ted Knight, for instance, who had left the organisation but had not broken politically or emotionally with Healy or understood what the organisation had come be to in 1980, or why.

The kitsch-Bolshevik WRP reclaimed the loyalty of Knight and others, and steered them to sell out the fight in local government against the Tories. The WRP rationalised retreat for them.[32] What should have been a registration of labour movement strength, in fact registered only the disorientation of the local

---

32. I had known and respected Knight in the far past. I like to think he would have had better instincts, without the WRP.

government left.

The WRP provided the right wing of the Labour left with strong "Bolshevik" credentials, by association, and from September 1981 with a new weekly paper, *Labour Herald*. Nominally edited by Ken Livingstone, Ted Knight, and another Lambeth councillor, it was printed at rates which let it survive on a very low paid circulation. It had "big name" contributors like the future Labour minister David Blunkett, then leader of Sheffield City Council. The WRP provided the working editor, a member of its Central Committee, Steven Miller.

Among the *Herald*'s contributions to the Labour Left were rancid antisemitic cartoons and comments on the Middle East. That is what the WRP was being paid good Arab petro-dollars to provide, among other things, and one motive for them founding *Labour Herald*. Thus Livingstone and Knight were holding hands with WRP leader Gerry Healy, who in turn was holding hands with Gaddafi and the Iraqi regime.

When Queen Elizabeth II opened a new Thames flood barrier in May 1984, Ken Livingstone was photographed bending at the waist to kiss her hand. A cartoonist would have had that image with Livingstone with a hand behind his back holding the hand of Gerry Healy, who was kissing a different part of his paymaster Gaddafi's anatomy.

It was both tragedy and farce, not separate but combined. The tragedy of a labour movement, so strong in 1979-81, which let itself be defeated without a serious fight until the miners' strike, which came very late in the day. And farce, in the behaviour of the local government left, so many of whom were in their own heads Bolsheviks, or at least allied with "Bolsheviks", that is, Gerry Healy's outfit.

The politics of the Left that faced Thatcher.

Between 1979 and the general election of 1983, which gave the Tories a crushing victory and set the scene for the left to be swamped in a Labour politics of politically empty anti-Toryism at all costs, the Labour left was strong, but inchoate.

The dominant Labour left ideas for an economic alternative to Thatcher's wrecking-ball Toryism were those of the "Alternative Economic Strategy", originating with the Communist Party. This was an amalgam of World-War-Two-style state-controlled "siege economy" and Stalinist models of planning, but linked to bourgeois-democratic liberalism.

For a very long time, "the left" had been a foreign-policy left, defined by its stand on the international concerns of the USSR. In other words, it was not really a Left at all. The so-named Second Cold War had been triggered by the invasion of Afghanistan at Christmas 1979. It coincided with the return if the Tories to power and the crisis of the Left. Important sections of the left backed the Russians, including some Labour MPs.[33]

Then as now, a lot of vocal Labour Party members were ex-members of the Communist Party or of revolutionary groups who had not changed their international politics or their prejudices. A large swathe of the left was still en-

---

33. Among the activist left groups, the only exceptions to support for the Russians in Afghanistan were the SWP and Socialist Organiser.

thusiastic for the Iranian revolution, long after its profoundly reactionary character had been made unmistakable. Some of the left — the Mandel Fourth International people — exulted in their "solidarity" with the Khomeinyite students who seized the US embassy in Tehran in November 1979 and held the American staff hostage there for 15 months.

In August 1980 the greatest working-class revolt against Stalinism since the Hungarian revolution of 1956 erupted in Poland. Workers seized the Gdansk shipyards. Ten million workers created the Solidarnosc trade-union movement, which was more than a mere trade-union, and forced its way into legality for 18 months. Benn and others on the left had an attitude of, at best, reserve or suspicion towards Solidarnosc. We saw the absurdity that Thatcher, who legislated against trade-union freedoms at home, and right-wing union leaders like the EETPU's ex-Stalinist Frank Chapple, who ran his union like a mini police state, "backed" Solidarnosc, while sections of the left were unenthusiastic or hostile. Arthur Scargill supported the suppression of Solidarnosc. Sections of the labour-movement right wing, too. The right-wing leader of the ISTC steel union referred in a speech to the leaders of the police-state "labour fronts" as his "trade-union colleagues" in Eastern Europe.

In the Falklands War of April-June 1982, a wave of chauvinism washed over Britain. The Tory government made itself popular by invading to rescue the British population of the Falklands/ Malvinas islands in the south Atlantic, seized as a political ploy by Argentina's tottering military dictatorship.

The press[34] used the occasion for a viciously personal campaign against Labour leader Michael Foot, whose instinct of less than wholehearted support for the war made him vulnerable in the chauvinist atmosphere. Meanwhile, large sections of the would-be revolutionary left came out in support of the fascistic Argentine military junta. For them, the war was not what it was, a freak event, but, on the Argentine side, a great anti-imperialist war of liberation.[35]

Benn for Deputy Leader, 1981

In April 1981 Tony Benn and his coterie decided that Benn would challenge the incumbent, Denis Healey, for the symbolic post of Labour Party deputy leader (the position carried no independent power). Prominent leaders of the Labour left, notably Vladimir Derer of the Campaign for Labour Party Democ-

---

34. The Thatcherite "coup" changed even the popular press for the worse. The *Sun* was until 1964 the *Daily Herald*, half-owned by the TUC. It remained politically and culturally the same for a while under its new name. Then it was turned into a yobbish caricature, with public schoolboys such as Kelvin MacKenzie, editor from 1981 to 1994, writing as they thought workers talked — backward, ignorant, and very reactionary workers. The *Daily Mirror* had been a good Labour-supporting popular paper, in its own way serious and would-be reader-elevating. It too fell in its intellectual level and its aspirations for its readers.

35. Again, the SWP (this was before it embraced kitsch anti-imperialism) and *Socialist Organiser* were the exceptions. And Militant, in an odd way. Militant "made the record" against the war in small print somewhere, but characteristically kept its head down and its mouth shut about it. Was it for or against Thatcher's war? Not the way to pose the question, comrades! What we need, as the alternative to war, is a Socialist Federation of Argentina, Britain, and the Falklands. Now? Immediately!

racy, criticised the decision.[36]

However, thereafter, until the vote at the end of September 1981, the focus of the entire left was on the deputy leader campaign. The leadership of the left was moved away from the representative Rank and File Mobilising Committee into Benn's "kitchen cabinet"

The deputy leader campaign split the harder from the softer left, among them Neil Kinnock. Benn got 81% of the CLP vote, but lost to Dennis Healey because of trade-union and MPs' votes.

In January 1982 the trade union leaders struck a formal compromise with the remaining Labour right. (Some of the right had quit Labour in March 1981 to form the SDP, which later merged with the Liberals to form today's Lib-Dems). A newly strengthened "soft left" emerged (the fey ones sometimes called themselves the "cuddly left" — but it wasn't with the left that they cuddled).

Michael Foot, who in the past had had some serious left-wing credentials, had been elected Labour leader (by the Labour MPs) in November 1980. He promised to lead a "crusade" against the Tories and against rising unemployment, which would reach 12% (3.3 million) in February 1984, but he did nothing like that.

He chose at the crucial point to turn on the left, denouncing it as hostile to democracy. He meant such things as mass pickets, defiance of the police, and active opposition to the anti-union laws which had been passed through parliament. The left, bit by bit, lowered its sights and turned to "anti-Tory" electoralism: increasingly, nothing mattered, not even anti-Tory policies, except getting the Tories out. When we finally got the Tories out, we got the Blair neo-Thatcherite, duck-egg-blue "Labourites" in.[37]

After the 1983 election, the Labour Party was led by Neil Kinnock, a soft left-talking man who had never, unlike Foot, had serious left-wing credentials. Kinnock was a mere windbag, progressively deflating. He would run helter-skelter down through the 1980s before the Tories. The Labour left revived, and Kinnock was put on the back foot, during the 1984-5 miners' strike, but after the miners' defeat Kinnock proceeded apace.

---

36. Derer was a Third Camp socialist: son of a Czech socialist MP, he came to Britain as a refugee around 1939, and was a member of the IKD (International Communists of Germany) in exile

37. The absurdity of anti-Toryism-and-don't-mind-the-politics was summed up much later in the slogan which Dennis Skinner, a once-left MP who had let himself by patronised by the Blairites and became something of a pet for them, raised in support of the Blairite candidate David Miliband in the 2010 Labour leader election: "Vote for the candidate the Tories fear".

The political anatomy of the left in 1979-85

What was the proper work of socialists in the 1980s? By way of Marxist education and agitation:[38]
 • to educate, multiply and group together the Marxists;
 • to bind them together in a coherent organisation, capable of both collective political thought and disciplined united action; and capable of knitting together the political and industrial fronts of the class struggle with a coherent battle of ideas against ruling-class ideas and for a consistently working-class world outlook;
 • to work towards building a rank and file movement in the trade unions;
 • to help organise a coherent class-struggle left in the Labour Party and trade unions;
 • to promote the class struggle day to day;
 • to work steadily towards reorganising the labour movement, by supplementing and replacing its existing bureaucratic structures by thoroughly democratic ones and its augmentation from the very large layers of unorganised workers.

Yet mostly, the neo-Bolshevik left danced to its own tune, choreographed to fit rhythms and priorities for its own ranks rather than those of the class struggle. What could they do? If they couldn't unite — and realistically, they couldn't — they should have created an ad hoc common front on a programme of active working-class self-defence, something like the Rank and File Mobilising Committee of 1980 set out to be. They needed to agree on goals in the fightback. To find ways to influence and educate the broader labour movement, the trade unions, the Labour Party, the unorganised workers, unemployed youth; to convince the working class and the labour movement to go with the logic of their own struggles. To organise common actions and dialogue about their differences. To organise as much as possible of the "Bolshevik" left into some sort of coherence and common action.

Inside and outside the Labour Party, the left was broadly as follows.

The Communist Party of Great Britain[39] still had over 20,000 members and large influence in the Labour Party. It consisted of a number of factions. Large sections of it were turning themselves into plain-and-simple liberals: in the early 1980s they would play the role of ideological ammunition carriers to Neil Kinnock and the Labour Party's treacherous soft left. Yet politically the Communist Party then was still altogether more serious than its descendant group round the *Morning Star* is now. It had condemned the Russian occupation of Afghanistan at Christmas 1979, when almost all the Orthodox Trotskyists either backed the occupation or refused to say that the Russians should stop their colonial war and get out.

There was a cluster of Trotskisant organisations which aspired to take the Bolsheviks as their model, guide, and political conscience.

---

32. As *Socialist Organiser* argued.

33. Not the odd little cult of dim political provocateurs that has that name now.

Tony Benn speaks on a Militant/ LPYS platform, early 1980s

The WRP

The WRP, as we have seen, became a factor in the local government left by way of *Labour Herald* and the collaboration with them of Knight, Livingstone, and others. By 1979 it was shrunken compared to the early 1970s, but still highly visible. With lots of money to throw around, it was publishing a daily paper that combined an impressive professional finish with very strange politics. By 1982 it had funked itself into not advocating a new Labour Government because that might provoke a military coup. The organisation was showing many signs of plain craziness, too.

As *The Club*, then the *Newsletter* group, then the SLL, and then the WRP, it had been far and away the dominant revolutionary organisation in Britain since the late 1940s. In the 60s and early 70s it was, in its size, energy, activities, and power of initiative, the nearest thing to a Trotskyist (Orthodox Trotskyist) party in the country. But it was a caricatural Stalino-"Bolshevik" organisation, a vicious sect run as a personal tyranny by Gerry Healy who, by the late 70s, was, arguably, mad, clinically as well as politically.

As it became crazier, denouncing almost everyone who disagreed with it as "police agents" or "spies", the organisation was overtaken by other groups on the left, in the first place the SWP, and then Militant. By the mid-70s the WRP's most important base was in the theatre. It recruited or influenced a large number of well-known players, directors, and managers.

In 1975-6 Gerry Healy had sold the organisation to the Libyan government and secret service. Over the subsequent years he received an immense amount of money, at least a million pounds, probably more. He also put the organisation to doing money-spinning jobs for the vile tyrants who ran Iraq, and entered into lucrative relations with other Arab states.

In return Healy offered the information-getting capacities of his organisation, and of the reporters and photographers of its paper *Newsline*, to his employers. It provided spy reports on Arab political dissidents in Britain, as well as making propaganda for the Libyan and Iraqi regimes. Some of those spied on no doubt paid dearly for it when they went home. *Newsline* (2 February 1979) publicly justified the killing by the state of 20 Iraqi CPers for organising politically in the army: one of them had five months previously brought greetings to a WRP conference! (bit.ly/nl-790202). The WRP's leading committee decided with one dissenting vote to approve the killings.

Mr Healy provided another service for his paymasters. His agreement with the Libyans obliged him (and his organisation) to provide information on prominent "Zionists" in business, entertainment, journalism and the arts. Here "Zionist" meant Jew (and readers of the denunciations of "Zionism" in Healy's press could not but be aware of it). Some of the WRP stuff was crazy Protocols-of-the-Elders-of-Zion type anti-Semitism, as when the paper proclaimed the existence of a "Zionist connection" stretching through the Thatcher government in Britain via the WRP's enemies on the left (of which it named Socialist Organiser) to the Reagan administration in the USA.[40]

The organisation would implode in 1985. One of the WRP's long-time leaders, Cliff Slaughter, denounced the group around Healy and the Redgrave siblings, Vanessa and Corin, as fascist in their attitudes. That was an exaggeration, but not very much of an exaggeration.

Militant

By the early 1980s the biggest Trotskisant, would-be Bolshevik, group was Militant. They had threaded themselves into the trelliswork of labour movement structures for two decades. They had controlled the Labour youth movement from 1969. They had grown in a strange symbiosis with the official Labour Party bureaucracy.

Politically, they were built around what they called their "perspective". They had two "perspectives", one for the rest of the world and one for Britain.

In their "world perspective", History, impatient at the slowness of the working class to resolve the epochal crises of capitalism, had generated a world of bureaucratic revolutions making non-socialist progress beyond capitalism, by way of Stalinist revolutions and military expansions. They called the Stalinist states what other Orthodox Trotskyists called them — "degenerated and deformed workers' states" — and added a special name of their own: "Proletarian Bonapartism". Not only the Stalinist states: Syria and Burma, so Militant's leading writer Ted Grant announced in 1965, were also "deformed workers' states" or cases of progressive "proletarian Bonapartism". Possibly Egypt too, or soon.

This vision was the same as the "Shachtmanite" picture of the forces in the world — only with the totalitarian new societies seen as "progressive" (which is all the "deformed workers' state" tag meant), and not the barbarism Shachtman saw in them.

These "proletarian Bonapartist" totalitarian states were immensely progressive. They developed the means of production where capitalism could not. Historically they bore roughly the same relationship to socialism as capitalism did in the Marxist view: they prepared the way, they developed the economic and social prerequisites of socialism, including a working class. The socialist

---

40. 9 April 1983: bit.ly/nl-830409. See Workers' Liberty 3/62, bit.ly/wl3-62.

world revolution was, for now, Stalinisation — a progressive totalitarianism spreading across most of the world in a series of bureaucratic revolutions.

It was a "first stage" of socialism — in which, alas, the workers were, in Trotsky's words of 1939 about the people in the territories annexed by Stalin, semi-slaves. This was Trotsky's nightmare of 1939-40 — and Shachtman's — but seen through the mists of a powerful hallucinogenic. Ted Grant and his disciples such as Peter Taaffe and Alan Woods turned Marxism into a comfort blanket for people who saw horrible things and were so demoralised that they became the self-deluding sycophants of the "historical process" — like religious people facing annihilation who let themselves believe that death is the start of a paradisaical future.

Militant backed Stalinism everywhere, "critically". For ten years, all through the 1980s, they backed Russia's colonial war in Afghanistan.

Twinned with the world outlook was a "British perspective". The labour movement was evolving incrementally, organically, towards an adequate commitment to socialism. The labour movement, proclaimed Militant, was too strong to be defeated or even to suffer severe setbacks. Major defeats were so improbable as not to be worth even considering.[41] The Labour Party would move unstoppably to the left, first to a mass "centrist" stage (like the Bevanite movement of the 50s), and eventually come under Marxist (that is, their) leadership. And then?

The vision of Stalinism making the anti-bourgeois revolution as the first stage of world socialist working-class revolution infected their idea of socialism in Britain too. They defined socialism as the "nationalisation of the top 250 monopolies" — by the existing bourgeois state. These nationalisations, if there were enough of them, would constitute "socialism". All this "Bolshevik" group's politics for Britain were a sloppy debauch of vulgar evolutionism and cheap — and silly — "optimism".

In the trade unions they did routine activity seeking — like the Communist Party — to capture offices and positions, and in those terms they were quite successful. By 1969 Militant was the only one of "the groups" still in the depleted Labour Party Young Socialists. They took it over and ran it for 17 years as a political school and breeding (and in-breeding!) ground, without disturbance from the Labour Party. One of their members was paid by the Labour Party to be its youth officer, and the LPYS had a seat on the National Executive, always occupied by another member. They coexisted with the Labour Party bureaucracy by retreating up the ladder of socialist abstraction, preaching nationalisation of the top 250 firms but avoiding conflict on current politics. They ran a very bureaucratic youth movement.

Politically the LPYS was on many issues to the right of the labour movement,

---

41. Back in 1966, when the prospect of a bourgeois assault like Thatcher's seemed remote to most people, this writer and others polemicised against Militant's argument that such an assault, or at least success for it, was impossible. "Against the argument that capitalism isn't always going to be basking in the sun you reply that any attempt to throw the workers out of the factories as the result of a slump, and create conditions where legislation would have a practical effect, would provoke such a revolt as would topple the system... Drunken optimism! That the workers will be forced to rise is inevitable — victory is not inevitable... The bourgeoisie too are very strong..." (*What We Are And What We Must Become*, www.workersliberty.org/wwaawwmb).

and of radical-minded young people, in the era of the "libertarian revolution" in sexuality and culture. Yet, in symbiosis with the Labour Party, the organisation grew. They won leadership and control of the labour movement in Liverpool, which had been the original heartland of the group back as far as the 1930s.

In the ferment on the Labour left after 1979 they were nominally or notionally involved in one initiative (the Rank and File Mobilising Committee). They stood aloof from the broader labour movement, concentrated on tilling their own factional garden at the edge of the broad field, made "socialist" (pro-nationalisation) propaganda, and climbed official trade-union ladders. Their young people thought themselves very left-wing because of the large number of industries they wanted to nationalise. During the miners' strike of 1984-5, when much of the left was out and about collecting money for the miners, they went on demonstrations with buckets to collect for the supposedly-upcoming "daily Militant", and got donations which the donors assumed were for the miners. That was the point. Militant showed that it was possible to be in the broad working-class movement — indeed, buried in the bowels of the Labour Party and the trade unions — and still be politically passive self-obsessed sectarians.

The group would half-collapse when their "world perspective" hit the buffers with the fall of the USSR and their "British perspective" with the blows dealt to the labour movement by Thatcher and the setbacks for the left in the Labour Party under Neil Kinnock.[42] For what concerns us here, their activity during the attempt of the labour movement to remake itself politically in the Benn surge and the 1984-5 miners' strike is most instructive.

Liverpool

Through Liverpool City Council in 1983-5, the avowedly Marxist and Bolshevik "Militant Tendency" had the leadership of a mass workers' movement.

For most of the 1970s, everywhere in Britain, labour was militant and on the offensive. The combativity of Liverpool's dockers was heroic. One of Liverpool's MPs, from 1964, was Eric Heffer, an honest working-class-loyal Third Camp socialist.[43] In 1973 the Liberals gained control of Liverpool City Council and held it for the rest of the decade. The Liverpool docks industry declined, so did shipbuilding. Chronic economic depression laid hold of the city. Against this background, Militant slowly grew to dominance over sections of the Liverpool labour movement. The pre-Windrush-era Liverpool seaport black community had been the subject of institutionalised racism for many decades: Militant never concerned itself with such issues. (In fact it would come into conflict with the community). Nor with the oppression of women (it campaigned against Mrs Thatcher with a placarded cartoon of a flabby airborne middle-aged female body with Thatcher's face, in a tight Wonder Woman costume, carrying a handbag, and the slogan "Ditch The Bitch").

But by 1983, when Labour regained control of the council from the Liberals,

---

42. It would take them some years after 1991 to register, or to admit, that "proletarian Bonapartism" had collapsed and was gone.

43. Who worked with *Socialist Organiser* until his death in 1990.

Militant had a big enough block of councillors to control events, behind a thin screen of "independent left-wingers", for example, council nominal leader John Hamilton. In mass-poverty Liverpool, conflict with the government was inevitable. Militant's representatives openly prepared for conflict, or seemed to, compensating for all the years of quiet burrowing with an orgy of public boasting and confrontation-rattling.

For a year, from May 1983 to May 1984, they organised mass demonstrations, pledged to resist the Tories. An opinion poll in May 1984, two months into the great miners' strike, showed that 55% of Labour voters were ready to back a local general strike against government cuts and government policies. Here were legions of class warriors ready to join the miners in confronting the Government. On any class-struggle overview of the best interests of the working class and the labour movement, in Liverpool and in the whole country, Militant should have worked for confrontation with the Government during the miners' strike, and linked up Liverpool with the striking miners. A battle then would have linked the cause of local government democracy and the Labour far left with the striking miners. It could have given a lead to other left councils. They had that option. They didn't have the nerve or conviction or overview of the class struggle to use it. In early July 1984, when the miners' strike was still at its strongest, Militant chose to do a short-term deal with the Tories. The Tories made some temporary financial concessions that let Liverpool off the hook, and won the Tories time to beat down the miners. Militant's leaders on Liverpool City Council, Derek Hatton and his friends — under the daily control of Peter Taaffe and other Militant national leaders — bought themselves safety from prosecution for exactly a year. Then, the miners beaten, the Tories came back to sort them out. And in the wake of the Tories came Labour leader Neil Kinnock and his friends to complete the rout of Militant.

The truth is that had Militant deliberately worked to demoralise and demobilise the Liverpool working class, it would not have behaved any differently.

Central to sect "Bolshevism" is an all-pervading organisational narcissism. Militant was within the trade-union and Labour Party organisations, but it was a closed-off sect, devoted to itself and its "perspectives", capable of such irresponsibility towards the open class struggle during the miners' strike as we have seen. Like a mechanical mouse, clockwork ticking away, it marched up to the top of the hill and then down again, looking neither to left nor right. Its dominant instinct was not attuned to the working class, the working-class movement, or the class struggle, but to its own narrow interests, as understood by very short-sighted people.

SWP-UK

The second in size of the "groups" was the Socialist Workers Party (SWP). That group had begun as a Third Camp breakaway from the Orthodox Trotskyists when the Orthodox backed North Korea and China in the Korean war. Very small in the 1950s, it grew in the 1960s, and was in the Labour Party and in the Young Socialists until 1967-8.[44] For most of the 1950s the group thought

---

44. The group then prided itself on understanding that capitalist imperialism had come to an end. They would only attain their present "anti-imperialist" pitch after 1987, when they switched to

The 1984-5 miners' strike was a watershed for the left.

of themselves as in the Bolshevik tradition. They switched dramatically in 1958-9 and proclaimed themselves "Luxemburgist", in the decadent-mythic sense of being anti-Lenin and anti-Bolshevik. "For Marxists in advanced industrial countries, Lenin's original position can serve much less as a guide than Rosa Luxemburg's, notwithstanding her overstatements on the question of spontaneity" (Cliff, *Rosa Luxemburg*, 1959).

From the mid-1950s to the mid-60s they had proclaimed it a principle to be in the Labour Party. Then progressively they cut themselves off from the labour movement by way of contrived and shifting definitions of what the labour movement was. First, in the mid-1960s, they redefined the labour movement as constituted only by its militant sections, in effect by the shop stewards' networks of the time. By juggling with words and arbitrary redefinition, they read the real labour movement out of their "labour movement". They proclaimed the need to "link" the militant "fragments" of fighting trade unionism, and offered themselves as the organisation to do it. At that point they began to switch towards seeing themselves as a self-sufficient, "revolutionary party", fully formed and needing only to gain more members, though it took them a while to understand what they were doing, to spell out and follow its logic.

In 1968 they officially rediscovered the need to build a revolutionary party, and proclaimed themselves "Leninist" again. Cliff's pamphlet on Luxemburg was reprinted, and the deprecation of Lenin, in favour of Luxemburg's attributed alternative, was replaced (without any modification of the preceding argument) by the following sentence: "whatever the historical circumstances

---

support Iran in the Iran-Iraq war. They were anti-Israel, but their hostility to Israel had not yet become a world outlook, with all its ramifications. The shift there can be dated to July 1986, when they published a pamphlet, much reprinted since, entitled *Israel, the Hijack State*, whose cover cartoon showed Israel as a mad dog on a leash held by the USA. Their later period of active alliance with Islamic clerical fascism (via the Muslim Brotherhood offshoot in Britain) covered the years 2002 to about 2007.

moulding Rosa's thoughts regarding organisation, these thoughts showed a great weakness in the German Revolution of 1918-19". In substance if not always in style, the SWP started more and more to ape the sectarianism and onanism of the SLL-WRP.

In the late 1970s they decided that the shop stewards' networks had become bureaucratic, and in effect they read the shop stewards too, or most of them, out of the labour movement. The labour movement was thus progressively redefined and step by step disqualified — in their heads — until all that was left was "the socialist alternative", that is, themselves. Tony Cliff said somewhere: if I'm the only man in the world I'm entitled to call myself "Mr Universe". He did. Except that he wasn't the only man in "the world", and with his wild-eyed and furiously ecstatic political onanism he took the organisation out of the real political and class-struggle world, and away from the to-be-or-not-to-be conflict which Thatcher forced on the labour movement in the first half of the 1980s.

The SWP shrugged off the threat of the Thatcher Tories in the 1979 general election. *Socialist Worker* said "vote Labour". The best-known SWPer, the journalist Paul Foot, told the *Evening Standard*: "For the next three weeks I am a strong Labour supporter. I am very anxious that a Tory government shouldn't be returned. and I shall be going around to meetings we are having telling everyone to vote Labour". An SWP leader explained:

"This [the election] is an issue we shouldn't get very excited about. In terms of the national alternatives we have to say 'grit your teeth and vote accordingly'... We will not involve ourselves in the nitty-gritty of election work, because we have a series of operations that we are attempting to mount which will absorb all the energies of our people".[45]

Tony Cliff published four volumes of biography of Lenin, in which he quoted great chunks of Lenin, but only what suited his own current political and organisational needs. It was more than a joke when someone said it was a political biography of Cliff written by Lenin. Now that the SWP were again Bolsheviks, Cliff was again the Lenin of his era.

The SWP stood aside from the Benn surge and the Labour Party. Cliff, with Paul Foot as amanuensis, reduced the significance of the Labour-left political movement to a silly little joke: "If you want to push a wheelbarrow, you don't sit in it". Translation: if you want to change the political labour movement, you do not participate in it. In 1979, as the Tories came to power and the Labour left erupted, the SWP put the coping stone on their turn away from the labour movement and towards exclusive focus on building themselves as "the party", a toy-town pseudo-Bolshevik "party" spinning on its own axis. That was the "theory of the downturn".

It perceived and interpreted the world around it according to an a priori schematism: "the downturn". That ruled out anything more than a small-scale, atomised labour-movement fightback. It was impossible. They knew! They read it off from generalities about "the period" in which nothing could be done except "build the party". In practice the SWP — together with the CP, though in a different style — preached universal defeatism in the life-or-death struggle

---

45. Duncan Hallas, interviewed in *Socialist Challenge* 12 April 1979

of the labour movement with Thatcher.

To the inglorious chronicles of kitsch Bolshevism it added an astonishing, epoch-defining display of sectarian obtuseness and irresponsibility. The "downturn" thesis licensed the SWP to recoil, like a snail from salt, into living in its own shell and augmenting it. "Build the party!" Build the shell! Barnacle Bolshevism! But it was a "Bolshevik" shell, festooned with quotations from Lenin. Politically, to switch the image, in relation to the Tory offensive, it was throwing in the towel, with a photo glued to it superimposing Cliff on Lenin.

When the miners in the great strike of 1984-5 launched what was the last great battle of the old-pattern labour movement, the SWP was disoriented for the first six months. Join or help the miners' support groups? No! That was just "left-wing Oxfam". Better to concentrate on selling more copies of *Socialist Worker* and try to recruit new members to the "revolutionary party". The SWP made appeals to Labour Party activists to "join the socialists", that is, to abandon the fight in the mass political labour movement and join those who preached that a general fightback was impossible or hopeless.

Didn't the miners' strike contradict the idea of an absolute downturn? Not at all! Rubbish! Tony Cliff pronounced: "The miners' strike is an extreme example of what we in the Socialist Workers Party have called the 'downturn' in the movement" (*Socialist Worker*, 14 April 1984). Wet is the new dry, white is the new black, gobbledegook is the new Marxism... Cliff is the new Lenin!

That he could do that — think that, let himself say it — and that he could get away with it, was proof, if proof were still needed, that he had turned the organisation into a politically dead sect.

Socialist Organiser

What would become AWL then produced the paper *Socialist Organiser,* initially as a broad Labour left paper. It was an Orthodox Trotskyist grouping, in the strongly but incoherently anti-Stalinist tradition of the James P Cannon of the early mid 1950s. It would break decisively from the Orthodox Trotskyist collective orientation by opposing and denouncing the Russian war in Afghanistan from December 1979, the event which triggered the "Second Cold War". Russia had been in bad odour with most of the left after the invasion of Czechoslovakia in 1968, to which even the Communist Party had objected. The new cold war shifted that backwards. Russia, including its colonial war in Afghanistan, was after all the progressive alternative to imperialism. CND became a mass movement again.

Socialist Organiser — perhaps too ambitiously, but under the compulsion of the situation the labour movement was in — tried to reorganise and unite the left. It raised the alarm against the Thatcher threat. With the 1978 Socialist Campaign for a Labour Victory conference it took the initiative in organising the Labour left against the Labour establishment, attempting to combine that with organising to fight the Tories. It also initiated the Rank and File Mobilising Committee that emerged in 1980.

The group itself was in a prolonged process of "revaluation of values". In theory it had James P Cannon's Zinovievite or Orthodox Trotskyist version of "Bolshevism", in which the organisation must always present a single policy to the outside world; political or political disputes, when they are allowed, must remain strictly internal and, usually, of pre-set duration; all initiative, even to allow or start or finish a discussion, is in the hands of a leading committee. But from the start it had in practice modified that so that the onus of proof was always to show specific reasons why there should not be open and public discussion on a particular issue.[46]

Most important had been the group's experience of the clerical-fascist revolution in Iran. There was a vast wave of sympathy and support outside Iran for the anti-Shah revolutionary movement. It was anti-imperialist and anti-West. Many currents combined in that movement. Democrats rebelling against dictatorship tended to cast the Shah as embodying dictatorship, and the mass movement hegemonised by political Islam as representing democracy.

Stalinists and Orthodox Trotskyists saw it as directed against the Shah's western alliances and therefore to the benefit of the "defence of the USSR". Among some Orthodox Trotskyists, politics had become so hazy and unfocused that they let themselves hope that "the Iranian (Islamic) revolution" would go beyond capitalism and inaugurate a socialist Iran, though in reality it had more in common with Hitler's reactionary revolution in 1933-4 than any other. The great swirling movements of millions of roused-up people — demonstrations, strikes, mass meetings — were an intoxicating spectacle to see and respond to, especially for revolutionaries who had learned not to ask awkward questions of any revolution that presented itself for their consideration.

There had been opposition to supporting the Islamic revolution within the group but it had been numerically insignificant. Events quickly showed, to people paying attention, that the leftist enthusiasm had been a stark expression of political bankruptcy: the Iranian clerical-fascist revolution was one of the most reactionary events of the 20th century. By the time the Iran-Iraq war broke out in September 1980, the group had sobered to the point that the advocates of support for Iran were now the numerically insignificant group. (The issue was debated openly, though pseudonymously, in the group's paper of that time, *Workers' Action*).

The Iranian experience prepared the group for the cold look it took at the Russian "expansion of the revolution" into Afghanistan, and for the Third Camp conclusions it drew.

No doubt *Socialist Organiser* committed stupidities and had inadequacies, including residual sectism. It was too slow, too careful and too timid in its political evolution. But its main limitation was that it was just too small. Its possibilities of growth were cut off by the bigger Trotskisant organisations and by

---

46. Its response to a number of events in the second half of the 1970s had been atypical of the Orthodox Trotskyist groupings. The terrible Stalinist regime in Cambodia, which in three years killed two million people, a quarter of the country's entire population, had, on the face of it, as much right as, say, China, to be characterised as a (deformed) workers' state. Some in the international Trotskyist current — the SWP-USA and its spread of co-thinkers in the Fourth International — ducked the issue with the judgement that Cambodia was entirely special and not a workers' state of any sort or degree. In fact, what happened in Cambodia was a variant, a repetition, intensified and condensed into a shorter time, of things that had happened in China and earlier in Stalinist Russia.

Ken Livingstone and Ted Knight hold up a copy of *Labour Herald*

the hostile reaction of many leftists against the "toy-town Bolshevik" groups. And *Socialist Organiser*, of course, like everything attached to the labour movement, shared the consequences of the defeats of the movement from 1979 onwards.

It was in Britain in 1979-83 as it was for the communists of Europe in 1919 — a matter of building fire engines when the fire was already raging. This time, in a political world where a number of vehicles labelled fire engines were rushing about, bells clanging and sirens wheezing. They were not fire engines, but toys with imaginary fire-hoses, driven by wild-eyed onanists who though they were somewhere else.

A large swathe of the Labour left had had involvement with or had observed the "neo-Bolshevik" organisations in action, and consequently rejected Bolshevism, including the proto-AWL's version of it. It was somewhat like the problem Trotsky described for the Fourth International in 1938: "The same generation which heard its [the Third International's] formation now hears us, and they say, 'But we have already heard this once!'"[47]

---

47. There was also the Mandelite organisation, the International Marxist Group (IMG), whose remnants have recently been sojourning, first in George Galloway's Respect, then in the Green Left faction of the Green Party, then in the short-lived Left Unity organisation launched by people quitting Respect, and now in the Labour left. The IMG was not very big, though active enough to be a visible factor on left-wing demonstrations and the like. Politically, it was a wildly unstable organisation, consuming itself in battle between a number of factions, some semi-permanent, kept together by a common adherence to "the International". It zig-zagged between calling on the left to break up Labour Party meetings in June 1970 (bit.ly/rm-700601) to joining the Labour Party, then pulling out, then jumping into the Benn current with a sycophantic book, *The Benn Heresy*, then... It lived, bred and bequeaths political confusion.

Left antisemitism

This was the time when absolute anti-Zionist antisemitism came to be a major aspect of the left. Israel invaded Lebanon in June 1982 in an attempt to destroy the armed PLO there. The PLO would recognise Israel's right to exist in 1988, and more explicitly in 1993, but in 1982 a state of no-quarter-given open war existed between Israel and the PLO. The delicately unstable balance of the multi-confessional Arab state of Lebanon had been shattered by the presence of strong armed Palestinian forces in the country: civil war had raged there since 1975.

Maronite Christian Arabs massacred many hundreds, maybe thousands, of people in the Sabra and Shatila Palestinian refugee camps, in September 1982. Overall control of that area was held by the Israeli army. You didn't need to be mortally hostile to Israel to attach some or most of the blame to those in overall military control. There were huge protest demonstrations inside Israel.

Among those who were already convinced that Israel was an illegitimate state without any right in principle to defend itself, hysteria fed on these events. *Socialist Organiser*, too — where advocates of Israel's right to exist side by side with a Palestinian state were still a small minority — was caught up in the "anti-Zionist" hysteria and scandalously fed the campaign against "the Zionists".

Between 1949 and Stalin's death in 1953 a series of show trials had been mounted in the satellite states of Eastern Europe in which many of the victims, leading Stalinists such as Rudolf Slansky in Czechoslovakia and Lazlo Rajk in Hungary, were accused as "Zionists" and enemy agents. The Stalinist press throughout the world echoed the "anti-Zionist" antisemites of Moscow, Budapest, and Prague.

The inner natures of Nazism and Zionism were symmetrically racist. The Zionists-Jews were "rootless cosmopolitans", part of a great international network of reactionary conspirators. Israel — which Stalin had at first backed, as a potential tool against the British Empire — was a spawn of imperialism and racism. It was a racist state.

In 1952 five Jewish doctors who had tended to the dwellers in the Kremlin were accused of poisoning and planning to poison the leaders. Stalin died before they could be brought to trial, and they were released. If Stalin had lived, a large-scale rounding-up of the USSR's Jews would probably have been the outcome of the "anti-Zionist" antisemitic drum-beating. And possibly Stalin would have taken up Hitler's cause and dealt out mass murder to Russia's Jews.

The poison that now warps the minds and politics of many people on the left originated in that Stalinist campaign. It was later rehashed in books published by Lenni Brenner in 1983 and 1984.

In 1976-7, as we have seen, the leaders of the steeply declining WRP sold its services to Muammar Gaddafi, Saddam Hussein, and the PLO as a propaganda outlet for them and a spy agency on prominent Jews ("Zionists") and on Arabs in Britain.[48] They published two big WRP pamphlets glowing with

---

48. Workers' Liberty 3/62, bit.ly/wl3-62

adulation of Gaddafi and Saddam Hussein. They could have been published by their regimes, and they certainly were paid for by them.

Part of the deal was that the WRP would make anti-Jewish propaganda. They salved their political consciences by saying Zionist when they meant Jew, but there was no mistaking the meaning. They saw "Zionist" conspiracies everywhere: for instance when a Jewish man, Stuart Young, was made chair of governors of the BBC, they roused a huge and cry in the labour movement against… Zionists.

The weekly paper, *Labour Herald*, which the WRP produced for Ken Living-stone and Ted Knight, echoed the WRP's hysterical "anti-Zionism" in articles and cartoons: one, in June 1982, showed Israeli prime minister Menachem Begin in SS uniform and giving a Nazi salute.

When in 1985 Livingstone went for and won the Labour parliamentary se-lection for Brent East, the WRP ran a vicious campaign against the sitting Labour MP, Reg Freeson, as "the Zionist Freeson", that is, the Jew Freeson. Even after the WRP collapsed, among much scandal, in 1985, Livingstone never dissociated himself from it. As late as 1994, he wrote a friendly intro-duction to a hagiographical biography of the WRP's Gerry Healy, who had died in 1989.[49]

The SWP too had a hardened animosity to Israel. The central figure in the group, Ygael Gluckstein (Tony Cliff), was in background a Palestinian-born Jew. He could say things that someone who was not Jewish could not get away with saying, still less get complimented for his candour and dauntless inter-nationalism. His wife Chanie, who was active in the group, and her brother Michael Kidron, who was for a long time a central figure in the group too, also had a Palestinian Jewish background.

Cliff's political evolution and history on this question can be traced in the files of the Trotskyist press. On the eve of war in 1939 he was involved in ex-changes about Palestine in the American magazine *New International.* At that point he was in favour of free Jewish immigration into Palestine.

In December 1944 the British Revolutionary Communist Party published an Open Letter from Palestinian Trotskyists (unsigned, but written by Cliff) to the Labour Party conference, urging delegates not to back a pro-Zionist motion from the Executive. As that text showed, somehow, living in Palestine, in the shadow of the Holocaust and in the course of it, Cliff had become a hardened opponent of free Jewish immigration.

It is tempting to explain this as neurotic guilt focused against Jews who had got to Palestine and like himself survived. One can speculate. But the hard po-litical fact is that some time in the period between the outbreak of war and 1944, he turned into a lifelong enemy of the Palestine Jews.

He left Palestine in September 1946. At a time when many Jews were still in displaced persons' camps in Europe, he had the resources to arrange a visa to go to Ireland as a student and to spend enough time in Britain to become part of the British Trotskyist movement.

Cliff's was a remarkable shift of opinion to make in the early 1940s and in

---

49. bit.ly/kl-gh

Palestine. In an interview in *Socialist Review* 100 (July-August 1987)[50], Cliff would say that he was wrong to favour free Jewish immigration to Palestine before the war. Considering the alternative — death — which those who escaped from Europe had avoided, that was to say that it would have been better if they had not escaped. Did he understand that? Did he mean to say that?

In the 1948 war, none of the Trotskyist organisations backed the Arab states. Nor did Cliff. He was silent on the question for many years thereafter.

An anti-Israel tone became dominant after the 1973 war, but the SWP's shift to absolute hostility to Israel becoming a world outlook, with all its ramifications, was as late as 1986-7[51].

---

50. bit.ly/cliff-55

51. There is a peculiar problem in establishing the political history of the SWP. The documentary evidence, and the gaps in it, can seriously mislead.

In 1958-9 the Socialist Review group, as it then was, republished as a duplicated pamphlet Rosa Luxemburg's 1918 criticism of the Bolsheviks. It took the text and the footnotes from the version put out by the Lovestone organisation in the USA in 1940, which would later also be republished by the University of Michigan Press, in 1961. The Lovestoneite pamphlet had registered the disbandment of what had been in the 1930s an important political element in the US labour movement, which had had important international links. Tony Cliff in Palestine had been a member of the equivalent group before becoming a Trotskyist. Socialist Review republished the pamphlet complete with Bertram D Wolfe's footnotes and with a short introduction by Cliff. That pamphlet, more than Cliff's later booklet on Rosa Luxemburg, was central to the group's self-transformation into "Luxemburgists", as distinct (so they saw it) from "Bolsheviks". As far as I can find, the pamphlet received no mention at all in their press, nor, as was customary then, did they advertise it in the Labour left weekly *Tribune*.

There is a copy in the British Library, but the only record of it from the time is a review by Walter Kendall in the Independent Labour Party the *Socialist Leader*: the ILP had been the "Luxemburgists" up to that point.

Similarly with the SWP's sudden call in 1992 for a general strike. You don't get an adequate picture of that from their press. It was a thing of mass-produced wall-posters and placards on demonstrations more than of their press.

Their pro-Serb politics during the Serbian state's attempt to ethnically cleanse Kosova of its ethnic-Albanian population: that is in their press, but the collection of pieces they afterwards published as a book on Kosova does not convey an adequate account of what they said and did during the 1999 conflict.

The left today

How does the left look today, measured by what it needs to do? The *Morning Star*, the paper linked to the Communist Party of Britain, circulates among a leading layer of the labour movement. Jeremy Corbyn used to write a column in it until his unexpected election as Labour leader in 2015. If there is a political organ of the Corbynites — Corbyn himself, Seumas Milne, Andrew Murray, and others — it is the *Morning Star*.

It shows that they are a Stalinoid formation. The strange difficulty the Corbyn group has in responding to the charge that they are antisemitic and clearing themselves of it is part of that. Corbyn is for two states, but he has difficulty in candidly declaring that advocates or supporters of the conquest and destruction of Israel by Arab nationalist or Islamic forces are antisemites. Some of the Corbyn inner circle do believe Israel should be destroyed. All of them had and have connections and shared platforms with people who do. The *Morning Star* coverage of the Israel-Palestine conflict is, at least, blatant incitement of antisemitism.

Nine trade unions are represented on the board of the "People's Press Printing Society", which publishes the *Morning Star,* and pay subsidies to the paper. In 2016 "The World Transformed", the large Labour Party fringe festival promoted by Momentum, was co-sponsored by the *Morning Star* and had the *Morning Star* prominently on sale within the event, while other left-wing papers were officially barred from selling there.

The *Morning Star* started in 1930 as the *Daily Worker*, a Communist Party daily paper. It was banned from January 1941 for pro-German "anti-war" advocacy during the Hitler-Stalin pact — August 1939 to June 1941 — and resumed publication in September 1942. It became the *Morning Star* in 1966. Its first words critical of Russia appeared only in its 38th year, and the CP's 48th, when Russia invaded Czechoslovakia to kill the experiment there in creating "socialism with a human face". Between 1930 and 1968 it supported and seconded everything Moscow did and said and hoped for. Between 1968 and 1980 it supported *virtually* everything Moscow did and said and hoped for. It remained generally supportive of the Russian and other European Stalinist regimes until they collapsed in 1989-91. A large bulk of its circulation was to Stalinist-ruled states, which thereby subsidised it.

Journalistically the *Morning Star* is a seriously bad effort, without political coherence, flair, élan, political fire, or much positive socialist conviction. To this day it retains something astonishingly like an attitude of support to Russia and to the totalitarian-capitalist fascistic regime (what else is it?) in China. For instance, it sides with Russia against Ukraine. This goes further than negativism towards the USA, though that, it seems, is basic. It is as if the *Morning Star* sometimes relapses and forgets what month of the political calendar it is, or as if those who publish it often do not know quite what they believe. And, positively, most of the time, they don't. Their comments on events are arbitrary and erratic. And, as in the Russia-Ukraine conflict, often reactionary. They are enthusiasts for Brexit.

In terms of its politics now, the *Morning Star*, like much of the ostensible left, is uncementing sand, pulverised by great upheavals and the all-reducing passage of the Stalinist glacier.

The *Morning Star* is the "left" shorn of positive principles or any coherent notion of what the socialism its publishers think they promote is and must be. A central characteristic of the ostensible left, more widely than just of the *Morning Star*, is that often it does not know, positively, what it is for. A viable new left will never be raised on the mildewed crumbs of old Stalinist politics and animosities.

There are a lot of people on the Labour left now — more than in the 1980s — who deny the need for any explicit socialist organisation. They argue that being individually active in the Labour left (and usually also in Momentum) is enough, and even that anything more ("paper-selling" and such) is "sectarian", unnecessary, maybe harmful. Sometimes they go along with, or part of the way with, the *Morning Star* current. Sometimes they do not; but they see no need for an activist across-the-board opposition to that trend.

Some of these are people who were active in the 1970s or 1980s — often in or around a Trotskisant group — and have now returned to political activity via the new Labour left. Their attitude is shaped by disillusion with the group they were active in or associated with. Others are younger, sometimes influenced ideologically by "NGO politics".

On the pseudo-Bolshevik left, the biggest force is the SWP. In 1979-85 the SWP remained a part of the rational left, albeit a very sectarian and cult-ridden part of it. They had, for example, resisted the "anti-imperialist" demagogy that led much of the left in 1982 to back Argentina in the South Atlantic war. All that changed in the second half of the 1980s.

In 1987, seven years after the start of the Iran-Iraq war, the SWP decided to support Iran. Reason: the Americans, you see, were on the side of Iraq. As they had been from the start...

In a world in which the bigger ex-colonial or ex-semi-colonial countries had achieved independence (Russia's sphere of control was the exception) and become important economic and political powers, the SWP rediscovered "anti-imperialism". It seemed a cheap (and usually meaningless) form of militant posturing. The SWP's hostility to Israel reached the screeching level of absolute anti-Zionism in 1986-7.

The cult of their Little Great Man, Tony Cliff, was airtight and windowless. The long-time preacher of the idea that in "the downturn" nothing much could be done then in 1992, eight years after the miners' strike, suddenly advocated a general strike against the Tories' decision to close the pits of the "working miners" — the scabs — of the 1984-5 battle. The SWP had scorned talk of a general strike during 1984-5, when it made some sense. Naturally, those few who resisted the sudden brainstorm were expelled from the ranks of Cliffite Bolshevism.

From about 1988 the SWP edged away from the "downturn" idea, at first vaguely announcing more "volatility". The SWP reoriented to seeking high union positions, where their members have become cautious and passive in the name of "left unity".

Tony Cliff died in 2000, at the age of 83. In 1999 the organisation started a

"Stop The War" campaign in support of a Serbian state that was attempting to kill or drive out ethnic Albanians in its colony, Kosova (93% of the population). What the chauvinism-drunk Serbs were capable of doing had been laid out before the eyes of the world in Bosnia. During the Bosnia war, as earlier in Serbia's war with Croatia, the SWP had remained its old self: aloof, refusing to declare any degree of support to the Bosniac Muslims, not even condemning the arms embargo which the West imposed on people fighting for their lives against opponents who controlled the armoury of the old Yugoslav state. Now it started a campaign in support of Serbia, against NATO. It was a gruesome example of the snares of mainly-for-posturing "anti-imperialism". They actively campaigned for a state engaged in attempted genocide, when the only issue in the NATO-Serb war was that attempt at genocide. (The war stopped when Serbia pulled its troops out of Kosova).

The "Stop The War" brand name was given a new coat of political paint and relaunched against the invasions of Afghanistan (2001) and Iraq (2003). On Iraq, when the "Stop the War" campaign was able to organise big demonstrations, the SWP entered into a civil partnership with the soft left Labour MP George Galloway.

Galloway never had even halfway serious left-wing credentials. He was an unreconstructed old middle-of-the-road Labour Party Stalinist who in the 1990s had become a freelance London spokesman for the quasi-fascist Saddam Hussein regime in Iraq. In the House of Commons he was jocularly, and not so jocularly, called the "MP for Baghdad Central". He had gained infamy in 1994 when he visited Baghdad with a delegation of European MPs to present Saddam Hussein with a pennant from admirers in the West Bank and, in front of BBC cameras, told the dictator: "Sir — we salute your courage, your strength, your indefatigability... We are with you... Until victory! Until Jerusalem!"

In 2004, together with this fragrant and dainty citizen, the SWP formed "Respect — George Galloway" (its official title on ballot papers). Those who in Kosova had frantically backed the people murdering Muslims en masse, now advertised themselves as "the best fighters for Muslims". They turned themselves into would-be communalist politicians. They publicly apologised for having defended Salman Rushdie, fatwa'd by the clerical-fascist Iranian regime in 1989 for writing a novel, *The Satanic Verses*.

When it came out in 2007 that Respect had taken a lot of money from a Dubai businessman, they gave it back. Respect subsequently split. The people who had run the Respect office and taken the money hived off from the SWP in 2010 and are now *Counterfire*: they run the Stop the War campaign and the People's Assembly.

Since Cliff's death the SWP leaders seem to have lost the confidence and elasticity required for any substantial reassessments. Their public profile is mediated mostly through an array of "broad" campaigns. In the unions they are cautious, distinguishing themselves mainly by what they call "political trade unionism", meaning, in their own words, such things as "using petitions, winning delegations to... anti-racist protests". That is sufficient? That is sufficient.

The SWP present themselves as enthusiastic supporters of Jeremy Corbyn, criticising him for his voiced objections to Brexit and to left antisemitism, which they explain as Corbyn conceding to the Labour right. They remain outside the new Labour Party on the grounds that "strikes and demonstrations" are necessary for progress as well as parliamentary politics, and who is more expert than the SWP at such activities?

They retain a fundamental conception of "the party" (themselves) as an organisational machine rather than an instrument for ideological education and political clarity.

The Socialist Party, after their debacle in Liverpool, declared that Labour had now become an ordinary bourgeois party. They themselves split and scattered in the early 1990s. Their venerable prophet, Ted Grant, discredited like his "perspectives" by the collapse of European Stalinism and events in the Labour Party, was expelled, together with those who sided with him. Grant objected to the turn out of the Labour Party[52] and at that point was not convinced that the progressive-totalitarian "degenerated and deformed workers' states" had collapsed. The SP exchanged its Little Great Man for a Littler Great Man, Peter Taaffe. Later in the 1990s the SP's main concentrations of membership, in Glasgow and in Merseyside, split away.

The SP revived a bit after about 2001. What the SP believes about the world now is not clear, at least not to this observer. For example, they have in theory pretty much the same position on Israel-Palestine as AWL does — two states — but they usually hide it. They retain the traits that endeared them to outsiders for so long, including the idea that you don't say what "the workers wouldn't understand" and might hold against you. Building on their support for British exit from the EU, they have extended that stance into an explicit opposition to free movement for workers across borders.

Since about 2000 the SP has held control of the PCS civil servants' union, in collaboration with the union's former soft-right.

It is much more usual than it was in the 1980s to find individual SP and SWP members openly sceptical about their organisation's policies. In terms of internal debate the regimes remain pretty airtight, the SP more so than the SWP.

The No-Party people

Across the valley, so to speak, from the toy-town Bolshevik shell-backed parties lived, in the 1980s, a lot of people who thought of themselves as Marxists but who had grown indifferent or hostile to any project of building a Marxist organisation.

This tribe, and it was quite an important component of the Labour left, marched or ambled, in so far as it expressed itself explicitly, under the idea: we will develop the influence of Marxism by promoting left-wing ideas in the existing broad labour movement, trade unions and Labour Party. No socialist organisation beyond the Labour Party and its coteries and careerist cliques was needed. The existing structures were sufficient.

This view was not often expressed in coherent argument, yet it was a most important current of thought or half-thought in the labour movement, the "po-

---

52. He seems earlier to have been a critic of the Militant's antics in Liverpool, which was under the direct guidance of Peter Taaffe.

sition" of numbers of ex-WRP, ex-SWP, and ex-Militant people who turned the sectarian fetish of "building the party" inside out, and of younger activists who took their cue from them.

This is an important question again now, in the days of the Corbyn surge. Routine labour movement activity is counterposed to the creation of a Marxist movement that is politically and organisationally independent, has a distinct job to do, and rhythms and short term concerns "of its own".

The structures and ethos of either the Labour Party or the trade unions can not substitute for the specific structures required for all-round Marxist — Bol- shevik — activity on the three fronts of the class struggle, the trade-union, the political front, and the battle of ideas.

You cannot meaningfully develop the "influence of Marxism" as a revolu- tionary force without building a revolutionary party. In the 1920s Trotsky thought that such a party, the Communist Party then, might slot into the ex- isting framework of a union-backed broad Labour Party. "The Communist Party will occupy the place in the Labour Party that is at present occupied by the Independents"[53].

At the end of the day, both formulas — spreading Marxism and campaigning to invigorate the labour movement in general, and building a revolutionary organisation — mean one and the same thing. At a certain point in the process they will have matched up and merged into one: a mass revolutionary party at the head of the broader labour movement. It is a matter of working out con- cretely at a given moment which is best of the possible ways the organised col- lective of Marxists, be they more or less numerous, can relate to an existing mass reformist labour movement and to ongoing working-class struggles.

More. The Marxists organise themselves so as to fight the class struggle on all three fronts — now. It is ridiculous to suggest that Marxists must wait until the movement is transformed before immersing themselves in the immediate class struggle, trade-union struggles, for example, and doing that effectively, that is, as an organised force. Equally ridiculous is the related idea that an or- ganised collective of Marxists able to act coherently as a fighting organisation is useless in the class struggle here and now. Or any idea that we can transform the labour movement apart from the class struggle.

In 1984-5, the miners' strike could have been won with solidarity action by dockers and other key workers, or had what remained of the local government left, notably Liverpool, opted to fight and coordinate its activity with that of the striking miners. A network of rank-and-file activists in key positions across industry, even if only a few thousand strong, might have won solidarity for the miners — that is, made the difference between possible victory and the all- too-real defeat for the miners, and for the whole working-class movement.

Who will build that sort of movement if not the Marxists organised as a mil- itant, distinct (for now) minority? If not now, when; if not us, who? Without revolutionary organisation we can only babble or maintain a preaching sec- tarian aloofness, muttering whatever self-consoling excuses we can foist on ourselves.

---

53. That is, by the ILP, which then, through Ramsey McDonald, Philip Snowden, and others, led the Labour Party: *Where Is Britain Going?* chapter 8

This is the answer to those who conclude from a bad experience with the kitsch-Bolshevik organisations that everything a small Marxist organisation does, beyond routine labour-movement activity, is futile and sectarian.

Moreover, individual Marxists naturally vary in knowledge, experience, and aptitude from person to person and area to area. One purpose of a Marxist organisation is to raise the level of the Marxism which the Marxists educate for to the highest level the collective can achieve.

The organisation needed to do the things that the Marxists must do, and only Marxists can do, has to be built over years of smaller struggles, in advance of the big struggles and crises. It has to sustain, and educate itself, by formulating and checking adequate collective responses to political events. That cannot happen without the continual interaction of the Marxist organisation with the working-class struggle and the mass working-class and other movements. A "Marxism" lacking embodiment in a militant organisation which strives for leadership in economic and political struggles would be like the clock with neither battery nor spring nor digits: an absurdist joke.

Two short quotations from basic Marxist texts sum up the Marxist position here:

"The Communists do not form a separate party opposed to the other working-class parties. They have no interests separate and apart from those of the proletariat as a whole. They do not set up any sectarian principles of their own, by which to shape and mould the proletarian movement" — Karl Marx and Friedrich Engels, *The Communist Manifesto.*

"To base one's program on the logic of the class struggle... these are the rules" — Leon Trotsky, *The Death Agony of Capitalism.*

Socialists who know the ABCs of Marxist politics do not wilfully try to "build the party" apart from the labour movement and the working class. But, equally, they do not sink the revolutionary group into the rhythms and norms of a labour movement which is not yet revolutionary and which at best involves only a minority of the working class. That is as much a recipe for suicide as the antics of the sectarians — by an overdose of tranquilisers rather than an excess of one or other of the sectarian hallucinogenics.

The labour movement now has another chance to transform, sharpen, and rebuild itself in politics. That transformation cannot happen spontaneously, as a spin-off of trade union class struggle. Nor can it happen as a by-product of political reform-socialist efforts (for example through the Labour Party under Corbyn). Unless the Marxists are strong enough to shape events we get fiascos and muddle and confusion as in the Bennite left of the 1980s. And catastrophic defeats and regressions. Marxists know that as well as evolution there is devolution, regression, defeat. That was true around 1980. It is true now. The politics of the Corbynites and their *Morning Star* stink not only of their own wrongheadedness but also, already, of the new defeats they may well bring down on the labour movement.

The last thing the working class needs is another pseudo-Bolshevik "revolutionary party". But it does need a democratic, rational, non-sectarian Bolshevik organisation to fight the class struggle against the bourgeoisie and for Marxist ideas and class struggle militancy in the labour movement. Again: if not now, when; if not us, who? We must build a democratic Marxist organisation, not a pseudo-Bolshevik sect counterposed, or half-counterposed, to the mass labour movement — a coherent three-front class-struggle Marxist organisation. We must do that, as slowly as necessary and as quickly as possible.

A Marxist organisation, to be effective, to be Marxist in any solid sense, must be an organisation where regular activity with the organisation and regular socialist self-education are conditions of membership.

It must have coherent, coordinated, planned, collective activity. It must have a structure of democratically elected and accountable committees and organisers capable of deciding and carrying through prompt political responses. It must supervise its members who gain official positions in the movement.

To proselytise, to promote its ideas, it must publish and distribute newspapers, pamphlets, magazines, books, leaflets, workplace bulletins, resolutions. No systematic development of Marxist politics, any more than of any other substantial connected body of ideas, is possible without systematically defining, studying, and criticising ideas in print. It must use the internet systematically, too.

These requirements for a Marxist organisation are liable, in today's left, to be censured as "sectarian". They are in fact part of what we must learn from the real history of Bolshevism in its great days. They are necessary today if the labour movement is really to be transformed.

Gregory Zinoviev wrote to the IWW in January 1920, as the Communist International was gathering its initial forces: "History does not ask whether we like it or not, whether the workers are ready or not. Here is the opportunity. Take it — and the world will belong to the workers; leave it — there may not be another for generations..."

The WRP Healyites and then the SWP turned into retrograde sects because they did not understand how the work of building the revolutionary party must be related to the already-existing mass labour movements, and to other socialists. And because they understood Bolshevism as a matter of building a near-depoliticised machine. Where their mirror-image "Marxists" sink — often without trace — completely into the existing labour movement, the sectarians conceive of "building the Leninist party" as a process more or less fully autonomous from the existing movement and even, sometimes, in practice if not in lucid theory, from the working class.

The idea that Marxists can be fully autonomous in relation to the working class and the labour movement is absurd. Yet some autonomy of the Marxists is essential. You cannot do all that needs to be done by way of the existing structures of the British labour movement alone, or by way of the existing mass consciousness of the labour movement. Both remove or minimise the creative activity of Marxists as an organised force in the present and future evolution of the mass labour movement. That is the real lesson which the experience and the work of the Bolsheviks passes down through a hundred years to our time and our struggles.

The organisational sectarians are sterile because they stand aside, mistaking political onanism for fruitful political activity; the others are sterile because they cling self-distortingly to the existing structures and become parasitically dependent on them, incapable of independent initiative. Passengers are not builders of new tracks and better engines! They fail to develop the sinews and muscles of an independent organisation in relation to the class, the class struggle, and the existing reformist labour movement. They fail to be what socialists must be: the representatives of the movement's future, active in the here and now to carve out that future. James Connolly said it well: "The only true prophets are those who carve out the future they announce".

Of course, the majority at a given moment has to set the politics and the goals of the organisation, and democratically elected and controlled officials have to be given authority to direct work day-to-day. If it is to have effect, the organisation must be made up of disciplined members, with the discipline not of the sect or the cult but of thinking Marxists who reason and argue and democratically arrive at the conclusions they carry out. Within that framework, without which the organisation would in practice amount to very little, there has to be full democratic freedom of opinion and freedom to express that opinion and organise people to fight for it in the organisation.[54]

Honour the Bolsheviks!

Crude parody of the Bolsheviks can be no proper part of the attitude and approach of those who aspire to learn from them. In his well-known passage about history repeating itself, once as tragedy, then as farce, Marx continues:

"... Luther put on the mask of the Apostle Paul, the Revolution of 1789-1814 draped itself alternately in the guise of the Roman Republic and the Roman Empire, and the Revolution of 1848 knew nothing better to do than to parody, now 1789, now the revolutionary tradition of 1793-95. In like manner, the beginner who has learned a new language always translates it back into his mother tongue, but he assimilates the spirit of the new language and expresses himself freely in it only when he moves in it without recalling the old and when he forgets his native tongue".

The Great Russian Revolution was not made, as its detractors and some of its politically numb-brained defenders try to insist, to engage in the fantastic experiments of social engineering from above by a totalitarian state developed by the Stalinists in Russia, China, Cambodia and other places. It was made to win the liberation of the working class. Bolshevism in politics was the fusion of Marxism and tremendous working-class militancy and creativity.

The Bolshevik party was not what the Stalinists and the bourgeoisie say it was. It was what Rosa Luxemburg said it was:

"Whatever a party could offer of courage, revolutionary far-sightedness and consistency in an historic hour, Lenin, Trotsky and all the other comrades have given in good measure. All the revolutionary honour and capacity which western Social-Democracy lacked was represented by the Bolsheviks. Their October

54. The November 1995 Workers' Liberty conference wrote into its constitution the long-existing right of people with dissenting views to publish these views in the group press. It broke the National Committee monopoly of political initiative inside the organisation by giving the Disputes Committee (or a percentage of the membership) the right to call a conference even if the NC majority did not want it.

uprising was... the salvation of the honour of international socialism".

Uncovering and telling the truth about the great working class revolution of 1917, is one of the irreplaceable things which socialists must do now. Not alone to do justice to the great women and men who made that revolution and to its leaders, in the first place, to Lenin — and to Trotsky, who died defending it against the Stalinist counter-revolution. Because knowing, studying, and understanding the Bolshevik Revolution is a major part of learning how to rebuild a viable working class socialist movement in the 21st century.

Those who forget the glories of the working class's past, or lazily let them be buried under mountains of ignorance and ruling class misrepresentation, diminish both themselves and the prospects we can plausibly point out to the working class about its future.

The Bolsheviks did not say the last word on socialism. If there is a last word, it has not been said yet. But they said much that socialists now need to heed, learn, remember, and work to apply in our conditions.

Whenever it comes to discussing the Bolsheviks or the rearguard Bolsheviks in defeat, the Trotskyists, I think of an Irish republican song from long ago. An old woman is discovered crooning to herself, ruminating about the revolutionaries of 50 or 60 years in the past, the Fenians. She sings a refrain: "Glory O, Glory O to the bold Fenian Men".

Some fell by the wayside
Some died mid the stranger
And wise men have told us
That their cause was a failure
But they stood by old Ireland
And they never shunned danger
Glory O, Glory O, to the bold Fenian men

In one of the verses she reflects: "We may have great men but we'll never have better".

Socialism will have great women and great men and great socialists in the future who will learn from the Bolsheviks (from their mistakes, too). They will succeed in doing more and doing it more permanently. But a big part of the socialist tradition that will inspire and educate them in doing this work is the heritage of the Bolsheviks and the October Revolution. Of the Bolsheviks it is appropriate to say what the Fenian woman in the song says about the revolutionaries of another age: Glory and honour to the Bolsheviks and their memory! Learn from them. And think for yourself.

Sean Matgamna

Under the banner of Marxism

In an age of apostasy

The Marxists are under tremendous pressure. The Second World War has come to an end but no socialist revolution has come to destroy capitalism. Where capitalism has been weakened or wiped out, it has been to the profit of Stalinism. Where Stalinism has profited, the working class and its vanguard have lost.

Before our very eyes, capitalism and Stalinism are preparing the Third World War[1], a war to the death to determine which system of exploitation and oppression shall rule the earth. The conflict between these two so mighty giants dominates all political thinking and all political life. The world is already so completely divided between the American side and the Russian that no other pole of attraction is visible. Because no other is visible, it seems that no other is possible, at least not before the conflict turns into a world-shattering explosion.

Where in this conflict is the proletariat, its independent program and goal, its own army and leadership? Program, goal, army and leadership of its own, it has none today. Docile or discontented, apathetic or restless, disheartened or disorientedly combative, it is everywhere under the thumb of drillmasters of imperialism and in its service. Nowhere does it fight its class enemy with a socialist consciousness. It fights only the "enemy of its enemy".

Where in this conflict is the revolutionary Marxian movement? It is everywhere small and isolated, weaker than it has been in decades. It has the ear of few workers and the allegiance of even fewer. While it discusses its problems, with earnestness and perturbation, it is assailed by the almost deafening roar of the two giants: "Enough idle talk about your proletariat and your socialism! Choose one of us or the other. Nothing else is real!" The pressure is not always blunt; sometimes it is bland and subtle. But it has never before been so great, and the movement is wavering under it. It is hard to admit these facts; it would be stupid to deny them.

The American Marxists are subjected to a multitude of pressures and difficulties. We suffer in general, like our comrades everywhere, from the defeat of the Bolshevik revolution, and the triumph of Stalinism, To think that we are unaffected by it in the United States because the American workers know or care very little about the Bolshevik revolution and Stalinism, is utterly superficial.

The vast majority of the American workers, including the most advanced among them, are steeped in the tradition of individualism, democracy, political freedom. Stalinist totalitarianism is repugnant to them. But in a general way, they identify Stalinism with Bolshevism and the revolutionary Soviet government, looking upon the former not as the exterminator of the latter but rather as its continuator and more or less legitimate successor. That is how they have been taught to think by the bourgeoisie, the labour leaders, and the growing

---

1. In the late 1940s, and decreasingly through to the end of the Korean war in 1953, people from many political viewpoints reckoned a Third World War between the USSR and the USA would probably or even inevitably come soon.

Auto workers on strike in 1945. After the Second World War there was a s great strike wave in the US.

crop of ex-revolutionists who specialise in providing "inside information" to prove the case. Consequently, these workers regard with suspicion or, at best, with doubt those groups which defend the Russian revolution and its principles even if they are anti-Stalinist.

On the other hand, a small minority of American workers likewise identify Stalinism with Bolshevism and regard it as its legitimate successor but, for reasons that are familiar to us, they are attracted to Stalinism. Even if from a different standpoint, these workers, too, regard with suspicion or, at best, with doubt those groups which are anti-Stalinist, even if they are defenders of the Russian revolution.

The decay of the revolution and the vast confusion created by the rise of Stalinism is therefore a double handicap to us. It requires a thorough political understanding of the problem, and patience and perseverance to overcome the handicap. Our movement is one hundred per cent right in going out of its way, so to speak, to distinguish and dissociate itself from Stalinism, its agencies and policies, in the ranks and the eyes of the working class, even if it entails courting (and — this we will continue to see to — overcoming) the risk of vulgar anti-Stalinism. Meanwhile, we progress only among that tiny sector of workers and students who grasp the polar opposition between Stalinism and Marxism and are ready to engage seriously in the fight for socialism.

We suffer also from the fact that the socialist revolution failed to win in a single European country after the Second World War. The American Marxist movement itself has been depressed as a result not only of the failure of a socialist victory in Europe — that includes even ourselves who entertained no stratospheric illusions on this score — but also of the failure of a strong and vigorous Marxist movement to appear in Europe.

Finally, there is the pressure of the mounting ideological war against Stalinism, carried out by the American bourgeoisie which, precisely because it is carried on by the bourgeoisie, is also directed against communism, against the revolutionary Marxist movement and not least of all against the working class, here and everywhere else. There has never been anything like it. Not even the war, military and political, that the imperialist world conducted against the Russian Revolution of 1917 can be compared with it. That war deceived very few of the people and won the support of even fewer. The Russian Revolution had comparatively little difficulty in defending itself against this war in the ranks of the working class. Even the conservative workers looked upon the great revolution with sympathy or friendly interest. It did not take too long to spread among them the truth, the simple facts, about what was actually going on in Soviet Russia. The bourgeoisie had at its disposal nothing but lies and calumnies. Now it is different. In its war against Stalinism, the latter has given the former such weapons as it never dreamed of having thirty years ago. What could the bourgeois press do then with its stories about Lenin as an agent of the Kaiser, about Lenin's Chinese and Latvian terror-troops, about Lenin's nationalisation of women? Very little, and not for very long; in a word, practically nothing. But just think of what this same press is able to do with its stories about Stalin's alliance with Hitler[2], about Stalin's GPU terror, about Stalin's nationalisation of slave labour, about all the cynicism, perfidy, knavishness, cruelty, the hideous oppression and exploitation of the mind and body of the people that characterise the Russian regime today.

One of the most important things it is able to do is to bring demoralisation into the ranks of the socialist movement, especially in the United States and above all in the United States.

When the American Way of Life and the Stalinist "Life of Joy" are contrasted, the difference is most impressive! No question about it: the American worker enjoys a far higher standard of living than the Russian, and he has incomparably more democratic freedom than the Russian, who has none at all. What is the historical origin of the difference? Its social causes? What is the relationship between the two? What is the trend of development for tomorrow and what determines it? Nobody bothers very much about these questions, nobody but a "handful of Marxian dogmatists." The only thing that counts is that there is the known and incontestable difference. The entire American bourgeois campaign is based upon this difference. From it flows the imperious demand: "Make your choice between the American way and the Stalinist way! And make damned sure that your choice is the right-minded one! Don't think for a moment that ideological pressure is our only weapon!"

Is there not, perhaps, another choice? ask the handful of Marxian dogmatists. Cannot mankind, leaving the path of Stalinist totalitarianism and American imperialist democracy, take a road that leads to socialist freedom, peace, and

---

2. In August 1939 Stalin and Hitler concluded a pact which, including through secret clauses, set the basis for the USSR to be the ally of Nazi Germany in the first part of World War Two, until Germany invaded Russia in June 1941. The USSR seized eastern Poland, part of Romania, and the Baltic states, and tried to seize Finland, under the terms of the pact.

abundance? The reply comes in a thunderous orchestral crescendo of editorials, sermons, lectures, moving pictures, radio broadcasts and speeches by Max Eastman[3]:

"No, no, and no! Capitalism is the only society in which freedom is possible. Capitalism has its weaknesses, it has its defects, it has its ugly sores. But here, where it is properly practised, it functions exceedingly well, on the whole. Where else do you have such a high standard of living and the possibility of its growing higher and higher? Where else can an office boy become President of the United States Government or United States Steel or both? Where else do you have so much democracy and therefore the continuing possibility of extending that democracy more and more? As for your socialism, it is a Utopia. The only socialism you ever saw is Stalinism. It is the only socialism you will ever see. It ought to be clear to you, as it is to us, that there never was any other socialism and there never will be. Even you have recoiled from Stalinism, but you don't understand it the way we do. With the best intentions in the world you are only Stalinists who have not yet taken power, just as Stalin is a Trotskyist who has taken power or a Leninist who has kept it. We have ample proof of that, with lots of facts, and lots of profound analysis. There are scores of books and pamphlets and articles by men who were once Utopians like yourselves but who have seen the light — Eastman, Hook, Gitlow, Oak, Lyons, Burnham, Budenz, dos Passos, Harrison, and any number, of other Smiths, Joneses, and Robinsons, every one, of them a scholar and an expert. Not one of them is a capitalist or an exploiter of labour, but they are all working for us now. Do not persist too long in the error of your ways. We do not even demand unconditional submission. If you want to make a criticism here or there, we are always open to constructive suggestions. We have any number of Labour Leaders and Critical Social Thinkers and we tolerate them. This is not Russia, thank God! But if you don't want it to become, another Russia, choose and choose the side of the just!"

The pressure is enormous; it comes from all sides; it comes in a multiplicity of forms; it is exerted everywhere in different degrees. And here it is more seductive and persuasive. Capitalism here is prosperous (more or less) and democratic (more or less). It has been a success. In a country like ours, a "practical" country which has never had much use for theory or generalisation, above all in social science, success equals good. The Marxian movement[4] has thus far failed to become a serious political force. In a country like ours, failure equals bad — to fail is to show that you didn't have it in you in the first place, that you have no future. In the looming battle between the two titans, who already trample over peoples and nations as though they were ant-heaps, it is not popular in the respectable circles to persist in revolutionary independence not only from Stalinism (mere independence from Stalinism in the United States is not only easy and cheap, but a sure road to safety and popularity if not to re-

3. Max Eastman had been an early defender of the Russian Revolution and of the Trotskyist movement, and translator into English of some of Trotsky's works. In the 1940s he became a free-market conservative.

4. By "the Marxian movement" Shachtman would have meant the two Trotskyist groups in the USA, the Workers Party/ISL of which he was himself a member, and the Socialist Workers Party led notably by James P Cannon.

The US Communist Party was founded as a genuinely revolutionary party. By 1949 it was Stalinist, but still attracted workers.

spectable acclaim) but also from reactionary American imperialism. It is not popular to persist in holding firmly to the socialist road and the socialist goal.

We will deal later with the *extremely significant and important* question of why the American bourgeoisie — that powerful bourgeoisie, so fat, so strong, so vaunting in its confidence — concerns itself to such an astonishing degree with promoting the speeches and writings of the "ex-Utopians" who concentrate on arguing that socialism is a failure or that Bolshevism is Stalinism and vice versa. We will see why this happens in a country where, apparently, capitalism is not in the slightest danger of assaults from a socialist working class or Leninist ideas or even from a substantial Stalinist movement. For the moment, it is enough to record that this hooligan ideological pressure against Marxism (the specifically American counterpart of the hooligan police pressure against Marxism in Hitlerite Germany) has yielded valuable results for the bourgeoisie. It has broken from any fidelity to Marxism and the socialist struggle a considerable number of our once prominent or well-known spokesmen and defenders in this country, above all among the intellectuals. And in breaking them from Marxism, with its high standards of intellectual discipline, its unquenchable passion for social justice, its tenets of cool objectivity in political judgement, it has also broken them as rational and distinguished men.

Every one of them would undoubtedly take offence at this, yet it must be said that there are firm grounds for comparing their cases with those of the Trotskyist and Zinovievist capitulators to Stalinism[5]. They are not the same people; they are not the same phenomenon. The two are, however — to employ a favourite phrase of Trotsky — symmetrical phenomena.

The capitulators to Stalin range from the most sordid and contemptible to the most saddeningly noble. Stalin made them "confess" so that they would

---

5. The United Opposition, led by Trotsky, Zinoviev, and others, was expelled from the ruling party in December 1927, and its leaders exiled to remote parts of the USSR. Soon after, Stalin turned "left" in economic policy, towards forced-march industrialisation and collectivisation of agriculture. From 1928, most of the leaders of the Zinoviev wing of the Opposition, and some of the Trotsky wing, capitulated to Stalin, and sought readmission to work in the government machine. Later, during the Moscow Trials from 1936, many of the capitulators were forced to "confess" as counter-revolutionaries, etc.

degrade themselves, disavow and defame their own revolutionary past and the past of the revolution. He needed their "confessions" in order to dig a deeper grave for socialist principle, the socialist road, the socialist struggle. How could he get such "confessions" from people whom Tsarist persecution could never budge? By appealing to something they still held sacred — the interests of the revolution, however terribly undermined. By convincing them of the hopelessness of the "old ideas" as a means of safeguarding what was left of the revolution. What were the "old ideas"? The mobilisation of the proletariat as an independent class force fighting consciously and in its own name for the socialist revolution. "Renounce Trotskyism! It is Utopian and therefore counter-revolutionary. Where is your world proletariat and its struggle for socialism? We alone can carry things through and only by our methods. If not for us, the world bourgeoisie will triumph over us all. Bear that in mind!" In substance, that was the Stalinist line which, in so far as ideological pressure was employed, impelled the capitulators to capitulate.

Our own "ex-Utopians" represent a symmetrical phenomenon. "Where is the socialist proletariat? Where is the struggle for socialism? Where is the socialist movement? All these things are — or were — sacred, but in any case they are not visible. What is left to fight for, actually, concretely? Some measure of democracy, and that is better than none at all. Who threatens it, actually, concretely? Stalinism. What can resist it? The revolutionary movement? Little or no sign of it. The proletariat? Let us not make an abstraction, a fetish, of the proletariat. It must have a leadership, a real one, not a desirable but mythical one. It will either be led by Stalinists, which is real, or by the bourgeoisie of American democracy, which is also real. We once thought that the socialist revolution would lead to democracy and socialism. Well, all it led to was Stalinism. We know what would happen to us and everyone else if Stalinism triumphed. That, above all and everything else, must be prevented. We don't care too much how it is done, either. This is no time for dogmatism or hairsplitting or any scientific refinements. To outstrip Stalinism we must travel fast. Travelling fast means travelling light. Overboard goes all our old baggage. We will concentrate on exposing and opposing Stalinism by speech, article and book, and we'll show we really mean it by joining with anybody and everybody else who opposes Stalinism. As an earnest of our intentions we will start by confessing the errors of our own past, thus warning off all others by displaying our own horrible example."

Basically, it would seem clear enough that this is the line along which the confessions and conversions of our "ex-Utopians" took place. (It is not without interest that Gitlow's book is entitled *I Confess*. Freda Utley's book is *The Dream We Lost*, Eugene Lyons' book is *Assignment in Utopia*, and Eastman's book is *The End of Socialism in Russia*.[6]) The Russian capitulators turned their backs reluctantly to socialism and surrendered to Stalinism. The American capitulators

6. Benjamin Gitlow was a former leader of the Communist Party USA and then of the CP splinter led by Jay Lovestone; Freda Utley was a writer who worked for the Russian government, escaped to the USA, and then after some brief contact with Trotskyists became a conservative; Eugene Lyons worked for a CP USA magazine, then the official Russian news agency, and then for a US news agency as its (pro-Stalinist) Moscow correspondent. He became a left-wing liberal in the mid 1930s and then moved to the right. For Max Eastman, see note 3 above.

did likewise, and surrendered to American capitalism. Stalin used Radek[7], until he murdered him, as his expert in attacking Trotskyism, that is, revolutionary socialism, proletarian internationalism, workers' democracy. Our capitulators, while pretty safe from judicial murder at the hands of the present White House, are used for pretty much the same end, even if not as directly and consciously as Stalin used Radek. The bourgeoisie — the thoughtful, intelligent bourgeoisie — and its press, its schools and lecture halls are at the disposal of the capitulators. Not only in the campaign against Stalinism, but also in the campaign against the revolutionary socialist movement and its ideas. Eastman, Hook[8], Lyons, and their imitators constitute an important division of the brain trust of American capitalism.

What makes them political experts from the standpoint of the bourgeoisie, is obvious. But what makes them political experts from the standpoint of the working class or even ordinary sincere democrats — to say nothing of socialists — is absolutely incomprehensible! Their principal qualification is that, although they teach in different classes, all of them belong to one and the same political school. It should be called by the cumbersome but accurate name of: "I-Was-a-Political-Idiot-but-now-I-Am-Smart" School.

Take Max Eastman[9]. He started out, in his courageous defence of the Russian Opposition, by showing that Stalinism had nothing in common with the revolution, that it was antipodal to socialism and was undermining it, that Trotsky and his comrades were the only loyal champions of workers' democracy and socialist progress. Then he decided to take Marxism gently to task. Its theory of the inevitability of socialism is a theory of fatalistic quiescence which does not provide for planned action by concerted individuals who understand that capitalism makes possible the fight for socialism and who are resolved to carry on that fight in good weather and bad. Lenin — there was a man after his own heart! Lenin, the scientist, the social engineer, the engineer of revolution! He was not content to sit in the British Museum and wait for socialism to come by its mystically dialectical inevitable self. He went out, organised a party of revolutionists by profession, worked out an engineer's plan to bring the revolution about, and proceeded scientifically and resolutely to work according to plan. The trouble with the revolutionary movement is that it is stultified with Marxism. If it really wants to get socialism, it has only to adopt genuine Leninism, as so clearly explained by Eastman. Then followed his Herculean wrestling matches with other giants like Hook and Burnham[10], where he tried

---

7. Karl Radek was a Marxist activist in the Polish and German movements before World War One, and then a prominent figure in the Bolshevik and Communist international leadership after 1917. He was a member of the Trotskyist opposition from 1923 until he capitulated to Stalin in 1929. He was forced to confess in one of the Moscow Trials, in 1937, and died in the labour camp to which he had been sentenced.

8. Sidney Hook was an academic philosopher; a member, briefly, of the American Workers' Party led by A J Muste; and a supporter of the Dewey Commission, which independently investigated Stalin's Moscow Trials charges against Trotsky and declared them unfounded. In the 1940s he moved to the right, though he continued to consider himself a sort of reform socialist until his death.

9. See note 3.

10. James Burnham was an academic philosopher who, like Hook, joined Muste's Workers' Party,

his best to pull them down from the clouds of dialectics and inevitability and pin them to the good earth of active revolutionary struggle for socialism. Socialism, he pleaded, is only desirable and possible, but it is not inevitable. To realise it, you simply have to organise and fight for it, as Lenin and the Bolsheviks did.

This struggle in favour of struggling finally exhausted him. Some people think and then write. Eastman wrote and then thought. Each new book implicitly renounced its predecessor. He ended with a book which introduced a slight change in his original position. Now, he informed his public: it is perfectly clear that socialism is neither desirable nor possible; and if you are imbecile enough to fight for it, the only thing that is inevitable is Stalinism. Stalinism is not the enemy of socialism, it is its true incarnation, its only possible incarnation.

The man who once chided the revolutionary movement for failing to understand the virtues of Leninism now appears before the convention of the AF of L[11] to chide the labour movement for failing to understand that the capitalist goose can keep on laying golden eggs for ever and a day provided the working class is intelligent enough not to kill it.

Upon hearing this new theory of perpetual motion, the labour bankers and businessmen at the convention burst into enthusiastic ovations for the new scientist and champion of labour's very best interests.

What is it, exactly, that makes him more of an expert on the subject than credentialled representatives of the Chamber of Commerce? Only one thing: his insistence upon the claim that practically all his life he has been a political idiot. That politics is practised by idiots of all kinds and of long standing, is a matter of common knowledge. But as a rule, the idiocy is subject to debate, and the politician in question is always ready to take the indignant negative. With our "ex-Utopians", it is different. They are downright anxious to convince the world that all their lives they have been political idiots and now they know it.

"But why does such a confession give you special letters of credit in politics? Wouldn't it be more decent if you hid yourself quietly for the next ten years without writing or uttering a word about the subject on which you were so frightfully wrong all your life? Why should anybody listen to you for one minute how?"

"Why? Because now I am really smart! Now I am really an expert! Now I am sure, of my ground — look how categorically I write!"

but, unlike Hook, went with Muste into a merger with the US Trotskyists in December 1934. He became a prominent writer for the new movement. In 1939-40 he sided with Max Shachtman and the minority who would form the Workers' Party in the dispute over attitudes to the Stalinist invasions of eastern Poland and of Finland. James P Cannon, the main leader of the majority faction in the Trotskyist movement, and Leon Trotsky, claimed that Burnham was setting the tone for the minority and leading it into capitulation to "petty bourgeois democracy". In fact Burnham quit the Workers' Party very soon after it was formed in April 1940, and soon became a prominent conservative publicist.

11. The American Federation of Labor (AF of L) was the majority, and more conservative, wing of the US trade union movement from its founding in 1886 through to when it merged in 1955 with the CIO, a more radical trade-union centre formed by unions expelled from the AFL in 1935. The merged AFL-CIO suffered a split named Change to Win, led by the SEIU and other unions, in 2005.

"But that's what you said yesterday and a year ago and five years ago and ten years ago, and each time you not only discovered that you had been an idiot but you insisted on communicating the fact to the still patient world."

"You don't understand a thing. You're a doctrinaire, a dogmatist, a hair split-ter, a Utopian. You're a typical Bolshevik — intolerant, arrogant, impervious to fresh argument and independent thinking. I know what I am saying and I insist upon being heard because this time I really know what an idiot I've been all my life."

Eastman is only one member of the School. Hook[12] is another — different from Eastman, as each is from all the others — but still a member of the School. His books and articles on what Marxism really is are not countless but they are numerous. Read them, if you can, one after the other. Hook is not as rol-licking and lusty and sword-direct in his confession of life-long idiocy as East-man is. He does the same thing in a more stately, dignified and academical way, but he does it. Each of his writings on Marxism, if it is not an outright re-pudiation of what he wrote before, is at least a mockery of it. He started out by explaining that Lenin's works showed an understanding of Marxism, on the whole, and that Kautsky's[13] works did not. He ended up (at least for the time being) by writing — just as firmly, just as authoritatively, just as confi-dently — introductory tributes to Kautsky's works. Withal, he remains the only man who really understands Marx and he is not the last to admit it. On theo-retical and political problem of the socialist movement, he is an increasingly awesome Authority, accredited to that post by years of implicitly acknowl-edged idiocy. To be sure, mistakes in politics are as common as daisies and God knows we have made our share of them. But there are mistakes and mis-takes, as the saying goes. We Marxist sinners are guilty of much, no doubt. But we have never based our claim to attention and respect on the fact that our lives have been an unbroken, all-sided, multi-coloured, high-vaulted and pos-itively ludicrous series of self-exposed and self-renounced political blunders. If we ever came to that, let us hope we will commit no public nuisance. We will have the elementary decency to retire from politics altogether and apply ourselves to such internationalist but less venturesome pursuits as stamp-col-lecting.

Or take Budenz, or Gitlow, or Oak[14], or any of a dozen others. With embar-rassing emphasis on personal detail, they write and speak of how, for many

---

12. See note 8.

13. Karl Kautsky, of Austrian origin but active in Germany, was the foremost writer of the Marxist movement from the 1890s through to World War One. Lenin esteemed him highly until 1914. Dur-ing World War One he took a "centrist" position, not supporting the war but urging that the move-ment eschew conflict on that question and wait until the end of the war would restore conditions for socialist cooperation across borders. He condemned the Russian Revolution as premature. Al-though he wrote a friendly obituary of Lenin in 1924, he denounced both Bolsheviks and Stalinists indifferently, equating them, until his death in 1938.

14. For Gitlow see above. Louis Budenz was a CP USA journalist and self-confessed Russian spy who became a right-winger, and an energetic collaborator of the FBI in targeting CPers and ex-CPers, after 1945. Liston Oak was active in the CP USA in the 1920s, was Director of Publicity for the Republican government in the Spanish Civil War, but after becoming disillusioned with Stal-inism worked for Voice of America.

years of their adult life, long after baptism, confirmation and attainment of majority, they were taken in by the communist or Stalinist movement as if they had no more sense than a cretin baby born of tested and registered cretin parents. With growing amazement, you read of their unperturbed membership in an organisation led by and largely composed of robbers, footpads, thugs, kidnappers, assassins, corruptionists, peculators, swindlers, confidence men, careerists, spies, union-wreckers, hedonists, rapists, cowards, adventurers, demagogues, hysterics, dreamers, fools, and rogues, rascals and scoundrels of every variety.

Alas, even an honest man can sometimes wander into a den of iniquity. But he has only to look around him and he takes to his heels. It may take him a few days or a week or two. If he is a slow thinker or has a cataract in his eye, it may take him a few months before he leaves. But these people were there for years and years. They not only saw all the things they write about (and a good deal they do not write about) but even contributed their not unappreciated mite to the universal foulness. Yet they stayed on and on and on.

Why? "You ask why? Simple! Because for years and years, more than we can remember, we were political idiots! We could not tell right from wrong, right from left, good from bad, top from bottom, inside from outside. Take our word for it, we know what we're talking about. This very minute we have manuscripts for more confessions which will wipe out the last scintilla of doubt about our idiocy."

"All right, we'll make an effort to read them too. But look here, even the most highly polished idiot in the world could not have overlooked these things for so long."

"Is that so? I'm surprised that you in particular should fail to understand how it happened. We were blinded to all that filth by our devotion to the socialist ideal, the socialist goal, the socialist principles, the cause of working-class emancipation. We were ready to do anything, to overlook anything, to further that aim. What fools we were! We didn't further it at all. We only brought it harm."

"Excellent! No more of those horrors for you and none for me. We'll build a clean and honest movement for socialism."

"Not now! I tell you, we've got to destroy Stalinism before it destroys us. I don't care who does it or how it's done, or whom I work with, or under, to get that job finished. I'm not a political idiot any longer. I know American capitalism is not perfect — you don't have to tell me that, you know. But it's the only power left that can handle Stalinism. You don't think for a minute that your tiny little socialist sect can cope with Stalinism?"

"No, as a matter of fact I don't. Not yet. We're not strong enough to destroy Stalinism. But we think that our road is the only one that can really lead to its destruction without undermining the working class and strengthening some other tyranny over the people."

"Road, road, road, road! That's all I hear from you people. I tell you again, when your house is on fire, you're crazy if you get finicky about how you put it out or who helps you do it." (The classical question of the burning house never fails to get dragged into the discussion, like that other question, so clas-

sical in California, about whether you would support the war if a Japanese soldier threatened to rape your sister.)

"I wouldn't be too finicky, either, except that I'd take care not to douse the
flames with wood chips, dry leaves, and kerosene, or give the hose to a man
who'll turn it on me. After all, firefighting is a science, with good sound rules
based on experience and…"

"Science! Rules! Experience! A lot of empty words. I thought Stalinism had
taught you something. You refuse to learn. You're still a doctrinaire, a bigot, a
theorist, a dogmatist, a Marxian theologian, full of the old and outlived answers, without a fresh idea in your head. I used to be a political idiot, but not
now. I must rush off to write a couple of articles for the *Reader's Digest* and the
New Leader[15], where people appreciate my work. We've got to smash the Stalinists. Why, those monsters actually believe the end justifies the means. They're
a menace. We must use any means to destroy them. Some day you'll understand that you're wasting your time with all this sectarian nonsense. Why, selling typewriters for a living, or even correction fluid, is more profitable than
what you are doing. Let go of me, and don't think I'm coming back."

Here, capitalism is not only rich but richer than ever. Only with the foregoing
well in mind is it possible to understand the case of those who have recently
oozed out of our party and into the void. We refer in particular to Goldman[16]
and Erber[17].

What an appalling display of irresponsibility they both make! What sadden-
ing examples of demoralisation!

Goldman looked upon himself as a political leader and a teacher. Perhaps
he still does. His services to the revolutionary movement are substantial and
serious; they cannot be erased — the last thought in our mind is to try. But
even the most glorious past cannot be used to justify or cover up everything.
Goldman organised and directed a political struggle inside the Socialist Workers Party [SWP]. It was a bitter struggle but there was no way to avoid it. Goldman assumed a heavy responsibility and discharged it in his own way. The
struggle culminated in the split in the SWP. Goldman and the comrades who
followed his leadership joined the Workers Party. In it, Goldman took his rightful place as a leader of the party as a whole, freely chosen for that position by
the entire membership.

---

15. *The Reader's Digest* was and is a big-circulation conservative "family" magazine in the USA.
Max Eastman worked for it from 1941. *The New Leader* was a magazine, originally associated with
the Socialist Party, later broadly liberal but in favour of purging CPers. Liston Oak (see above) became its editor.

16. Albert Goldman was a leading member of the US Trotskyist movement from 1933 to 1949. He
sided with Cannon and the SWP majority in the 1940 split, and until 1943 was one of the leading
writers for their viewpoint. In 1943-6 he, together with Felix Morrow, Jean van Heijenoort, and
others, went into opposition in the SWP. In 1946 he and some others left the SWP to join the Workers Party. He quit the Workers Party, and quit politics, in August 1948.

17. Shachtman's whole text is written as a response to a letter of resignation from the Workers
Party by Ernest Erber in September 1948, available online at bit.ly/erber-docs. Erber had joined
the Trotskyist movement in the mid-30s and was a leading figure in the Workers Party between
1940 and 1948. After resigning Erber made a career as a town planner and was only marginally
active in politics, considering himself a sort of social democrat.

The question of our attitude toward the Marshall Plan arises[18]. Goldman presents his own view inside the leadership; he presents it to the membership orally; he presents it in our press so that our friends everywhere may be acquainted with it. An extensive discussion ensues and in the final decision on policy, Goldman's view is rejected. So far, everything is normal. Then arises the question of our attitude toward the socialistic candidates in the [1948] presidential elections. Goldman, and not he alone, proposes that our party give its support to one of the three candidates, Thomas[19]. In the debate, his view is discussed as attentively as any other. The party rejects it, however, and decides that we recommend a vote for Thomas or either of the other two socialistic candidates, Dobbs or Teichert. Explosion! Moving with unusual speed, Goldman resigns from the party. Just like that!

Why? Has the party betrayed socialism or the working class? Has it abandoned its program? Has it suddenly adopted a decision on a question which is in flagrant and catastrophic conflict with everything it stands for? Or perhaps the difference of opinion on this question is, for Goldman, only the culmination of a long and fierce struggle he has fought in the party over a whole series of vital controversial questions, the straw that has broken the camel's back? No, no, and nothing of the kind. Is it then possible that a man of Goldman's political experience can really be so exercised over this one question, which is, after all, one way or the other, a pretty episodic trifle, that he must quit his party on account of it?

After all, a political party, good or bad, tight or loose, is not a movie theatre that you stroll out of with a yawn or thrash out of with a snort when the newsreel doesn't please you. But Goldman is not afflicted with such a sense of responsibility. He does not make the slightest attempt to persuade either the party leadership or the membership. In a fit of petulant demoralisation, he explodes and quits. A charitable person may say reprovingly: "Do not be so harsh. Soften your fanaticism with the milk of human kindness. After all, it is not the first time he has flounced out of an organisation. Be a little more tolerant." He would then have to reply hastily: "No, no, it is not intolerance we feel, but apprehension. If he acts that way a few more times in a few more organisations, people may eventually get the idea that he is not to be taken seriously as a political person because he does not take his political allegiances seriously. And such an idea, although clearly erroneous, may end up by impairing his political prestige. That's what we are apprehensive about." That is what we would answer. But how many people would believe us?

The case of Erber is even more lamentable. His abilities and contributions to the party were often valuable and properly recognised for what they were. Also recognised was the fact that he fell deeper and deeper into the habit of undermining his own value, often to the point of negating it entirely: his im-

18. The Marshall Plan was a US program of economic aid to war-shattered Western Europe between 1948 and 1952, designed to combat Stalinist influence. Goldman and others wanted the Workers Party positively to support the plan; Shachtman and the majority rejected that view.

19. Norman Thomas was the candidate of the traditional social-democratic Socialist Party in the 1948 presidential election. Farrell Dobbs was the candidate of the "Orthodox Trotskyist" SWP, and Edward Teichert of the "De Leonite" Socialist Labor Party. Thomas gained 139,569 votes, Teichert 29,244, and Dobbs 13,613.

pressionism. Impressionism in the political movement is not an altogether bad quality, and this was often proved in Erber's case. It made him sensitive to new situations and new problems, and in itself this sensitivity is certainly a plus.

But if it is not balanced by theory (generalisation from significant experience), scientific discipline (the employment of tested methods and standards), and intellectual firmness (resistance to the clamour of the prejudiced, the ignorant, the weak and — in politics — the reactionary, and an understanding contempt for the pressure they seek to exert upon you) — if it is not thus balanced, the plus easily becomes a minus. Events make their impression on you like winds on a loose-legged weathervane. You veer, you reel, in a stiff blow you collapse, and then you are not even any good for telling the direction of the wind, or rather you are carried off by it. If Erber is not evidence of the sure upshot of unbalanced impressionism, then there never was any and there never will be.

Erber grasped long ago the existence of a problem that others also grasped, sooner than he or later, more perspicaciously or less. Trotsky often referred to it in these terms: the crisis in society is the crisis of revolutionary leadership. How shall we bridge the gulf between the over-ripeness of capitalism for socialist reorganisation and the unreadiness of the proletarian revolutionary movement to carry out that reorganisation?

In the past [Erber] at least had some sort of grip upon Marxism and it kept him in the revolutionary movement. In the last year or so, this grip became looser and looser. There was no lack of warning to him against the demoralisation that was assailing him. He was reminded of what he once knew pretty well, that nowadays above all others the revolutionist must steel himself and steel others against the slings and arrows of the times. The warnings and reminders did no good. His disintegration could be watched week by week the way you watch a man expiring on his deathbed.

Is there, then, not something wrong with the movement itself, not with him — since it could not hold him in its ranks? Indeed there is. We do not even dream of denying it! The movement is small and weak and isolated.

Marxism is on the defensive. The forces arrayed against us are multiple and mighty. They are arrogant or persuasive, bestial or subtle, but always persistent and insidious. Small as we are today, they devote no small amount of attention to us, for their good political instincts make up for what they lack in thought-out understanding of what we represent. They pour down upon us a steady trickle of acid. What, after all, is so surprising if this trickle dissolves the elementary lime deposits that make up the mental bones, the political bones of some of us?

Has this thought been invented for the special purpose of explaining away Erber "after the fact" and "exempting" the movement from "its responsibility"? Surely even the arrant sceptic must be moved by the mountain of contrary evidence. We adduce but one item. It was written not so long ago, in July 1944.

"The Socialist editor who recently, in defending the victims of the Minneapolis frame-up, referred to Trotskyists indulgently as somewhat naive people who still believe in the *Communist Manifesto* as originally written hardly realised the historical significance of his statement. Yes, Trotskyists are the only people to whom the great document of Marx and Engels remained a living program.

"The two decades of struggle against the current proved an unbearable strain to most of our fighters, both in the ranks and in the leadership. No need to refer to the capitulators in Russia or the many stalwarts of this early Comintern in Europe who disappeared from political life. Suffice it to note how few of the pioneers of the Communist League of America[20] still remain. It is difficult to estimate how many thousands have passed through the organisation in the last 15 years. (Among them a host of very able men like Muste, Burnham, Hook, Spector[21], and others.) No more severe tests could be devised to test the optimism of a movement than those to which we have been subjected from one field of working class defeat after another. Only those with a broad historical vision, a firm grasp of theory and life lived in close personal connection with the movement and the problems of the workers could survive."

And further: "This decay of class feeling and militancy was an inevitable result of the decay that penetrated the entire social organism. No class, above all not one so basically rooted in the productive process as is the proletariat, can base its politics upon the status quo of a rotting society without beginning to rot itself. The proletariat could only save itself in a revolutionary struggle against the status quo.

"But only a tiny segment of the working class understood this and was willing to wage such a fight. More accurately, it was not even a segment of the class but only an ideological grouping that consciously expressed the historic aims of the working class and identified itself with the most advanced program and revolutionary traditions of the working class.

"This core of irreconcilables was all that was salvaged from the revolutionary years. They remained all but immune to the all-pervading decay of the times."

No great skill is needed to guess the author. It is Erber, the Erber of yesterday. Despite the somewhat heavy pathos, every word he wrote is true — true then, true now. The past struggle did prove an "unbearable strain" to many. Indeed it did and does. Only those with a "broad historical vision, a firm grasp of theory and life lived in close personal connection with the movement and the problems of the workers could survive" these most severe tests. Yes indeed, not only in 1944 but in 1949 too. Even the core of irreconcilables that remained true to the fight for socialism was not altogether exempt from the effects of the decay of society (for it lives and breathes in capitalist society) or the decay of the working class (for it also lives and breathes in the working class). No, not altogether exempt. We will be just as exact as Erber was five years ago: the core "remained all but immune" to the all-pervading decay.

For proof, in 1944, Erber pointed to his host of very able men. Spector folded his tent one night and stole away into a silence he has never broken. Burnham

---

20. The first US Trotskyist organisation, launched in 1928.

21. Maurice Spector was the founder of the Trotskyist movement in Canada, in 1928. Later he was active in the USA. He quit the movement in 1939 and was later a member of the reformist Socialist Party.

collapsed in a stupor from which he was aroused to a simulacrum of political life only years later by the vision of de Gaulle. Hook gradually wended his penitent way back under the roof of his father, Dewey. Muste returned to his knees before the Father of all Fathers.[22]

It did not occur to Erber — how could it? — to find or even seek the cause for these (let us use a polite word) departures in some misplaced commas in our principles and program or in some awkward sentences. He found the cause where it was and is: in the all-pervading decay of the times. That is where we find the cause also for his... departure. We are not immune; we are only "all but immune." The acid of decay trickles down dissolvingly upon the soft-boned and loosens their grip upon Marxism. For proof today, we point to Erber.

"Hold on there for a minute!" cry out some of our friendly critics whom we do not need to identify further. "There are at least a couple of things wrong with what you have written thus far. In the first place, you are giving the impression that all is well with your principles, your program, your traditions, your tactics, and that in the face of a fact that not even you deny, namely, the Marxian (or Leninist or Trotskyist) movement has not progressed but, if anything, it has declined, it is weak, it is isolated, it is without serious influence. You are giving the impression that everything in your old, old arsenal must be left intact, nothing added and nothing subtracted, and that in the face of the fact that with these weapons you have not gotten very far. That's the point!

"In the second place, you have been talking about these people pretty much in psychological terms — their lack of character, their lack of firmness and endurance, their lack of understanding, their personal deficiencies and the like. After all, you are not confronted with a psychological but with a political problem. And after all, even if all those who left the movement did suffer from the personal shortcomings you ascribe to them, wouldn't it still be necessary to deal objectively with what they write, with their criticisms, that is, to deal with their ideas independently of the personality of the authors, on the basis of the merit or demerit of the ideas themselves? That's the point!"

This outcry is familiar but it does not seem to us to be very valid.

We are not inseparably attached to our Marxian principles or to our party, in the sense that we live for their sake. We are inseparably attached only to that conquest by man of nature, to that emancipation of man from all forms of physical, mental and spiritual slavery which will assure the fullest flowering of the human personality. There and only there is our goal and our reason for associated existence. The social relations that make up capitalism (not simply the fact that a capitalist has more money and a worker less) stand in the way of realising this goal, so we combat it. Stalinism is our enemy for the same rea-

---

22. For Spector, Burnham, Hooke, Muste, see above. In 1948 Burnham had done and published a sympathetic interview with André Malraux, an ex-leftist who was now an adviser to Charles de Gaulle. De Gaulle had been president of France in 1944-6 and by now led a strong right-wing political movement, the RPF. De Gaulle would retire from politics in the early 1950s but, after a "soft" military coup, return as president between 1958 and 1969. Muste, a former preacher, returned to the Church after quitting the Trotskyist movement. Sidney Hook in later life faded out the philosophical differences he had developed in his radical years with his teacher at university, John Dewey, an advocate of "pragmatism" as a philosophy.

son. The socialist revolution is indispensable for realising this goal, so we work for it.

This work we perform with our instruments. The instruments are our ideas (principles, program, tactics). The "we" are those who agree on the goal, agree that these are the instruments for attaining it, and seek to keep them clean and sharp and to perfect themselves in their use. The "we" make up the revolutionary party at any given time. To get us to abandon these instruments, it is only necessary to offer us better ones. We will not hold on to what we have simply because the proffered alternative is brand new or very old. We will hold on to them, however, and fanatically, if the alternative is false and inferior. We will certainly hold on to them if *no* alternative is offered.

Our ideas did not come into full and hermetically-sealed roundness overnight. They have developed for a hundred years and more. With our own mortal abilities we have tried to continue that development. Those that did not bite smoothly and deeply into the material they were applied to, we modified as needed or discarded altogether, sometimes later than others did and sometimes sooner. You need to have only a hat-tipping acquaintance with our party to know this. Yet, with all this, we have at no time found it necessary to abandon the fight for socialism, or the fundamental ideas of Marxism, as the only worthwhile weapons in the fight. We can even add that it is only by means of these very ideas that we have been able to explain intelligently and intelligibly the reasons why these ideas have not yet triumphed and have even suffered prodigious defeats.

We have not yet solved every problem in the realm of our own ideas. We are ready to discuss every proposal or criticism, but that does not mean that we are hopeless bigots because, in the discussion, we defend those views we hold to be correct. We are ready to accept any proposal or criticism — new or old, and regardless of its source — but only after good arguments have been made to establish its validity.

Let us say we are trying to get far off the ground. Science and experience tell us that the best way is by means of aircraft or perhaps a rocket. To the best of our ability, we build one machine after another. Some of them are very promising; others take off and crash with the pilot. It is discouraging! But we persist in our ideas, and each time these ideas explain to us our preceding failure. Someone comes along and says: I have a better idea. The way to get to the moon is to dig a great big hole in the earth. We are just a little sceptical. He does not convince us. Off he goes, muttering: Visionaries! Utopians! Bigots! Fanatics!

The analogy is loose and primitive but the point is not too obscure. Our actual experience in the defence of Marxism for the past quarter of a century is that in practically every case our critics wanted us to dig a great big hole in the earth. So, with this freely acknowledged reservation, which we hope will not be condemned too harshly, we reiterate our readiness to listen, to discuss, and to change.

The ideas of the critic have their own status, too, and merit consideration "in their own name." That would be even more evident if the same ideas were presented anonymously. We have enough confidence in our own ideas to de-

Lenin during his last illness

fend them openly and directly against any that challenge them. We note briefly that the same cannot be said about Erber. That is disclosed by his whole procedure.

Not once did he even try to submit the ideas contained in his new document to those with whom he had worked and exchanged opinions throughout most of his political life. Toward the end, he made a couple of mumbled references to the fact that he was thinking about some basic questions, that he was changing or had changed his mind about "democratic centralism" or "the war question". To think, to change your mind, these are God-given rights. But just what he was thinking, just how he had made up his mind, he did not find it necessary to inform his comrades and colleagues. He was content to leave them in the helpless position of guessing. That too is a man's right. Erber abused it to the limit.

At long last, he came to a conclusion, if that is what you can call his document. He submitted it simultaneously with his resignation from the party. What about discussing his views first? Oh no, not that! He dropped the document into our laps and fled like a man who had screwed up all his courage and then shot his wad in one super-audacious act.

He threw his rock into the window and took to his heels like those who throw rocks into windows and take to their heels. He has no confidence in his own views. It was safer to nibble away at us from a distance, in a closed-off corner. Our critic is discreet.

If the author would not come to grips with us, we can come to grips with his work. If we lacked enthusiasm for the task to begin with, we must admit that it has come and grown considerably in the course of performing the task, on about the same principle as the appetite grows with the eating. If the first reaction, upon reading the document, is downright repugnance, the second is consolement, and both for the same reason. If, after all the reading and cogitating which the author claims, this document is all he could produce as the criticism of our principles and tradition, then things are not so bad with us. It is certain that if we get no heavier blows than this, we will manage to... survive. Erber's document is really so lamentably porous that it is not worth a minute's attention so far as settling accounts with him is concerned. But in so far as it offers the opportunity to re-state our principles, to demonstrate their validity again, to reaffirm the fight for internationalism and socialism, it is worth treating, Before taking adieu from our late friend, we will consider the questions with which he concerns himself: the Marxian theory of the state; the nature and unfoldment of the Russian Revolution.

Lenin and the Marxian theory of the state

Erber starts with the Russian Revolution and follows with Lenin's concept of the state, to which he counterposes the Marxian concept as he understands it. We have no particular objection to that order and hope that none will be offered if we reverse it. Erber begins with an acknowledgement: the theory of the state is of key importance for socialists. The attitude toward this theory is the bedrock difference between Leninism and reformism (and "centrism"[23]), since it determines the road to power. Lenin's theory cannot be accepted "without following the strategy and tactics outlined in the basic documents of the early Comintern." This acknowledgement must not be lost sight of for a single moment. What is demonstrated on this score cannot settle every question before the socialist and labour movement. But it can settle one very important point.

If it can be shown that Lenin's theory of the state is in conflict with Marx's theory, then the strategy and tactics of the early Comintern which follow from Lenin's theory certainly do not follow from Marx's, and therefore the rejection of Lenin is not — or is not necessarily — a rejection of Marxism (Erber rejects Leninism in the name of Marxism.) However, if it can be shown that Lenin's theory is nothing but a restatement of what Marx and Engels taught, then the rejection of the former amounts to the rejection of the latter, and the repudiation of the strategy and tactics of the early Comintern amounts to the repudiation of Marxism, that is, of the proletarian struggle for socialism. (It should not be necessary to add that when we speak of the "basic documents of the early Comintern" we are not speaking of this or that comma, word, sentence, paragraph, or page, but of what was fundamental. Of precisely that which distinguished the revolutionary Marxists of that time from the reformists and — without Erber's condescending quotation marks — the centrists.)

Erber rejects Lenin's theory of the state because it is — hold fast, now! — "simplistic, crude and static," dogmatic, false, and in conflict not only with Marx and Engels but also with historical and political reality. How a theory that is so patently gross and preposterous could have fooled an eye as critical and a mind as luminous as Erber's for so many, many years is hard to understand. But that's another matter which relates far less to eyes and minds than it does to feet in flight, and we pass on.

Why is Lenin's theory of the state crude and simplistic? The minute this not entirely inappropriate question is applied to Erber's document, you see how completely he has immured himself in a fog so dense that it is almost impossible to find a landing place in it without the aid of radar equipment. To accept

23. In the socialist movement in World War One, the "centrists" were those who rejected both the mainstream leaders' support for their "own" governments' war efforts, and the revolutionary left, people like Karl Kautsky (see note 13). In later usage the term "centrist" came to describe left-wing groups wavering between reformist and revolutionary politics.

or reject Lenin's concept of the state, you ought to know what it is and be able to express it. There are two good reasons why this should be easy, even for Erber.

The first is that Lenin's concept is not cloudy, complex and dynamic but thanks to years of patient and finally successful explanations made to Erber by Kautsky, Martov, Algernon Lee[24], and the late President Roosevelt — crude, simplistic and static. The second is that Lenin not only devoted at least one work specifically to this subject — *State and Revolution* — especially since that is precisely and particularly the work in which Lenin seeks to show in painstaking detail, the view held by Marx and Engels. Doesn't Erber quote from this work? No. Not even one paragraph? Not even one. A sentence, perhaps? No, no sentence. At least a line? No, and not even a word. The only document from which Erber does quote is what Lenin wrote in "his"[25] theses for the Second Congress of the Communist International on "Communism, the Struggle for the Dictatorship of the Proletariat, and Utilisation of the Bourgeois Parliament".

Why this document, and particularly the quotation which deals almost entirely with the question of parliamentarianism, is chosen against the work which Erber himself regards as "the textbook of the Leninist school," is a conundrum which only its fogbound author can answer. We will leave the question of parliamentarism for later and try to stick to the general and basic question which Erber started to describe as the one of key importance. How are we to learn what Lenin's theory was, at least in Erber's view? We have to grope around in the fog until we come to something that has some corporeal feel to it. It is not easy. Careful search discloses two more or less specific references, and that is all. We will cite one of them briefly, not in order to get it out of the way, but because adequate treatment of it belongs elsewhere. "Lenin could conceive of only two states — a bourgeois state and a proletarian state. The former directed the economy in the interests of capitalism, the latter in the interests of socialism.

"Had Lenin not been blinded by the simplistic notion that a state is merely an instrument of one class for the suppression of another, he may have conceived of the possibility that the state apparatus could free itself from class control, wield its power over the economy against both classes, and constitute itself the new ruling class, based upon state ownership of property."

You have to read this three times under a light before you can believe it was actually written, Erber, on the hunt for proof of Lenin's simple-mindedness, provides us with perfect proof of his own light-mindedness, Lenin, you see,

---

24. Algernon Lee was a veteran reformist leader of the Socialist Party of America. Julius Martov was a leader of the Mensheviks, and specifically of the left wing of the Mensheviks. For Karl Kautsky, see note 13.

25. Shachtman's note: The Theses were not written by Lenin, but in all likelihood, by Bukharin. The only work of Lenin from which Erber quotes was not even written by Lenin. What does a little slipshodness in such matters amount to? He reads and doesn't know what he is reading. He writes and doesn't know what he is writing, So, he also quotes without knowing whom he is quoting. But we grant Lenin is politically responsible for the Comintern's theses.

could conceive of only a bourgeois or a proletarian state. Lenin's theory was simple-minded. Why? Listen carefully now; Lenin did not foresee the bureaucratic-collectivist, the Stalinist state, which is neither bourgeois nor proletarian! He did not foresee it because his theory ruled out such foresight in advance. It "blinded" him. Lenin lived right up to and including the year 1923, plus a few weeks in 1924. How was it possible for him to fail to foresee the non-proletarian, non-bourgeois Stalinist state? What excuse can conceivably be found for his ridiculous simple mindedness? Erber, severe, relentless, unsparing, in other words, dressed up like a real theoretician, refuses to let the quality of mercy drop like the gentle rain. He himself is distinctly of the school of Marx and Engels, whose unapologetic champion he makes himself. Engels — there is the man who really had the right theory of the state. With it, as every child knows, Engels, in contrast to the crude, simple minded and static Lenin, did foresee the Stalinist state without waiting till 1924 to die, because, as every schoolboy knows from the mountain of evidence in Engels' writings, he did not conceive of only two states. It should of course be added, in a lame attempt to apologise for Lenin, that Engels also had the advantage of being equipped with a very high-quality crystal ball, a ouija board, and an astrologer's chart found in a royal Egyptian tomb. At least, we must assume they belonged to Engels, for where else did his present champion and heir get his theoretical instruments?

If we had Erber's ruthless standards, we fear, we would have to paste the label of simple mindedness not only on Lenin but fix it also on Marx and Engels. We would add to what Erber says about Lenin the same thing that was said about Marx and Engels fifty years ago by Erber's predecessor, Eduard Bernstein.[26] But we would be unable to stop there We would have to admit that there were some thinkers outside the Marxian movement who did "foresee" the Stalinist state, and were therefore not simplistic or crude.

Has Erber ever heard of Herbert Spencer or Hilaire Belloc and the "Servile State"? If he has not, he will. Has he ever heard of any of the rest of the anti-socialist schools of thought, down to our Robert Michels and our present day Hayeks, von Mises and the editors, politicians, priests, and turncoats from socialism whom they inspire[27]? The very heart of their teaching is this: A bourgeois state is possible but the "socialist state" never was, never can, and never will be a workers' state because the victory of the "socialists" must inexorably mean the triumph of a tyrannical bureaucratic class which crushes bourgeoisie and proletariat alike; and the existence of the Stalinist state is proof of this and

---

26. Eduard Bernstein was one of the leading writers for the German Marxist movement in the 1880s, when it was illegal and its central publications had to be produced abroad. Bernstein lived in London and worked closely with Engels. From about 1896 he became a chief writer for a "revisionist" strand in the movement which argued that the politics of Marx and Engels should be "revised" and made more reformist and less militant.

27. Herbert Spencer was a late 19th century British sociologist and philosopher, an advocate of free-market economics and "social Darwinism" who argued that "all socialism is slavery". Robert Michels was an early-20th-century ex-socialist academic, later a supporter of Mussolini, who argued that all socialist movements would fall prey to an "iron law of oligarchy". Friedrich von Hayek and Ludwig von Mises were 20th century writers developing ideas similar to Spencer's.

proof of the fundamental invalidity of Marx's theories.

But really now, is Erber's criticism of Lenin identical with their criticism of Marx? Of course not. The latter is, however, the not too distant cousin of the former. We sincerely hope that Erber does not discover this relationship and draw the plain conclusions from it, as Eastman and others already have. But his document does not provide very much solid ground for such a hope.

We will leave Erber's first "specific" reference to Lenin's theory for a moment, and proceed to his second and only other reference: "In Lenin's view the bourgeois state was now stripped down to its real function, as nothing else but a machine for the suppression of the working class by the bourgeoisie, of the mass of toilers by a handful of capitalists." This is too much and too little for Erber. This twenty-five-word quotation which sums up Lenin's theory of the state is just too crude and simplistic — even static — to be accepted by a refined and subtle intellect. He rejects it, and accepts in its place the theory of Marx and Engels, particularly Engels. Why the latter in particular? Because in his hunt, he has discovered, late in life, Engels' famous letter to Conrad Schmidt (October 27, 1890)[28], from which he quotes with such confidence, frequency and a feeling of relief that for a fleeting moment a dozing reader might get the impression — which is patently ridiculous — that Erber has so much as a glimmer of comprehension of what he is quoting. Before we cite this letter which Erber so imprudently selected out of the voluminous writings of Engels and Marx on the state, we can establish beyond question the clear, unequivocal views that both these great socialist thinkers put forward at all times.

"The state," wrote Engels in his new introduction to Marx's *Civil War in France*, "is nothing more than a machine for the oppression of one class by another, and indeed in the democratic republic no less than in the monarchy." Where does this differ in one single respect from Lenin's view which Erber quoted, except to strengthen it by the additional emphasis on the fact that the state is no different in the democratic republic than it is in the monarchy?

Lenin, we recall, was "blinded by the simplistic notion that a state is merely an instrument of one class for the suppression of another." Who originated this simplistic notion, who blinded Lenin, if not the man who wrote that the state "is nothing more than a machine for the oppression of one class by another"? What is the difference between the two? That Lenin says "merely" and Engels says "nothing more than"? That Lenin says an "instrument" and Engels says a "machine"? That Lenin says "suppression" and Engels says "oppression"? That Lenin had a short beard and Engels a long one? That Lenin died in Russia and Engels died in England? We have given all these possibilities the consideration they merit, and we have concluded that none of them gives us the answer. The right answer, charitably formulated, is: Erber is a muddlehead.

But after all, it might have been a chance formulation — crude, simplistic and static — that Engels let slip into an introduction to a brochure by Marx,

---

28. The letter by Engels cited by Shachtman advises German socialists of the time to beware of simplistic and exaggerated versions of "economic determinism", and to take account of the relative autonomy of politics, culture, and ideology. Conrad Schmidt was a German socialist, later a reformist.

but which did not represent his refined and profound and true view. All right, we will try again. On January 24, 1872, Engels writes to Theodor Cuno: "While the great mass of the social-democratic workers hold our view that state power is nothing more than the organisation with which the ruling classes, landlords and capitalists have provided themselves in order to protect their social prerogatives, Bakunin maintains that it is the state which has created capital, that the capitalist has his capital only by favour of the state." (A depressing thought strikes us at this point: if that is what Engels said about Bakunin, what would he say about Erber?)

"Our view," says Engels. If we may guess, that means Marx and Engels. But isn't that view pretty simplistic? Hmm. Then Lenin was simply plagiarising from Engels. Hmm. Well, thinks Erber, if Lenin can plagiarise from Engels, why can't I plagiarise from Bernstein? That's not an easy challenge to take up.

On April 18, 1883, Engels writes about the state to Van Patten in exactly the same vein and in almost the same words:

"The main object of this organisation has always been to secure, by armed force, the economic oppression of the labouring majority by the minority which alone possesses wealth."

But why do we need private and posthumous letters for this miserable job of grabbing for quotations which are familiar to all students of Marxism, knowledge of which should be a sine qua non for everyone who presumes to write on the subject? Didn't Engels write a whole book on the subject of the origin of the family, private property, and the state?[29] Didn't he stick firmly by the views he set forth in it in everything he wrote subsequently, including his preface to the fourth edition written only a few months after the letter to Conrad Schmidt that has so overwhelmed Erber? Isn't the text available, in all modern languages, including English? Yet, Erber finds it possible to write about the state, and about Engels' theory of the state, without so much as a mention of Engels' classic work, let alone citing from it. After all, why should he? Anyone who can discuss Lenin's theory without citing *State and Revolution* would be untrue to himself and his method of exposition if he quoted *The Origin of the Family* while discussing Engels' theory. Since we ourselves prefer the old-fashioned way, we will quote from Engels:

"The state is the result of the desire to keep down class conflicts. But having arisen amid these conflicts, it is as a rule the state of the most powerful economic class that by force of its economic supremacy becomes also the ruling political class and thus acquires new means of subduing and exploiting the oppressed masses. The antique state was, therefore, the state of the slave owners for the purpose of holding the slaves in check. The feudal state was the organ of the nobility for the oppression of the serfs and dependent farmers. The modern representative state is the tool of the capitalist exploiters of wage labour."

And further: "In most of the historical states, the rights of the citizens are differentiated according to their wealth. This is a direct confirmation of the fact

---

29. *The Origin of the Family, Private Property and the State*, published by Engels in 1884.

that the state is organised for the protection of the possessing against the non-possessing classes."

And still further: "The aggregation of civilised society is the state, which throughout all typical periods is the state of the ruling class, and in all cases mainly a machine for controlling the oppressed and exploited class."

Engels' theory, and Marx's, remained to the end the one they succinctly advanced in their earliest works, in one that is so well known that Erber must have heard of it at one time or another — the *Communist Manifesto*.

"The executive of the modern state is but a committee for managing the common affairs of the whole bourgeoisie".

Under socialism, the state will die out because there are no class distinctions to maintain. The state — the "public power" — "will lose its political character." And what is this power, according to the *Manifesto*? "Political power, properly so called, is merely the organised power of one class for oppressing another."

In every case, almost word for word and comma for comma, Lenin's formula is identical with the one invariably employed by Marx and Engels. It is not the basic theory of Lenin that Erber has abandoned in his demoralised flight from revolutionary socialism, but the basic theory of Marx and Engels.

Still, what about the letter to Conrad Schmidt in which "the theoretical basis for such understanding was laid down by Engels many years before Lenin began his study of the state", in which "Engels gave us an insight into the relations of the bourgeoisie and the state apparatus from which we can understand such politically diverse trends toward statification as the New Deal and the Nazi state"?[30] The letter has been familiar to students for some time now, first in Sidney Hook's publication, then in a slightly revised publication in *New International*[31] and later in the Stalinist edition of the Marx-Engels letters. What we have to deal with is an Erberian rediscovery of America, an authentic one because he too does not know what continent he has actually landed on.

"There is a *reciprocity* between two *unequal* forces (wrote Engels); on the one side, the economic movement; on the other, the new political power which strives for the greatest possible independence and which having once arisen is endowed with its own movement. The economic movement, upon the whole, asserts itself but it is affected by the reaction of the relatively independent political movement which it itself had set up. This political movement is on the one hand the state power, on the other, the opposition which comes to life at the same time with it."

Then, skimming lightly over Engels' polemic against Dühring[32], he snaps up

---

30. The New Deal was a name for policies developed in the 1930s depression by the US administration of Franklin D Roosevelt, which gave an increased role to the state via public works programs and deficit spending.

31. *The New International* was the magazine of the US Trotskyists from 1934. The title was continued by the Workers' Party and the independent Socialist League from 1940 to 1958.

32. The passage quoted just before this reference is from Frederick Engels' book *Anti-Dühring*, a polemic against Eugen Dühring. Dühring, an academic and writer, claimed to have a whole philosophy of socialism alternative to Marx's, and for a short time had some influence in the German

another morsel: "... the exercise of a social function was everywhere the basis of political supremacy."

The conclusions which Erber draws from these thoughts are enough to raise your hackles. Here is another of those all too frequent cases which show that your politics are not unrelated to the functioning of your intellectual equipment. A state of political demoralisation can so disarrange your grey matter as to prevent an intelligible thought from filtering into it. This is not a harsh but really the kindest observation that can be made: he doesn't understand what he is reading.

With the "insight" which the quotations from Engels give him, Erber works out these novel ideas: The officers of the primitive organised communities "developed into the economically dominant class," freeing themselves from the control of the community and dominating over it by virtue of the increase of the productive forces which added to their, the officers', social functions. These officials "continued to serve a social function but they now had an additional function — a strictly class function — to preserve the new class division against attack from the exploited class." This new class function required a special apparatus, armed men assisted by material appendages, prisons, and repressive institutions of all kinds. "It is this apparatus that Marx and Engels called the state". But he emphasises, even when it had "the added function of an organ of repression," the social function of the state not only did not end, but "on the contrary, it could maintain its supremacy only as long as it carried out this social function." All right. Next: If the state, under capitalism or under any other class society "takes steps to keep society from disintegrating, regardless of the needs of the economically dominant class," it can survive, and only then. But "whenever the state safeguarded the needs solely of the ruling class at a time when those needs ran counter to those of the economy, i.e., when the state no longer fulfilled its social function, it lost its 'political supremacy'; it was overthrown. The French Revolution is the classic example of this." But that was only the classic example, you see. "Not all feudal states were overthrown. The feudal-monarchic bureaucratic state in Germany adapted itself to the needs of the economy and ruled on behalf of the bourgeoisie, while keeping them at arm's length from the state apparatus itself. (This was the real content of Bismark's policy.)"[33] You get the point? The non-classic but nonetheless real example shows that the feudal state did not have to be overthrown: it could be transformed by the feudalists themselves or by their smartest, Bismarckian, representatives who could continue to rule for and without the bourgeoisie by "adapting itself to the needs of the economy." Do you follow so far? Then we will go ahead one step further.

What made possible this "adaptation to capitalist society by feudal states"? Erber has no secrets from anyone: it was made possible by "the peculiar nature

socialist movement although his views were nationalist and antisemitic.

33. Otto von Bismarck was the chief political leader of Prussia, and then of the newly united German state, from 1863 to 1890. His "Bismarckian" policy was to adapt the state to the economic needs of growing German industrial capitalism while maintaining the political hegemony of the semi-feudal landlord class.

of the bourgeoisie." Which is what? It is "a class that confines itself to economic functions and demands only that the state be in friendly hands, that is, that the political power is not used to obstruct, but to facilitate the economic operations of the bourgeoisie." That is not how it was before capitalist society rose. "In slave society and under feudalism, the holders of economic power occupied themselves directly with the exercise of political power. In contrast, the bourgeoisie occupied itself with business and left the administration of the state to those who, in time, made a profession of it... The result was the development of a state apparatus which, while under the influence of the bourgeoisie in greater or lesser measure, was not under its complete control." At this point, Erber shows that he knows very little about feudal society and nothing at all about slave society. He would have the devil's own time proving that especially in the latter the actual owners of slaves always exercised political power directly. But let's skip over this piece of simplistic and crude ignorance and get along with his peculiar bourgeoisie and the state which is under its influence "in greater or lesser measure" but "not under its complete control."

"The modern bourgeois democratic state, based on universal suffrage, is an exceedingly complex mechanism" and it "defies simple definition." Very good, even the lowly amoeba "defies simple definition." What would be a complex definition? "If it is in instrument for the bourgeois domination of the working class, it is also an instrument for the workers' struggle against that domination. To add to the complexity, the bourgeois state is forced to intervene in the economy against the resistance of the bourgeoisie and develops a momentum of its own which conflicts now with the one and now with the other of the two basic social classes." The complexity of the bourgeois democratic state is not at an end. Neither is the state of confusion of Erber. He is really working up to something. "Its (the state's) reciprocal action upon the economic base can often be opposed to the interests of the economically-dominant class. In periods of social crisis, and most acutely in periods of revolutionary (or counter-revolutionary) change, the state can be wrenched loose from its economic roots and used against the economically-dominant class, with the result that the state, in turn, undergoes vast internal changes in the process. This is above all true under capitalism."

The fog is beginning to lift, isn't it? Outlines of something are becoming discernible. First we have a state. It has two functions. One is a social function, which is rather noble. Pasted on to it, like a dirty and ignoble plaster, is a class function. If the state carries out a purely class function, it won't last long — certainly not if Erber has anything to say about it. It can survive only if it carries out its social function, which is not, we hope you understand, a class function. The capitalist class has only a class function, which is not — this you also understand, we hope — a social function. The worst of it all is that this state, even when bourgeois democratic, is an instrument for the domination of the working class. But — lift up your hearts! Erber is riding to the rescue, his head bouncing steadily in the saddle of a quotation from Engels. All is not lost, for the state is also an instrument for the workers' struggle against bourgeois domination. The bourgeoisie has a certain influence in the state ("greater or lesser measure"), but thank God, it doesn't have complete control. If Erber arrives on time, maybe we can wrench the capitalist state loose from its capitalist economic roots like a plucked potato, use it against the dominant capitalist class, and introduce vast internal changes in the process. Word of honour, it is possible, "above all… under capitalism."

A little more patience now, for we don't have much further to go. Erber is not content with grey theory alone. He piles evidence upon evidence, straight from the rich mine of living history. He shows, by one example after another, that wherever the bourgeois democratic state was based upon an extensive electorate, "it became an arena in which the other classes fought to bend the powers of government to their own purposes." Every group has a fighting chance — to get its interests taken care of by the bourgeois-democratic state, to "bend" it to its purpose.

And such historical erudition, so impressive, so overwhelming, so conclusive. There's been nothing like it outside a high school textbook for years. All about the struggles of Hamilton and Jefferson, Biddle and Jackson, the Republicans and the slaveocracy, Johnson and Stevens[34]. And not only struggles between sections of the propertied classes. When labour comes on the scene, it too shows that it can bend the state to its purposes: the Clayton Act, the Norris-LaGuardia Act, universal suffrage in Belgium and Austro-Hungary (even though it was a monarchical and not a democratic state, mind you!) And then the New Deal itself, which "remains incomprehensible from the Leninist theory of the bourgeois democratic state." And earlier historical evidence than that, even more erudite and overwhelming, which can be found only in Erber and in high school texts on civics and history (Second Year course), jam-packed with the Athenian Democracy, the Roman Senate, the cantonal democracy of the Swiss mountaineers (no mention of William Tell, but only for lack of space), the Hanseatic League, the Dutch Republic of the merchant families, Hamilton, Madison and Jay, and Girondins and Jacobins, and anything else your heart desires. It is all there. All that, plus the amazing and Leninistically-incomprehensible New Deal, proves it to the hilt.

Proves what? Proves that the state loses its political supremacy if it "operates solely to safeguard the interests of the bourgeoisie." We know that it hasn't lost it. Why? Hold your breath now: "This has not happened to date because the states of the capitalist world have taken measures to fulfil their social func-

---

34. In the first period of the USA after the War of Independence, Alexander Hamilton was the main proponent of a strong central government taking some economic initiative, and Thomas Jefferson of a decentralised free-market regime. Nathaniel Biddle was the president of the Bank of the United States (a forerunner of the Federal Reserve as central bank for the USA) from 1822 to 1836; Andrew Jackson was the US President who cancelled the Bank's charter. The Republican Party was at the time of the US Civil War the party of the opponents of slavery. Thaddeus Stevens was an advocate of Radical Reconstruction in the southern states after the Civil War to secure equal rights for ex-slaves; Andrew Johnson was the US President who halted and reversed Radical Reconstruction. The Clayton Act of 1914 was a law to limit monopolies in industry; the Norris-La Guardia Act of 1932 stopped bosses getting injunctions against routine workers' industrial action. For the New Deal, see above.

tion." Hold your breath just one more minute and you'll finally find out what this social function — not the "purely class" but the social function — is: "To protect society as a whole from economic dislocation and general impoverishment, the state has intervened increasingly in the economic sphere, to regulate the economy in order to secure its more normal operation."

There, we've quoted it at last and we feel that an enormous burden has rolled off our shoulders. Isn't your imagination electrified by the magnificent vistas opened up before mankind by the possibility of the bourgeois state ripping off the plaster of its purely class function and going about its proper business of fulfilling its social function? Aren't we all endlessly lucky that the bourgeoisie is a peculiar class which only has influence but not control in the state, and that the proletariat can bend the state to its own purpose? Aren't we favoured by Nature and the Almighty himself by the fact that the bourgeois state can adapt itself to the needs of the economy, despite the selfish resistance of the selfish bourgeoisie, and proceed with its social function of protecting society as a whole — which means all of us, them as well as you and me — and of regulating the economy, which selfish capitalists are braking, so that its more normal operation is secured? Intoxicating thoughts!

Intoxicated with them, Erber breaks out into lyrical song about his state and its social functions. "The trend toward statification of production, foreseen by Engels and observed by every prominent Marxist theoretician since, is nothing else but the effort of the state to fulfil its social function. It has succeeded in fulfilling it to a remarkable degree despite bourgeois opposition... The state is adapting itself to the needs of the economy."

The temptation to shout Hurrah! for this highly commendable state which not only makes an effort but succeeds in a remarkable decree in protecting society as a whole, is overcome only by the sickening and humiliating thought that at one time Erber was allowed to edit a magazine which calls itself an organ of revolutionary Marxism. After fighting off the wave of nausea, we re-read that all this was foreseen by Engels and observed by every prominent Marxist theoretician since.

What did Engels foresee? The continuing tendency toward what he calls in his *Anti-Dühring* "the conversion of the great organisations for production and communication into joint-stock companies and state property, which shows that for the purpose of controlling the modern productive forces the bourgeoisie can be dispensed with." Then Engels adds: "All the social functions of the capitalists are now carried out by salaried employees. The capitalist has no longer any social activity save the pocketing of revenues, the clipping of coupons and gambling on the Stock Exchange, where the different capitalists fleece each other of their capital." it appears that in Engels' view the capitalists did not merely have a "purely class" function but "also" a "social" function.

As we shall see, this is entirely correct, and it is likewise correct that Erber has been splashing around in a semantic puddle. But according to Engels, who foresaw, what is the *social* significance of this state intervention into the economy? "But the conversion into either joint-stock companies or state property does not deprive the productive forces of their character as capital. In the case of joint-stock companies this is obvious. And the modern state, too, is only the organisation with which bourgeois society provides itself in order to maintain the general external conditions of the capitalist mode of production against encroachments either by the workers or by individual capitalists. The modern state, whatever its form, is an essentially capitalist machine; it is the state of the capitalists, the ideal collective body of all capitalists. The more productive forces it takes over, the more it becomes the real collective body of all the capitalists, the more citizens it exploits. The workers remain wage-earners, proletarians. The capitalist relationship is not abolished; it is, rather, pushed to an extreme."

First, then, the modern state, "whatever its form" — be it democratic or monarchical — is an essentially capitalist machine, the ideal collective body of all capitalists. This is such a shockingly crude, simplistic, and static definition, so much like Lenin's even though it was written by Engels, that Erber refuses to dignify it by quoting it. Second, this state exists in order to maintain the conditions of capitalist production against encroachment by the working class or by individual capitalists. The social function of the state is precisely its class function. The latter is not an "addition" to the former, like an extra suit of underwear which the state puts on when one class influences it and takes off when another's influence is greater.

The class function of the bourgeoisie is precisely its social function. The function of the bourgeoisie is the organisation, direction, and control of production and exchange, the development of the productive forces, the preservation (and extension) of the national framework ("the fatherland") within which the solid basis of this production and exchange exists. This is its social function. It can fulfil it only as a class function, on the basis of the capitalist exploitation of the proletariat and on no other. When Erber talks about the bourgeoisie not having a social function but only a purely class function, he just shows that he doesn't know what is meant by the words he is whacking away at in his puddle. He seems to think that a class function is not at the same time a social function. Against such ignorance you feel pretty discouraged if not helpless.

In the course of fulfilling its social function, which is also its historical function, the bourgeoisie develops the productive forces to the point where they come into violent conflict with the limits placed upon them by the capitalist mode of production itself. Crisis! Collapse! Widespread destruction of the productive forces! Stagnation or retrogression! In the "old days", the production cycle was resumed by the operation of the "natural processes" and without any significant intervention by the state. The capitalist class emerged from the crisis by itself, as it were.

But the sharper and deeper and more convulsive the crises become, the less capable the capitalists show themselves of resolving them on a capitalist basis. The "economic movement" of which Engels writes to Schmidt is here the movement of the productive forces which "press forward with increasing force to put an end to the contradiction, to rid themselves of their character as cap-

ital, to the actual recognition of their character as social productive forces."
Correspondingly, the capitalists are forced more and more "to treat them as
social productive forces, *in so far as this is at all possible within the framework of
capitalist relations.*" (This last phrase of Engels', underlined by us, is especially
important.) By speeding the centralisation and concentration of capital by
means of corporations, trusts, and the like, the bourgeoisie continues to fulfil
its social function — Erber's mortifying ignorance and confusion to the contrary notwithstanding.

But the more severe the crises, the harder the productive forces "press forward... to put an end to the contradiction" — the more helpless are the individual capitalists to deal with the economic problems. This applies even to the
biggest of the individual capitalists or of the aggregations of capital. Increasingly, the problem must be dealt with by the "committee for managing the
common affairs of the whole bourgeoisie" — the state. Increasingly, it is "constrained" to intervene in the economy.

Erber knows that and every child knows it. The question is, why and how it
intervenes, what function it fulfils in its intervention. The state intervenes because, by its very structure and character, it is in a better position to recognise
the social nature of the productive forces and to treat them as such, than is the
individual capitalist. It finds itself obliged to take over the management and
control of more productive forces, and in some cases or for certain periods of
time even the direct ownership. *That* is what Engels calls taking over "the social
functions of the capitalists," and he is right. But the modern state, even in its
democratic form, is the state of the capitalist. It takes over more and more those
social functions which decaying capitalism makes it difficult or impossible for
this, that, or the other capitalist to perform. But the more it takes over, "the
more it becomes the real collective body of all the capitalists, the more citizens
it exploits," explains Engels. The modern state, being what it is, intervenes in
the economy not only in a capitalist *manner* but in the interests of the capitalist
class *as a whole* (which means, more and more, the most powerful and monopolistic sectors of the capitalist class) and above all in the interests of *maintaining
the shaky foundations of capitalist exploitation itself.*

When Erber, however, speaks of state intervention in the economy to protect
society as a whole, he is not just talking like a liberal, at least not like a half-educated liberal, but like the most vulgar and backward of liberals. There is *no
such thing* as "society as a whole" and there has not been [for] hundreds and
thousands of years. That Erber can even use such language shows he has
reached new shores where a different tongue is spoken from that of Marx and
the socialist movement. There *is* such a thing as *capitalist society,* divided into
classes with conflicting class interests. The modern state can and does intervene to protect capitalist society *as a whole,* nothing else. That is its *social function.* It can fulfil it only on the basis of the *class* society in which and for which
it performs this function. Indeed, by performing this social function, the class
relationship of capitalism is not abolished, "it is rather pushed to an extreme."
So much for the Engels who foresaw and who gave us an insight plus a key to

an understanding of the state.

To quote from Engels has more than one value, among them the important one of finding out what he really thought and wrote and advocated, the important one of keeping our theoretical lines from being fouled and distorted by anyone who is capable of testing the patience of paper. But there is no need to stop at quotations from Marx and Engels or Lenin. We can go on to see how their theories — and Erber's — stand up in the light of social reality.

Outside the Stalinist countries, what is *the* outstanding example of the increasing intervention of the state in the economic sphere? *The German Hitlerite state*! There indeed was a first-class example — because it was carried out on so intensive and extensive a scale — of how "the states of the capitalist world have taken measures to fulfil their social function." Was it not under Fascism that we saw most clearly that [the] "trend toward statification of production" which our fogbound critic of Lenin tells us "is nothing else but the effort of the state to fulfil its social function"? Why doesn't Erber give us a few pages, or at least a few words, about how that state intervention worked "to protect society as a whole from economic dislocation and general impoverishment," how it worked "to regulate the economy in order to secure its more normal operation." Even a few words on this subject would make interesting, if not instructive, reading. For example, a few words on how the New Deal led the USA right into the Second World War, which was a splendid model of protection from economic dislocation and impoverishment and of securing normal economic operation. We grant that the war was only a trivial episode, but that is why we propose that only a few words be devoted to it.

Erber tells us that a study of Marx and Engels does not provide a "finished answer" and that "fifty years of history since the death of Engels provide us with materials on this subject infinitely richer than that on which the founders of scientific socialism had to base themselves." All right. Isn't the experience of the Nazi state rich enough material for Erber? Or is it too rich? In either case, he has nothing to say about it. He has space to tell us that there was once a Hanseatic League[35] (which we must admit is a fact) and that the British bourgeoisie eventually conquered Parliament but not without the Cromwellian interlude (which, with a yawn, we must also admit is a fact), but no space to deal with the Nazi state. Yet, if our memory does not betray us, the Hanseatic League and even Cromwell came before Engels died while Hitler came afterward. More's the pity. We confess to a somewhat morbid curiosity about how our cat would have walked around this bowl of hot porridge.

But enough of this talk about the Nazi state for the time being. It was on the whole a decidedly unpleasant experience. Let us take instead a more attractive bit of infinitely rich material, one which makes the bosom of every forward-looking American heave and quiver with patriotic pride — the New Deal. Two reasons qualify it for Erber's consideration: it occurred after Engels died and it is "incomprehensible from the Leninist theory of the bourgeois democratic

---

35. The Hanseatic League was a confederation of merchant guilds and cities which dominated Baltic maritime trade from the early 13th to the late 16th century. Centred in Lübeck, at its zenith it stretched from London to Novgorod in Russia.

Some were over-impressed by Roosevelt's "New Deal".

state." Erber writes about the New Deal as if it were his most crushing end unassailable argument against Lenin. Actually, it is just about the sorriest section of his sorry document — not *the* sorriest section, for the competition is strong there, but just about. What is the refutation of Lenin's theory that the New Deal represents?

"...in the midst of capitalism's worst crisis, one that shook it down to its very depths, the bourgeois democratic state in the United States passed as much pro-labour legislation in a few years as took the European workers decades to win." We are duly impressed, but not more. In a few years, democratic American capitalism, richest in the world, more powerful, industrially, financially and in resources than any country in the world has ever been, clothed in vast layers of economic fat, more advanced technologically than any other land — gave the working class as much pro-labour legislation as took the European workers decades to win, but which they did win *long, long before* the Americans, in the backward and relatively backward countries of Europe, including the Germany of the Hohenzollerns, the dual monarchy of Austro-Hungary and half a dozen other countries. In this field, then, the United States finally and belatedly caught up to Western Europe. That is indubitably an achievement, but on the face of it, it would seem to be a boast that would impress people in Arkansas more than it would people in England and Germany. In any case, we are too obtuse to have gotten the point as yet, so far as Lenin's simple-mindedness is concerned. We must read further.

"Roosevelt had a relatively free hand vis-a-vis the bourgeoisie. He took over at a time when widespread state intervention in the economy was the only means of restoring social and economic stability. Such intervention was firmly resisted by the big bourgeoisie, despite their weakened political power, which reached its low point in the 1936 elections. As a counter-force to the bourgeoisie and as a mass base for themselves, the New Deal bureaucracy encouraged and facilitated the organisation of labour, especially the CIO. With the power of labour increased, and the power of the bourgeoisie weakened, the two fundamental classes were more evenly balanced and the 'Bonapartist'[36] position of the state enhanced.

---

36. In Marxist discussions, the term "Bonapartism" — after Louis Bonaparte, nephew of Napoleon, and ruler of France from December 1848 to 1870 — means a regime where the state machine gains large day-to-day autonomy from the ruling class, while still serving its fundamental and historical interests.

"Yet this 'Bonapartism' was unlike the regimes to which Marxists have traditionally applied the term. These regimes were invariably dictatorial and relied upon police measures. The New Deal, on the contrary, was overwhelmingly popular with the masses... The New Deal encroached upon the freedom of property and expanded the freedom of organised labour. While the bourgeoisie organised the Liberty League[37] to defend the 'freedom of property' against state intervention, the rights of labour to bargain collectively were written into law and enforced through the National Labour Relations Board."

We learn also — for the first time, of course, which is why we read with such tooth-tearing yawns — that a contradictory trend was at work in the New Deal: a process of regimentation affecting all classes, greatly enhanced police powers, a huge state bureaucracy. But once admitted, it isn't really too much to worry about.

"Yet this process had no adverse effect upon political liberties. Even the war brought no serious curtailments of civil liberties. To the extent that restrictions took place in some fields, gains were registered in others. It is necessary to conclude that, to date, there is no evidence that state intervention in the economy has resulted in the curtailment of *political* democracy... If state intervention remains purely bureaucratic and unaffected by the labour movement, its expansion will pit the workers' economic struggle increasingly against the state. The bureaucracy will seek to defend its own interests by restricting the freedom of the labour movement. Even here the attack will be against labour's economic rights, since an attack upon political liberty must necessarily affect that of all other strata of the population and confront the state with a united people. Unless a mass fascist movement appears, the conversion of the democratic state into a police state can only be a slow process punctuated by struggles. The likelihood is that it would erupt, at some crucial point in a violent struggle.

"But there are no grounds for believing that state intervention must inevitably take a purely bureaucratic form. Such a perspective is valid only on the assumption that the working class will not rise above a trade union level and will prove incapable of entering the political arena, as an independent class force to fight for its own program. The latter development cannot but have a tremendous impact upon the government's role in the economy. This is what Engels meant when, in dealing with the effect of politics upon economics, he wrote; 'This political movement is on the one hand the state power, and on the other, the opposition which comes to life at the same time with it.'"

Read these lines over again, if you have the intestinal stamina, for such a job. "It is necessary to conclude that, to date, there is no evidence that state intervention in the economy has resulted in the curtailment of *political* democracy... But there are no grounds for believing that state intervention must inevitably take a purely bureaucratic form." You might say that "it is necessary to conclude" that these lines could be written only by an extraordinarily clever person who sat down with the deliberate purpose of seeing if there were any

---

37. The American Liberty League was a right-wing group formed in the mid-1930s by big-business opponents of the New Deal.

limits to the amount of confusion that could be racked into a few lines. Obviously, this is not the case with Erber.

How explain, then, that he did write these lines? Have we not had, and precisely during that period in world history when Erber was still alive, the results — including the results in the field of political democracy — of the most extensive and intensive state intervention in the economy in Germany, Italy, and Japan, to say nothing of the organisation of the economy by the Stalinist states, to say nothing of the intervention into the economy of all the states in wartime? Are these known results what Erber means by "there is no evidence"? There is indeed evidence that state intervention need not "inevitably take a purely bureaucratic form", and that evidence is provided *exclusively* by the intervention into the economy of the revolutionary workers' state of Russia after 1917. But Erber is certainly not talking about that; he is talking about bourgeois states. Can he adduce one serious example of intervention by such states which was not or is not bureaucratic, which did not and does not increase the ranks and the power of the state bureaucracy, be it in the fascist or the bourgeois-democratic states — one serious example, just one, to prove that he has given any serious thought to what he has written? How can he write this tedious trash about bourgeois state intervention not resulting in any serious curtailment of civil liberties — we ask again — in the light of the clear and pronounced trend that operates in present-day capitalism as a whole? The only answer is: his thinking has been corrupted by the extraordinary and exceptional power of American imperialism. The ghost of Roosevelt and the ghost who succeeded him can congratulate themselves on the new "Marxist" scalp they have added to their belt.

Of all the important bourgeois states, the United States is the only important example which Erber can point to as one in which there has been widespread intervention into the economy under an out-and-out bourgeois regime without "serious curtailments of civil liberties." We can grant that without hesitation, all the more readily since we pointed it out and explained the reasons for it at the very beginning of the war (when Erber was in a sweat piling up evidence of a speedy police-state development in this country!).

A serious thinker, writing about such a fundamental question, could not and would not confine himself to the mere recording of this fact. He would attempt to relate it to other relevant facts. Once stated, he would have to ask himself why the state intervention led to such different political (and economic) results in the United States and for example, in Germany. Does the difference lie in the nature of the "German soul" perhaps, or the chemical makeup of the Italian, or those commonly known qualities which enable any red-blooded American to lick any ten Japanese? Does it lie in some mysterious and mystical inability of German capitalism to resist the laws of capitalist degeneration and an equally mysterious and mystical inability of these laws — to make an impression on American capitalism? If Erber is aware that this is precisely the problem that any serious political scientist or sociologist, to say nothing of a Marxist, must deal with when he writes about such things as the New Deal, he is very successful in concealing it.

American capitalism is subject to exactly the same economic and political laws of development and degeneration as German capitalism — exactly the same and no different. But as with all such laws, the results of their inexorable operation depend at any given stage upon the concrete materials within or upon which they operate, upon the "resistivity" of these materials. To overcome the social crisis in Germany in 1933, on a capitalist basis, required fascism — given Germany's poverty and economic wreckage, given the means and resources she had at her disposal, given her position in the capitalist world, given the state of the working class, given the relationship between the contending classes. The social crisis that assailed American capitalism at the same time was caused by the same fundamental contradictions of capitalism. But to overcome it in this country in 1933, still on a capitalist basis, did not require fascism. For that, the fantasmagorical motley of the New Deal was sufficient — given America's immense wealth, the means and resources at her disposal, given her world position, the state of her working class, given the relationship between the contending classes here. When Hitler said that democracy is a luxury of rich nations — he showed a better grasp, if not of Marxism then of economic and political reality, than is revealed in all of Erber's incoherent mumbling.

The Roosevelt New Deal showed that its champion was a much wiser and abler bourgeois statesmen than, let us say, Hoover. The former served the capitalist order which the latter's policies were imperilling. Time and again, Roosevelt pointed out that, under the given circumstances, his was the only course for restoring the capitalist system including capitalist production and profit. The results proved his case: time and again, Roosevelt pointed out that in order to save capitalism from chaos, particularly "revolutionary chaos," it was necessary for the state to "prime the pump" for the bourgeois economy[38] and to grant to the American working class the much-belated economic concessions which their fellow workers in the advanced European countries had obtained in struggle years and decades before. If he succeeded with his "priming" and his concessions to a far, far greater extent than any bourgeois statesman in Europe during the same period, it was not because the latter were politically stupid or inept in comparison with him.

It was not because the bourgeois state is anything else but an instrument for preserving capitalism and suppressing the proletariat. It was because American capitalism was and still is a "peculiar" capitalism in that it has resources — layers of fat around it — that no other capitalist country possesses or can afford to dispense with. From these layers of economic fat, accumulated for the bourgeoisie by the working class, Roosevelt pried loose a few slices for the

---

38. Roosevelt's policies were in part an implementation of the ideas of the British economist J M Keynes. Keynes wrote, in an open letter to Roosevelt in 1933: "as the prime mover in the first stage of the technique of recovery I lay overwhelming emphasis on the increase of national purchasing power resulting from governmental expenditure which is financed by loans and not by taxing present incomes". In a slump, argued Keynes, that public spending would prime the pump: it would have a snowball or multiplier effect, generating enough economic activity from otherwise idle workers and shut-down factories to more than pay for itself.

working class and a slice or two for the growing mob of bureaucrats needed for the operation.

Naturally, the bourgeoisie howled with pain at this unaccustomed operation; naturally, the working class shouted with glee at this contrast with Hooverism. But Roosevelt, true to his ungrateful class, and the Roosevelt democratic state, true to its capitalist nature, stood firm, more or less, and continued with its operation, more or less. Roosevelt's class conscience and the class nature of the Roosevelt state both remained intact.

To keep the bulk of the fat on the capitalist body, to add more fat to the capitalist body, and to save the body itself from forces that could threaten its complete destruction (as an intelligent bourgeois statesman could see from the revolutionary experiences to which European capitalism was being subjected), it was wise and necessary to throw a few slices to the working class.[39]

Did that operation bring Roosevelt and the New Deal state into conflict with certain sections of the capitalist class? Of course! Did it even bring them into conflict with big sections of that class? Certainly! Did it force them to seek support in the ranks of the working class? To be sure! But what does that prove about the validity of the Marxist-Leninist theory of the state?[40] Capitalist society could not exist for ten minutes without the capitalist state. That holds true not only because the main contending classes would, in the absence of a state power to regulate their conflict, immediately tear each other and therefore society as a whole into bleeding shreds, but also because each capitalist and group of capitalists would tear one another to bleeding shreds in the unrestrained and unregulated struggle for the greatest share of the total surplus value, for the greatest power.

The state is necessary not only to regulate the conflict between the classes

---

39. Part of the background to the 1939-40 dispute in the Trotskyist movement over Stalin's invasions of eastern Poland and Finland, which led to a split in April 1940 and the launching of Shachtman's Workers' Party, was Trotsky's belief in his last years that capitalism was in an absolute impasse. Fascism and the New Deal were both desperate and doomed expedients. That idea inclined him to see Stalin's relative economic successes as proof that the USSR's economic system retained some more progressive character. It also inclined him to think that in the world of the time the USSR was so acutely unstable, so finely balanced between a capitalist restoration which could only take fascist form and a new workers' revolution, that it could not really be characterised as an ongoing system at all. See *The Fate of the Russian Revolution, vol.1*, pp.59-64. In the 1940s Shachtman and his comrades corrected those ideas inherited from Trotsky, though in 1949 they still thought of capitalism as in deep decay.

40. Shachtman's footnote: *The New York Times* recently showed how well a bourgeois can understand the "social function" of the bourgeois state when it intervenes in the economy. In an editorial (January 8, 1949) on the economic program of Truman, who pointed out that we "may require the temporary exercise of selective controls in our economy", the *Times* chides those critics who, because of their opposition to a given proposal for state economic intervention (concretely, Truman "threatening the steel industry with subsidised government competition if it doesn't expand as fast as Washington thinks it should"), fail to grasp the significance of state intervention as a whole. The editor says: "If experience is of any value as a guide in such matters, there are times, we think, when government has a positive responsibility to save the free enterprise system, not from its enemies but from its more zealous and uncritical friends." The *Times* editor understands the problem a thousand times more clearly than Erber. The state must step in, despite the resistance of capitalists, in order to save capitalism, for the capitalist class!

(in the interests of the economically dominant class — not, as Erber so loosely puts it, the "wishes" of the capitalist class), but also to regulate the conflict within the capitalist class which is united, as a rule, only in opposition to its mortal enemy, the proletariat. This inner-capitalist conflict, the state also regulates (seeks to regulate) in the interests of the economically dominant section of the capitalist class (monopoly capital), coming into conflict with it only when the promotion of the immediate interests of this section would endanger the preservation of the entire class and its social system.

Neither the Hitlerite state, on the one side, nor the New Deal state, on the other, defied the Marxian definition of the state. Both can be understood only in the light of the teachings of Marx and Lenin. In both cases, it was shown that the modern state is the "committee for managing the common affairs of the whole bourgeoisie" — not of this particular Morgan or that particular Thyssen, of the Liberty Leaguers or of the German-Jewish capitalists, but of the bourgeoisie as a whole. In both cases, the economic and social power of the big bourgeoisie was strengthened. In both cases — the "New Order" of Hitler and the "New Deal" of Roosevelt — it culminated in the economic, political and human massacre of the Second World War. For that war, both states turned to the production of cannon and the consequent fortification of the strength and power of monopoly capital. But if, in doing so, Germany had to forego butter and the United States still had butter (which we use here as a popular synonym for everything from real butter to "civil liberties"), that proves nothing against Lenin's theory of the state. It proves only that a poor capitalism can produce cannon only by foregoing butter, whereas, a rich capitalism can produce cannon without foregoing butter altogether. And it also proves that this wealth, which underlies all that is peculiarly American in political Philistinism and peculiarly American "exceptional" chauvinism, is still dazzling enough to blind weak-eyed people to the fundamentals even of American capitalism, to twist their thinking machinery, and to twist and subvert the political allegiance of American revolutionists. With this acknowledgement, cheerfully made, we pay our tribute to the power of American chauvinism and express our contempt for the ex-Marxian Trilbys[41] whom it hypnotises into intellectual paralysis.

Erber has, therefore, discovered, with shattering effect upon himself, that the bourgeois state can come into conflict with the bourgeoisie, and that the working class can extract economic and political concessions from the bourgeois state, especially from the bourgeois-democratic state. It is an important discovery. At any rate, it was important when it was first made, a century or two ago. We truly regret that in the wearisome daily routine of party work, we neglected to tell Erber about it and left him to re-discover it for himself. It is somewhat disconcerting, we admit, to think now that the De Leonists[42] are suffering

---

41. In an 1895 novel by George du Maurier, Svengali, an hypnotic rogue, seduces, dominates, and exploits the heroine Trilby, and makes her a famous singer.

42. De Leonists: the Socialist Labor Party (see above), sectarian relicts of the early 20th century Marxist writer Daniel De Leon.

from Leninist simple-mindedness in their opposition to struggle for reforms from the bourgeois state; that the early Communist Party movement in this country, which had a similar attitude, was likewise Leninist in its malady. But we are relieved to think that the Stalinists of today are not Leninist, since they do call for all sorts of reforms which they demand from the bourgeois state, and that both the Socialist Workers Party and ourselves, who since our inception have made various demands on the bourgeois state for economic and political measures in the interests of the workers, are also free from the Leninist curse. We are also very happy to read that the only two sinister examples of our sectarian Leninist simple-mindedness on the state which Erber records was our failure to demand of the Los Angeles School Board that it refuse to let G L K Smith[43] speak in a public school (which would have given us a powerful basis for demanding that we be allowed to speak there), and that we did not propose a very practical alternative to trial of the Nazis by the Nuremberg Allied court. (This infuriates him so much that he forgets to offer his own "practical" alternative).

Our muddlehead is not yet finished with this question. Blue in the face, he insists that the workers should try to and can get reforms — from the bourgeois-democratic state. (Why only from the bourgeois-democratic state? They can even get them from outright totalitarian bourgeois states. Erber does not seem to know that under Hitler, the German workers received quite a few economic concessions, including improvement in their conditions, particularly so far as cheap housing is concerned, that they never had under the Weimar Republic, at least not to the same extent. We shudder to think of what would happen to Erber if someone were careless enough to tell him of a few facts like this!) And these reforms can be obtained by parliamentary means, he insists further. That is what Lenin's theory precludes and excludes. The reader will recall Erber's long quotation from Lenin's theses on parliamentarism at the Second Congress of the Communist international which were not written by Lenin. We will not say that Erber is dishonest, because we want to be kind as well as accurate. He has temporarily lost his sense of proportion and his sense of context. We will deal with the question of parliamentarism in more cogent reference in another chapter. Here, only a few words.

Every real student, every responsible investigator, can understand Lenin's position on this question without difficulty. We are not speaking of malicious political enemies, superficial journalists, catch-me-on-the-fly impressionists and professional exposers of Bolshevism, but of real students. The latter know that Lenin not only favoured the most militant struggle for a genuinely representative parliament in Russia, (the struggle under Tsarism for the Constituent Assembly, about which more later), but that he strongly supported participation even in the Tsarist pseudo-parliament, the Duma. They also know that Lenin devoted a classical polemic against the "left sickness" of ultra-radicals in the Comintern who opposed parliamentary activity and participation.

Against what was Lenin, and the early Comintern, fighting in the Second

---

43. Gerald L K Smith was a fascist leader in the USA from the 1930s through to the 1950s.

Congress theses? Against the Kautskyans and reformists of the world who were committing a double sin against socialism: first, in their advocacy of the purely parliamentary road to working class power in general; and second, in their advocacy of this view *above all in those days*. And what were they? They were the days when an unprecedented revolutionary crisis was sweeping over all of Europe; when civil war between proletariat and bourgeoisie raged in one country after another; when power lay in the streets, very often, and bourgeois parliamentarism had collapsed to the point where it could be propped up only by the degraded social-patriots[44] in the labour movement; when the idea of the Soviet form of government was in the mind of millions of militant Europeans workers (even in far, far-off Seattle and Winnipeg — yes, American Seattle! — Soviets, workers' and soldiers' councils were being spontaneously established!)[45] and the watchword of Soviet power was on their lips; when parliaments were the last trench of the bourgeoisie.

The issue was not, as so often in our own day, the defence of bourgeois democracy from fascist totalitarianism, but the social-democratic defence of parliamentarism and capitalism from revolutionary proletarian power. Erber once knew this so well that he went out of his way (quite rightly, too) to insist on including this thought in one of our party's political documents (our resolution on the "National Question" in Europe, 1943). Now he has the familiar political disease of amnesia.

Once the big post-war revolutionary wave had ebbed, it was necessary for the defeated revolutionary and working-class movements of Europe to regroup. (Lenin's theses were, of course, primarily though not exclusively concerned with the situation in Europe). Accent had to be shifted considerably to "parliamentary" activity. Lenin was not only aware of this, but at the Third Congress[46], along with Trotsky, he led the struggle against the contrary-minded ultra-leftists and even threatened to split the International if they prevailed.

That did not change Lenin's fundamental position on parliamentarianism. First, parliamentary activity that is not subordinated to the daily class struggle of the independently mobilised working class is parliamentary cretinism. Second, parliamentary government is the ideal form of bourgeois rule over the working class and it can and must be utilised to the full with that basic fact in mind. Third, parliamentarism is not and cannot be the governmental form of working-class rule, inasmuch as that requires the Commune-type of state. Fourth, with the sharpening of the class struggle and its development to rev-

---

44. Revolutionary socialists during World War One and after referred to the reformists who backed their "own" governments in the war as "social-patriots".

45. Workers in Seattle formed a General Strike Committee which acted as a "virtual counter-government for the city" during their general strike in February 1919, sparked by a dispute in the shipyards. Workers in Winnipeg, in Canada, formed a Central Strike Committee during their general strike in the city in May 1919, sparked by union-recognition disputes in the building and metalworking trades.

46. The Third Congress of the Communist International, in 1921, adopted the ideas which would become known as the "united-front approach", but against great resistance from many delegates. Lenin demonstratively declared himself to be on the "right wing" at the congress.

olutionary situations and civil war in one country after another the struggle of the working class is necessarily transferred outside of the parliamentary field, and parliaments offer less and less possibility for even minor reforms for the working-class, they become more and more an obstacle in its revolutionary struggle. Finally, the peaceful, organic, parliamentary road to socialism, in the light of all historical experience, is an illusion.

These five points of Lenin's views are sufficient for the moment. A refutation of these views ought, one would think, to deal with the points on the basis of the experience accumulated in Lenin's time or since Lenin. Not a word about this from Erber. He quotes from the theses, and with his customary empty pomposity, considers them disposed of without further ado. But just to make sure, in case the idiocy of Lenin does not communicate itself directly to the reader, he points out one of its horrendous consequences:

"The view that it is their state, that we make demands only to expose it, that we expect nothing from it, that we will 'utilise' it since the bourgeoisie is stupid enough to permit us to, creates a frame of mind in our movement which is alien to the workers of a political democracy and isolates us from them. This approach robs the Leninist of a moral basis for his struggle in a democratic arena, and consequently, makes his agitation devoid of the moral indignation over violations of democracy by the bourgeoisie."

As Engels would say, that's enough to give you epilepsy! We do not believe that the bourgeois state is a genuinely democratic state. We believe that even the most democratic bourgeois state is still a form of rule of the bourgeoisie over the proletariat. Therefore, how can we be morally indignant if the bourgeois state outrages the working class? This argument is so stupid, from every conceivable point of view, that it is hard to believe even a demoralised man can be very happy with it. It is not only stupid, but old and hoary, and unlike wine it has not improved with age.

Let us take a man like John Chamberlain[47]. He is an honest liberal, an honest democrat, a political thinker, with a full understanding of Marxism which he must have acquired while reading the *Communist Manifesto* on a subway train one night. Like Erber, he is not a blind New Dealer. In fact, he is a bit critical of it. Unlike Erber, he does not claim to be a socialist, because, you see, if you start with socialism you end up inevitably with Stalinism. Like Erber, he writes a book on politics. Like Erber, he deals with the question of the state, the very first words in his book being almost a dead copy of Erber: "No book on politics can be worth its salt if the author lacks a clear notion of the origins and evolving nature of the state." Like Erber, he has no such clear notion. Like Erber, he does not care for our attitude to parliamentarism which, like Erber, he traces to our theory of the state. Listen to what he says — he, Chamberlain, not Erber:

"The Marxian may protest his belief in the temporary uses of parliamentarism, but he will inevitably be less interested in making the democratic forms

---

47. John Chamberlain was an American journalist, left-ish in the 1930s and a participant in the Dewey Commission which independently investigated Stalin's charges against Trotsky, who moved to the free-market right in the 1940s.

work than he will be in proving that they can't work beyond a certain point. The Marxian is committed by his theory to a use of parliamentary institutions and free speech not as something good in themselves or because they are needed instruments of people whose love of freedom transcends purely material interests, but merely as a sounding board or a theatre for propaganda looking eventually to the strict racket of the proletarian dictatorship. In his fight to preserve and deepen democracy the Marxian is beaten before he starts. For by the very nature of his philosophical assumptions he is psychologically unprepared to fight for democracy beyond the point where his strict racket succeeds in capturing the State. He is by definition interested in class power, not in freedom or democracy."

This appears in Mr Chamberlain's *The American Stakes*, an excellent title which pairs perfectly with Erber's American "Marx", published nine years ago, in 1940. While our approach robs us of moral indignation, we have enough of it left to denounce Chamberlain's outrageous plagiarism from Erber. In this anticipatory plagiarism, however, he does not try to separate Marx from Lenin, because, as we said, he is after all, an honest liberal. But that apart, he simply "repeats" Erber's criticism, sentence for sentence and almost word for word. What is quoted above is only one sample. There are so many others in Chamberlain's chapter on the state that for a moment you get the feeling that Erber must have copied wholesale from it, which is of course untrue, for Erber is a Marxist, isn't he? And an original thinker to boot, whereas Chamberlain is anti-Marxist and not very original. If we didn't feel fairly sure that there are two persons involved, here, we would suppose that there is only one. It is not Erber — God forbid! — but Chamberlain, who then adds: "… if the theoreticians continue to popularise the belief that the State belongs solely to the capitalists, they will be encouraging, not revolt, but a dangerous passivity. The only sound slogan is, 'It's your State!' Under that slogan you can raise a fighting army or the defence of rights. Under any other slogan you merely encourage the growth of cynicism."

But away from Chamberlain, the plagiarist, and back to Erber, the original and seminal thinker. We have never believed that moral indignation is a substitute for objective political analysis and political conclusions and we do not believe it now. But to say that our views robs us of "moral indignation" is, if we may repeat the harsh word, stupid, just plain stupid.

Erber has some dubious praise for the IWW[48]. The extreme view which most Wobblies had of the state in general, and of the democratic bourgeois state in this country in particular, is pretty widely known. Did this view rob them of "a moral basis for their struggle in a democratic arena" when the state, clubbed, framed, imprisoned and even shot them down? Was the anarchist campaign to save Sacco and Vanzetti[49] from democratic judicial murder "de-

48. The IWW, Industrial Workers of the World, or Wobblies, were a revolutionary trade-union movement, organising mainly precarious and immigrant workers neglected by the older unions. The IWW's heyday was between 1905 and 1914.

449. Nicola Sacco and Bartolomeo Vanzetti were Italian-born American anarchists who were put to death by the US state in 1927 despite a long campaign, in which Shachtman himself was active,

void of moral indignation" because of their view of the state? Or still better and still clearer, if someone says the Hitlerite state is "their state," the state of antisemites, of anti-Marxists, of anti-democrats, of Fascist assassins — which it undebatably is, or was — is he robbed of moral indignation when he carries on a struggle against the wholesale murder of Jews and Marxists and democrats by the state which he says is organised to murder Jews and Marxists and democrats?

If Erber is not saying that, then he is saying that because Marxists (or Leninists, as he prefers to call them) believe the bourgeois state is a bourgeois state, they have no basis for fighting against the murder of Sacco and Vanzetti or the lynching of a Negro and nobody should take their fight seriously. In that case, he is saying something reactionary as well as stupid.

However, we will restrain ourselves and "for the sake of argument" admit that it is not "their state." Whose is it? Is it "our state", the state of the working class? Of course not! Do you think Erber is a total blockhead? He doesn't say that for a minute, and to ascribe it to him would be downright calumny. Right is right. It isn't their state and it isn't ours. It is not the bourgeoisie's or the proletariat's. The state, Erber reminds us, is what Marx and Engels called "armed men" with material appendages like "prisons and repressive institutions of all kinds." Presumably the armed men put the workers into the prisons when the bourgeoisie exerts the greater pressure on the state and the bourgeoisie goes into the prisons when the workers exert the greater pressure. "The bourgeois democratic state is amenable to various class pressures." If the New Deal hasn't proved that to you, you are pretty hopeless...

It may have taken the New Deal or even post-New Deal reflections for Erber to learn that working-class struggle and pressure can gain many reforms and concessions from the bourgeois state, democratic or autocratic. The political facts of life come to him slowly. We do not complain about that. Every man has his rhythm. But we suggest that we learned this some time back, almost immediately after discovering that two plus two makes four. Among others, we learned it from Lenin.

In his two best-known classics on the questions, *State and Revolution* and the Anti-Kautsky[50], Lenin just doesn't bother with repeating the veriest commonplaces of the Marxian movement. His two works are devoted to the question of the state in connection with the revolution, with the direct struggle for power, and not in connection with the daily political struggles of the working class in non-revolutionary periods. What those struggles are, how they are to be conducted, what they can accomplish — all that he takes for granted. He is assuming that his readers are at least half-educated socialists to whom fifty pages to prove that two plus two makes four would be a bore.

In fact, with reference to Kautsky, Lenin impatiently writes, in the first pam-

---

against their conviction on murder charges in a trumped-up trial in 1921.

50. By "The Anti-Kautsky", Shachtman refers to Lenin's late-1918 pamphlet *The Proletarian Revolution and the Renegade Kautsky,* a defence of the 1917 Revolution and the Bolshevik government against Kautsky's denunciations.

phlet: "Opposition and general political struggle are beside the point; we are concerned with the revolution." In his second pamphlet, he apostrophises Kautsky as follows; "It would not be amiss for you to know that 'opposition' is a concept that belongs to the peaceful and only to the parliamentary struggle, i.e., a concept that corresponds to the non-revolutionary situation, a concept that corresponds to a situation marked by an absence of revolution." Lenin knows all about the state being "amenable to class pressures," all about the usefulness of bourgeois democracy to the struggle of labour and socialism. But he also knows that what he was called on to deal with was the fundamental question over which a tremendous controversy had broken out — the state and revolution. It is there that Lenin made his exact contribution to the debate — not by revising but by reviving the Marxian teaching on the state in the light of the experience of the Russian Revolution. It is this contribution that has to be attacked, and the dull page upon page to prove that in the rich and wonderfully peaceful United States workers can get reforms from the bourgeois state and that the bourgeois state collides with bourgeois opinion is "beside the point" and only gives you a headache.

Even Erber begins to feel a little uncomfortable about not having gotten to the question that bothers him at the outset. Trotting ever so gently on the extreme ends of his toenails, he writes toward the end of his document: "It may be argued that even if the bourgeois democratic state is amenable to various class pressures, this does not prove that it can be 'pressured' into solving any basic problems in accordance with the workers' needs and certainly, could not be taken over by a working class parliamentary majority to establish socialism.

"It is quite true that it has not been proven. However, neither has it been disproven. No one can say over what road, or combination of roads, the struggle for power and socialism will lead. What is necessary for us is to establish that both Scheidemann's road[51] and Lenin's roads are failures. Both must be rejected."

He squeezed it out at long last! Congratulations! Is the parliamentary reform of capitalism the road to socialism?

Not exactly a brand-new question. It's as old as the Marxian movement itself — even older. It's been debated thousands of times, by thousands of people in and out of the movement, in dozens of countries, in tens of thousands of books and articles and speeches, under the most diversified of circumstances. The debate now has a good hundred years of living historical experiences of all kinds, and in all countries, to base itself on — not just books and theories and abstractions, but a whole spectrum of actual experiences.

Erber plunges into all this, without a stitch of clothing on him, without impediments of any kind, naked as a baby, with a "fresh approach" that is highly commendable. He dives in again and again; he swims; he floats; he darts to

---

51. I.e. the reformist road. Philipp Scheidemann was a leader of the German Social Democratic Party, and Chancellor of Germany at the height of the revolutionary upheavals after World War One, from February to June 1919.

the bottom, stirs up the mud, lifts up all kinds of stones to find what he's look-ing for; zooms to the top again, tries out every stroke, floats on his back and on his belly.

At last he comes out of the water, just the way he went in. "What did you find? What did you learn? What do you think?" asks the anxious throng that immediately surrounds the fearless venturer into the depths. "Is there anything to the reformist road of Scheidemann after all?" He sits down slowly, sinks his chin on his hand, tightens his brows, focuses his eyes to an impressive faraway look, draws a deep breath, mumbles and sighs for two solid hours, and with oracular majesty proclaims: "It is now clear to me that two plus two makes four!"

Tremendous excitement and elation from the throng! "But what about Schei-demann's road — is it right?"

Another deep breath, another sign, another proclamation; "It is quite true that this has not been proven."

Fine, splendid, very statesmanlike, very categorical, no shilly-shallying there! "Then Scheidemann's road is wrong?"

The chin sinks deeper, the brows grow sterner: "However, neither has it been disproven. Both Scheidemann's road and Lenin's road were failures. I finally saw that when I hit six fathoms. Both must be rejected."

There is a slight murmur of disappointment in the throng, but hope has not yet fled. "Tell us, oh Wise One Who Has Risen from the Depths, what road shall we take to get out of this infernal mess of ours?" The noble head is lifted, the eyes acquire a post-infinity stare, you can virtually see the fine wheels of the mind churning at a furious pace as The All-Revealing Revelation is uttered: '"No-one can say over what road, or combination of roads, the struggle for power and socialism will lead."

"No one? Not even you?"

"No, no one, not even me." This need not be told to the end. Any reader can easily imagine the ensuing outburst of joy and enthusiasm in the throng, the pervading feeling of new morale in the struggle that courses exhilaratingly through their blood, the renewed determination they feel for launching the as-sault on capitalism, the profound gratitude they feel toward the Almighty and toward Nature for vouchsafing them so luminous a thinker, so inspiring a leader, such a man among men.

"It hasn't been proved, but then again, and however and nevertheless, and on the other hand, and to give all sides a fair shake, and to put it clearly, neither has it been disproved."

"But the experience in Finland, in Germany, in Austria, in Italy, in Spain[52]. Do they not prove something about the reformist road, oh Wise One Who Drips Water?"

"What does that prove?" says the Wise One impatiently, shaking off some more water. "Only five countries! Why, there are more than fifty countries be-

---

52. Between World Wars One and Two, Finland, Germany, Austria, Italy, and Spain all had social-democratic or radical-liberal governments which failed to secure lasting reforms and fell before fascist or right-wing reaction.

longing to the United Nations alone! When I say it hasn't been proved, I mean it hasn't been proved to the hilt. Has it been proved in San Marino? In Ecuador? In the islands of the Polynesian Archipelago? Of course not! Then how can we be so Leninistically dogmatic? Just because thousands of people have died from a stiff dose of potassium cyanide, does that prove it is necessarily fatal, or that it was more than a coincidence? Have we any conclusive proof that it was administered properly, by diplomaed medical men, under controlled laboratory conditions, or that the victim did not actually die of boredom? Thousands of people — what is that to a scientist? Why, there are four hundred million people in China alone who have not been tested with potassium cyanide. The sampling to date is too small, and there is always Ecuador.

"I am a scientist, not a dogmatist. I say, don't I, that Scheidemann's road has not been proved. But I also say it has not been disproved. And don't I add that I don't know what road to take, or to recommend? That shows how modest. I am, how humble I am, how ignorant. I am honest enough not to pretend that I know the answer, as you conceited folk do. That may not appeal to everyone, but I know some people it will appeal to: those who do not want to do anything in the struggle, or who have grown tired in the struggle and who envy and admire me! And thank me for giving them a totally scientific reason for not doing anything."

Having rejected the reformist road and the revolutionary road, Erber can no longer remain in the Workers Party. "Were I a member of a broad, Marxist educational society, without a program, without a line, based upon more than one historical tradition, where all views have equal status, the struggle for my ideas [what ideas?] would have an educational significance aimed toward the crystallisation of a programmatic grouping at a later stage."

Let us help out Erber with a free and sincere suggestion: Form your educational society! You will find plenty of somewhat faded material for your membership. Title: League of Know-Nothing Socialists. Alternative title: Loose Alliance of Associated Muddleheads (L.A.A.M). Program? No program. Line? Likewise none. Size? Broad. Historical tradition? More than one and take your pick. Views? All men are created equal and your views are as good as mine if not better. Purpose? Crystallisation, but not now. Meetings? On call, whenever a member gets an idea, or thinks someone else may have gotten one. Rules and regulations for meetings: First half hour devoted to proving that Lenin's road has been disproved. Second half hour devoted to proving that Scheidemann's road has not been proved or disproved. One half hour of silence, devoted to earnest searching of the soul. One quick minute of mutual questioning, for the sake of formality, to see if any one has yet found the right road to socialism. No collection taken to pay for the light in the hall, since none has been shed. Singing of the familiar League (or Alliance) song, first stanza beginning with: "Ignorant and muddled are we, with nothing to say or do." Adjournment. All that takes less than two hours, leaving enough time to see a late movie.

We will cheerfully provide Erber with the names and addresses of enough prospective members for foundation purposes. We are even ready to go further and arrange a definite division of labour with Erber and his army of friends. If anyone comes to us saying that he is for socialism, we will try to convince him that the revolutionary road is the only road by which it can be realised; therefore he should join our ranks. If he says, at the end of our discussion, that he is not convinced, and that he wants an organisation that takes the reformist road, we will unhesitatingly give him the address of the Social Democratic Federation (if he is over 60 years old) or of the Socialist Party (if he is under 60)[53].

However, if our arguments prove unconvincing and he says he wants an organisation that knows how socialism cannot be achieved, but hasn't the slightest notion of how it can be achieved, we will thank him for his patience and recommend him to the ministrations of Erber's League. We do not insist on reciprocity. What could be fairer?

Upon reflection and closer reading, however, it occurs to us that one sector of this division of labour may well be superfluous. Erber's rejection of Lenin's road is clear and categorical. The same cannot be said about his view on Scheidemann's road. His rejection of it is purely verbal and confined to the mere statement in the quoted passage. Read a little further, and you see that Erber has accepted Scheidemann's road, that is, the road of reformism, the road of parliamentarism, the road of class collaboration in industry and in government. His rejection of this road is pure camouflage (conscious? unconscious? That is beside the point). The camouflage consists in falsifying the classical reformist position. Why Erber rejects the revolutionary road, he tries to explain in page after page of stertorous gasps. But why does he reject Scheidemann's road? (We assume that Scheidemann stands not for the individual, but for the "road" which he symbolises.) On that, Erber gives us one single sentence, not less but also not more: "Social Democracy came to view the struggle entirely as a parliamentary one, with the economic organisations of the workers limiting their role to improving the lot of the workers under capitalism, until a parliamentary majority would introduce socialism from above." And how, in contradistinction, has the Wise One Who Drips Water come to view the struggle?

"In the last analysis, the strength of the working class is, therefore, economic. Its political strength has real meaning only in terms of its economic strength. The latter cannot be given to it by socialist parliamentarians nor can it suddenly appear under the conditions of the highest tension of the class war. What is wrong with a 'purely reformist' participation of the workers in measures of economic control in a non-revolutionary period? The organised power of the workers must seek means of asserting itself in economic controls on the level of the department, plant and industry. Without such economic power as its base, the political victories of the workers rest on a very frail foundation. This exceedingly important concept was one of De Leon's contributions to Marxist

53. Reformist socialism in the USA had split in 1936, with the more right-wing Old Guard, headed by trade-union leaders, splitting away from the Socialist Party of America. The SDF and the SP reunited in 1957.

thought."[54]

As you can see, the Wise One is back at his muddle-headed march in two directions. The political strength of the workers has real meaning only in terms of its economic strength. Bull's-eye! What follows? What could follow except a muddle? The achievement of Reuther's GM program[55] would have been a tremendous stride forward for the workers... And the IWW concept of workers' management of industry was not altogether wrong... All right, all right. Where is the muddle?

What the revolutionists of all shadings — Leninists, Luxemburgists, even militant syndicalists[56] — have always talked about is the workers' management of industry as a function of their ownership of the means of production and exchange. Short of that, we have talked about workers' control of industry, with the bourgeoisie still owning property and managing industry, but this control only as the direct transition, in a period of sharp class struggle, to ousting the parasites altogether. But we have always rejected the sharing of control or management of industry by the two irreconcilably hostile classes as a long-lasting stage in the development of the struggle.

Why? Out of some abstract dogmatism? Not at all! Our position is only a generalisation from experience in the class struggle. We are for the workers, through their organisations, fighting at all times for certain rights and privileges in industry, which safeguard and improve their economic and social position, strengthen their self-confidence and weaken the power of the exploiter. We are for factory inspection, for reducing the work week, for increases in pay, for workers' control of hiring and firing, and the like. But we are not for joint worker-employer committees or boards in industry, any more than we are for joining together the AFL and the Chamber of Commerce, the CIO and the National Association of Manufacturers.[57] What are these "joint labour-management councils"?[58] First of all, they are based on the conception of the "essential identity of interests" between two classes whose interests cannot be reconciled, which means their very basis is false and rotten. Second, they are one of the means whereby the workers' representatives in the councils are called upon to police the workers without being able to police the employers, that is, to enforce upon the workers the decisions which, at bottom, only the employer makes or can make by virtue of his ownership of industry. Such joint councils

---

54. De Leon argued that by building industrial unions the workers could enable themselves to take control in each industry, and the socialist revolution would consist of that industrial change simultaneously with political revolution.

55. In 1945 Walter Reuther, leader of the United Auto Workers, announced a "GM program" to force General Motors to concede a 30 percent increase in workers' wages and a freeze on car prices, and to open the company's books to the union.

56. Militant, or revolutionary, syndicalists, believed that a socialist revolution could be made by trade-union action alone, by trade-unions taking over industries and the political state being thus collapsed, without political-party action.

57. The AFL and the CIO were the chief US trade-union organisations. See above, note 11.

58. I.e. "purely reformist participation of the workers in measures of economic control in a non-revolutionary period", Erber's recommendation from a few paragraphs earlier.

exist and can exist only on the basis of the mutual understanding that owner-ship requires and legitimatises profit, that the industry must be managed in such a way as to assure profit. And since profit can be realised only by the ex-ploitation of workers (or is that too a simple-minded Leninist dogma?), the guarantee of profit is the binding framework within which any management — "pure" "bourgeois, "joint labour-employer" or "joint government-em-ployer" — can operate under capitalism.

That is why, wherever such "joint councils" have been established (with the enthusiastic or enforced agreement of the bourgeoisie), they have operated for the benefit of the capitalists and against the interests of the workers. The work-ers and their representatives are required to take the responsibility, singly and collectively, for the anarchy of capitalist production which they cannot over-come so long as the bourgeoisie runs the economy and so long as the working class does not organise and control the economy in its own name and its own way.

"Times are bad. The company is not making any money. Things are all wrong in the economy." The revolutionist replies: "If that is the pass to which things have been brought by our 'managers', they are social bankrupts. They cannot keep industry going properly. They cannot keep the workers at a decent stan-dard of living. Throw them out! We have enough ability to organise industry properly by ourselves." The reformist, the labour bureaucrat, the class-collab-orationist, the champion of labour-management committees, replies: "We must keep this industry going. We must work harder and make fewer demands. We may even have to take a wage cut, because if this industry goes out of business, we are wiped out of our jobs. We have seen the bosses' books — they are really up against it."

Is it such "joint labour-management councils" so dear to the reformists, to our own Gompersists[59] and their imitators in the AF of L and CIO today, that Erber has in mind when he writes that the power of the workers "must seek means of asserting itself in economic control on the level of the department, plant and industry"? That is the only conclusion we can come to from the most careful reading of his muddy prose. In that case, he does not differ from the social democratic reformists! He has falsified the Scheidemann conception! His description of how the "social democracy came to view the struggle" is one-sided and a misrepresentation, and we do not hesitate to come to the defence of the Scheidemanns against their misrepresenters. "Strict" social-democratic theory declared that just as it was possible and necessary to win parliament to socialism by gradual penetration, so it was possible and necessary to win in-dustry to socialism by joint factory councils with the employers.

Just as a coalition with the bourgeoisie was necessary on the political plane, so it was necessary on the industrial plane. The social-democrats were the orig-inators, particularly in Scheidemann's fatherland, of the joint labour-manage-ment councils in industry, and exactly and precisely on the basis of Erber's not very novel idea that the power of the workers in politics must be coupled by

---

59. "Gompersists", after Samuel Gompers, leader of the AFL (see note 11) from 1895 and 1924.

"asserting itself in economic controls on the level of the department, plant and industry."

What, if not that, has been the "pro-labour" side of the argument in this country, for decades, in favour of all the "joint labour-management" schemes? What, if not that, has been meant by all the "industrial democracy" schemes of the labour leaders, reformists, liberal muttonheads and enlightened capitalists in this country? We used to know what planet Erber inhabited. We are not so sure today.

It was through their "joint councils" that the German Social Democracy finally discovered that, in the decay of capitalism, labour is called upon to be not the grave-digger of capitalism but, as it was put, the "doctor" who seeks to cure the ill. If Erber is simply ignorant of these familiar facts, he should stop abusing the social-democrats. If he is simply ignoring these facts, it is only in order to present his acceptance of class collaboration in a "less Scheidemannistic" light. It may be said: This is too strong. We reply: it is not strong enough, and here is proof of that.

Erber criticises "our demand that the MRP be ousted from the MRP-SP-CP government during the crisis of 1946 (in France), instead of demanding the ousting of the CP."[60] (Erber does not indicate that he agreed with us at that time and now considers that he was in error. That is not new. In *none* of his writings has Erber ever acknowledged that he made any political error — we repeat, in none! As he was told years ago, he suffers from the offensive malady which the Germans call *Rechthaberei*[61] and therefore always writes with a pained expression of injured innocence on his face.) Then he adds: "The minority, which opposed the slogan of 'CP-SP to power' dared not even think of the latter alternative, and therefore had no political proposals whatsoever."

This is not the place for rehashing our 1946 discussion. Only one thing is interesting. According to Erber today, the correct political course at that time would have been to demand that the Stalinists be ousted from the coalition government, that is, to demand a coalition government of the Social Democracy and the Popular Republican Movement (MRP)!

There is your true un-Leninist socialist policy! Erber may not know the whole road to socialism, but he knows his intermediate stations. He is not a Scheidemannite. Oh no! He's just a Kautskyan, which means he differs from Scheidemann like one banana from another. The road to socialism lies through a coalition government with the MRP, that is, with the party of the bourgeoisie. And not just any old party, but the party of the Vatican, the party of social

---

60. Between 1944 and 1947 France's "provisional governments" were coalitions of the Socialist Party, the Communist Party, and the MRP, a Christian-Democratic party. In a government crisis in 1946, the Workers' Party majority raised the slogan for the ejection of the bourgeois MRP from the government and a coalition government of the SP, the CP, and the biggest trade-union confederation, the CGT. A minority of the WP, including Hal Draper, Irving Howe, and Al Glotzer, disagreed, arguing that the CP would dominate such a SP-CP-CGT coalition, and the CP had to be assessed primarily as an agency of the Stalinist government of the USSR, rather than as an odd version of a traditional reformist party based in the local working class and the local labour bureaucracy. For the debate, see bit.ly/wp-sp-cp.

61. Rechthaberei: a "know-it-all" attitude.

Catholicism, of clerical political obscurantism, beloved by the Lord and by Washington.

A man must possess a stupendous amount of triply distilled effrontery to be able to advocate a Social-Democratic-MRP coalition government in the name of socialism, and in the very same document in which he associates himself, to even a minute extent, with such people as the Debs, Haywood and St John of the IWW, Daniel De Leon, and Rosa Luxemburg. If any one of them were alive today, the best that Erber could expect is that they would hang themselves after seeing who took their names and toward what end.

Erber is for collaboration between the working class and the bourgeoisie in industry. Erber is for coalition governments between the working class and a bourgeois party in parliament. That all this separates him sharply from Marx and Lenin is self-evident. In what significant way he considers himself separated from Scheidemannism, i.e., from traditional reformism, is a puzzle that is insoluble in political terms. It must be unravelled by students of psychology, which is not really our province. We still think there is room for a League such as we have recommended. But it is now doubtful if there is room in such a League for Erber, provided it adheres firmly to the basic principle which alone justifies its existence: "We do not know what to say or do." If he is admitted to membership, then only on the ground that he himself does not yet know that he has become a perfectly house-broken social-reformist.

Maybe he doesn't know that; but at least he knows something about Engels' letter to Conrad Schmidt[62], doesn't he? Therewith, we are at the last point in this chapter. No, he doesn't know anything about Engels' letter; he doesn't understand it. All you can say about him is that he remains faithful to the virtue of muddleheadedness. So, back again we go to Engels.

In this letter, as in several others written toward the end of his life, Engels deals with greater preciseness than ever before with the question of the interacting relationship between economy and politics, between substructure and superstructure, between economic development and "force". He found it necessary to introduce a balancing corrective into the popular understanding (misunderstanding) of the Marxist theory. As he wrote in another letter of the same period: "Marx and I are ourselves partly to blame for the fact that younger writers sometimes lay mere stress on the economic side than is due to it. We had to emphasise this main principle in opposition to our adversaries, who denied it, and we had not always the time, the place, or the opportunity to allow the other elements involved in the interaction to come into their rights."[63] (The phenomenon is a familiar one, and applies no less to Lenin than to Marx or... Trotsky.)

The letter to Schmidt is occasioned by an anti-Marxist criticism by the German, Paul Barth, who charged that the Marxists "deny any and every reaction of the political, etc., reflexes of the economic movement upon the movement itself." "He is simply tilting at windmills," jeered Engels. There is an important

---

62. See note 27.

63. Shachtman quotes here from Engels' letter to Bloch of 21-22 September 1890: bit.ly/eng-bl.

interplay and reciprocal influence exerted by the economic movement and the political power which has a relative independence from the economy, but "it is the interaction of two unequal forces." The state power can speed the economic development by moving in the same direction. It can stand in the way of the economic development, in which case it can do it great damage or be itself smashed. The same thoughts are expressed, even more trenchantly, in the chapter on "The Force Theory" in Engels' *Anti-Dühring*[64]. Let us supplement the Schmidt letter with an instructive quotation from the *Anti-Dühring*, which will make it even clearer that Erber just doesn't know what to make out of the material he stumbles over:

"… After the political force has made itself independent in relation to society, and has transformed itself from its servant into its master, it can work in two different directions. Either it works in the sense end in the direction of the normal economic development in which case no conflict arises between them, the economic development being accelerated. Or, force works against economic development; in this case, as a rule, with but few exceptions, force succumbs to it.... But where — apart from cases of conquest — the internal public force of a country stands in opposition to its economic development, as at a certain stage has occurred with almost every political power in the past, the contest has always ended with the downfall of the political power. Inexorably and without exception the economic evolution has forced its way through — we have already mentioned the latest and most striking example of this: the Great French Revolution…

"That force, however, plays another role in history, a revolutionary role; that, in the words of Marx, it is the midwife of every old society which is pregnant with the new, that it is the instrument by the aid of which social movement forces its way through and shatters the old, fossilised, political forms — of this there, is not a word in Herr Dühring. It is only with sighs and groans [Engels is speaking here, rest assured, of Dühring, not of Erber] that he admits the possibility that force will perhaps be necessary for the overthrow of the economic system of exploitation, unfortunately, because all use of force, forsooth, demoralises the person who uses it."

In Erber's scholarly and erudite survey of human history (wherein the Hanseatic League and the Long Parliament find the place due them), he comes at length to the French Revolution. It is the "classic example" of how the state (feudal) lost its political supremacy when it "safeguarded the needs solely of the ruling class at a time when these needs ran counter to those of the economy." "However," notes the observer whose vigilant eye misses nothing, "not all feudal states were overthrown. The feudal-monarchic-bureaucratic state in Germany adapted itself to the needs of the economy and ruled on behalf of the bourgeoisie, while keeping them at arm's length from the state apparatus itself." From which follows at least one "suggestive" conclusion:

"The possibility of adaptation to capitalist society by feudal states was based on the peculiar nature of the bourgeoisie."

---

64. For *Anti-Dühring* see note 31.

The "economic movement" under feudalism was the tendency to replace agriculture and handicraft by machinofacture, land by capital, production for use by production for profit, the shut-in primitive market by the world market. Force was represented by the political power of the feudal state and its possessors, the feudalists. The economic movement was represented by the advent of machinery, by merchant and lending capital, and by its possessors, the young, mixed bourgeoisie. In France, the state power persisted in fettering with outlived feudal bonds the unfoldment of the economic movement. The revolution, which also represented force, a force opposed to the state, overturned the state, cut the bonds and, under Napoleon in particular, the economic movement, capitalist production and exchange, was vastly and rapidly accelerated not only in France but elsewhere in Europe. The young bourgeoisie took clearer form, and consolidated itself and its social power, under the protection of the new bourgeois state. Right!

Just as right is the story of the German development. There, the gifted and ruthless statesman Bismarck led Germany to national unification by breaking the power of the particularist nobility, consolidating a German nation, and promoting the "economic movement." He promoted a capitalist development of Germany under a *capitalist* state (*not* under a feudal state), but without yielding the political power to the very young and very cowardly German bourgeois upstarts, who were terrified at a reproduction of the Great French Revolution because a new force — the militant proletariat — was already on the scene threatening to make the revolution... permanent. So, while it is not as simple as it appears in Erber's version, the fact is absolutely indisputable that for Germany to move from feudalism to capitalism, the violent overthrow of the feudal regime by the new class, the bourgeoisie, was not an indispensable requirement. From which our innocent one shyly suggests, does it not follow that to move from capitalism to socialism, the violent overthrow of the capitalist state by the new class, the proletariat, may also prove to be not an indispensable requirement?

One by one, your hair stands on end! Reasoning by analogy is always risky, but at times necessary and enlightening. But there has to be some basis for the analogy in the first place. To establish such a basis in his case, Erber would have to show, by his Bismarckian example, that the *bourgeoisie*, the young, rising and oppressed class which was the authentic bearer of the new social system, began its climb to power *by infiltrating the feudal state*, exerting more and more "political pressure" on the feudal state, forcing it more and more to "adapt itself to economic needs," and gradually, after a while and without a revolutionary overthrow of the feudal state and the feudalists, taking command of the state. Once he showed this, he might have at least one square inch of ground to stand on in saying that, correspondingly, the *proletariat*, the young, rising and oppressed class of today, can climb to power by infiltrating the bourgeois state, exerting more and more "political pressure" on it, and so on and so forth, until it has converted this state into its own instrument for the inauguration of socialism.

But he not only does not try to show that this is what happened in old Germany — he insists that something quite different happened. It was the representative, not of the bourgeoisie, but of the old feudal caste, Bismarck, who brought about the social transformation in Germany. And the young, rising bourgeoisie? Elsewhere — without any thought about what he is writing, without any thought of the need of tying together what he writes on one page with what he writes on another — Erber tells us: "The German bourgeoisie was fated never to establish its control over the state apparatus; its control passing from the Junker domination to reformist labour domination to Nazi domination." This is exaggerated, but it will do. Now then, since it is generally understood that modern Germany has been a bourgeois state for several decades, we must conclude that the other end of the analogy would have to look something like this:

Under capitalism, some Bismarckian representative of the bourgeoisie will use the bourgeois state in such a way as to adapt it to the needs of the economy. He will see to it that the bourgeois state does not "operate solely to safeguard the interests of the bourgeoisie". He will see to it that it fulfils a "social function". He will direct the transformation of capitalist society to socialist society, just as Bismarck directed the transformation of feudal society to capitalist society. Then, to make the analogy real and complete, Erber will tell us a hundred years from now that "the American proletariat was fated never to establish its control over the state apparatus."

Socialism without socialists! Socialism without revolution! Painless socialism! Testimonials by Fredrick Engels certified by Ernest Erber, produced on request! Capitalist teeth extracted with Bismarckian laughing gas! Come and roll on the floor with hilarity!

But, bless our soul, all this bears frightening resemblance to the new theory of some Trotskyists and ex-Trotskyists: just as Bismarck and Napoleon brought capitalism into existence without giving the bourgeoisie any political power to speak of, so Stalin and the Stalinist bureaucracy are destroying capitalism and bringing socialism (a workers' state) into existence without giving the proletariat any political power at all.[65] Be it said in extenuation of their madness that they have not tried to make poor Engels responsible for their "theory."

For all the difference between these two theories, there is one thing basic to them both: Neither one of them understands *the* specific characteristic of the proletarian revolution, of the socialist transformation of society. Both theories are *reactionary*.

Erber writes: "The possibility of adaptation to capitalist society by feudal states was based on the peculiar nature of the bourgeoisie." He picked up this cloudy notion in one of his daydreams. What is "peculiar" about the bourgeoisie is the characteristic it shares with *all* preceding ruling classes: it is an exploiting and oppressing minority which requires a state power to protect its social domination from assaults by the exploited and oppressed majority. It is, like its predecessors, a wealth-possessing and property-owning class. The state

---

65. Shachtman is referring here to the ideas of Isaac Deutscher and those who thought like him.

A poster from the days of Lenin's and Trotsky's Third International

of antiquity, the slave state, could be transformed into a feudal state without *necessarily* undergoing a violent revolution and destruction. Why? Because this social transformation could take place without the slave-owners *necessarily* being destroyed, without the slave owners *necessarily* being expropriated of their wealth (and therefore of their social power). The slave-owner either was an owner of land already, or the wealth accumulated for him by his slaves enabled him to become a landowner and therewith a part of the new feudal ruling class. Where the old state power did not "run in the same direction" as the economic movement, it was generally speaking destroyed. But it also could run in the same direction; and the old ruling class could become the new ruling class or part of the new ruling class.

Similarly, in the case of the transformation of the old feudal states into capitalist states. They did not *necessarily* have to undergo a violent revolution before the transformation could be completed (although if it did not undergo one, the transformation was never as "classic" and "complete" as in the case of eighteenth and nineteenth century France). The wealth and power of the feudal lords did not *necessarily* have to be expropriated and destroyed for the transformation to capitalist society and the capitalist state to take place. The economic wealth and the political power accumulated by the English feudal lords under feudalism were the means by which they perpetuated themselves — *in the form of capitalists* — in the course of England's transformation from feudalism to capitalism; which is how the roll of the British capitalist class today includes so many distinguished and aristocratic noblemen, up to and including authentically titled *capitalist* Dukes. The same holds true of Germany, where the roll of the capitalist class included not only the sons of authentic old bourgeois (merchant capitalists and loan sharks) but also sons of the no less authentic ex-feudal Prussian Junkers. Similarly in Japan and in any number of other countries.

To achieve the transition from feudalism to capitalism, from the feudal states to the capitalist states, it was not necessary to destroy wealth, to expropriate it, to destroy its possessors. It was only necessary to establish the supremacy of capital over landed property, of capitalist economy over feudal stagnation, of the capitalist nation over feudal dispersion and particularism. That revolution could be and in some countries was achieved "organically" (more or less), in others violently. But more important: that revolution could be achieved in some countries by the peasants, artisans, incipient proletarians, and petty-bourgeoisie; in other countries by the bourgeoisie itself; and in still other countries by the feudal caste itself.

The historically-outlived class was able in many cases to adapt itself without much difficulty to the social transformation, i.e., to the bourgeois revolution. It could do it by becoming part of the new class which was the principal beneficiary of the new social order. It could adapt itself even without becoming part of the new class, simply by remaining a landlord class, which, while not strictly speaking a capitalist class, shares in the surplus value extracted by the capitalist class. (Landlords are not part of a "pure" capitalism; they are a class "handed down" from feudalism; and since almost nowhere have they been wiped out by capitalism, the latter continues to exist side by side with this "feudal remnant" to the present day.) Erber does not give the slightest sign that these thoughts ever occurred to him.

How do matters stand with the socialist transformation of society? Here is exactly where the term "peculiar nature" is appropriate. The proletariat is the only class in all history that comes to power not for the purpose of oppressing or ruling over another class but in order to abolish all classes, itself included. It cannot consolidate its power, let alone move on to socialism, without abolishing all privileges and all class wealth, including all private property in the means of production and exchange. Some proletarians may not yet know what their historical mission is on this earth. But the bourgeoisie of the entire world does know — including all the ex-feudal capitalist aristocracy of Italian and Swedish Counts, German Barons, French, Japanese and Georgian Princes, and British Dukes. Hence, their fierce, unrelenting, intransigent, irreconcilable hostility to the socialist revolution or, to use a salon term, transformation. In comparison, the opposition of the feudal lords to the capitalist transformation was insignificant.

Take the matter from another angle. The *class* which is the bearer of the new economic movement under feudalism is the bourgeoisie. It alone can bring this movement to full unfoldment. But capitalist society and the capitalist state, as we have pointed out, does not necessarily have to be brought into existence *by this bourgeoisie.* It can be brought into existence by feudalists and even by slave-owners; it can be brought into existence by popular revolution or by bureaucratic measures from above; it can be brought into existence by the political power of the bourgeoisie or without that power, by depriving the bourgeoisie of such power; it can be brought into existence democratically or by blood and iron — because all it needs basically to assure its existence is the supremacy of capital, of private ownership of the means of production and exchange.

The class which is the bearer of the new economic movement under capitalism is the proletariat. Its "peculiar nature" is that it is a socialist proletariat. It alone can establish a socialist economy. But the socialist state — more accurately, the workers' state — without which all talk of a socialist society remains talk and empty talk, such a state cannot be brought into existence by slave-owners, by feudal lords, by capitalists, by Sidney Webb[66] bureaucrats, by trade-union bureaucrats, or by Stalinist bureaucrats. The proletariat is the only class in history (see how peculiar it is!) that nobody but the proletariat itself can emancipate.[67]

Being peculiar, it is unlike the bourgeoisie in another respect. For the triumph of capitalist society over feudalism, the bourgeoisie does not have to have political power. But for the triumph of socialist society over capitalism, the proletariat must have political power, a state of its own. No political power for the proletariat, no democratic rule by the proletariat, means: no socialism. Only by means of this political power — its own state — can the ruling proletariat carry though its socialist economic transformation. It cannot even begin to carry it out within the womb of capitalist society the way capitalists carried out their economic transformation within the womb of feudal society.

Erber quotes from Engels' letter to Schmidt to show that Engels understood that the state power can greatly influence the economic movement. "Force (that is, state power) is also an economic power." Erber is as honest as the day is long. But he was either in a hurry or else he was anxious to spare us the trouble of reprinting too long a document. That is the only explanation we can make for Erber's having decapitated the quotation. But poor as we are, we still have enough paper to print an extra sentence, especially if it was written by a man whose conception of the state made such an illuminating impression on Erber's mind. Here it is, head as well as trunk: "Or why do we fight for the political dictatorship of the proletariat if political power is economically impotent? Force (that is, state power) is also an economic power." We think that with its head restored, it presents a more rounded picture.

But after all, cannot the dictatorship of the proletariat be achieved by purely parliamentary means, without violence? And can it not be exercised through

---

66. Sidney Webb was a leading figure in the Fabian Society, a group of socialists in Britain who explicitly argued that socialism could best be achieved by gradual permeation of capitalist society.

67. In the first clause of the Rules of the First International (the International Workingmen's Association) in 1864, Marx wrote: "the emancipation of the working classes must be conquered by the working classes themselves". That statement has been much repeated by Marxists as a keystone of our politics.

a democratic parliament? Didn't Kautsky once explain that there are more workers than capitalists? Cannot the former get their parliamentary majority democratically and rule over the tiny-dying-off capitalist class? Is it not — don't be dogmatic — just barely possible? Haven't you forgotten that while the bourgeoisie had no democratic political rights under feudalism, the proletariat, at least in some capitalist countries, does enjoy democratic political rights which it can use for impressive parliamentary victories? Isn't Erber speaking, after all, of bourgeois democracies?

Let us see how far we get *once we have granted this*. Just to spread the greatest happiness among the greatest number of Lenin-killers, let us say: Abstractly, we can conceive of a special combination of circumstances under which a genuinely and resolutely socialist proletarian party — not just a party of reformist bureaucrats — can obtain an electoral majority and a majority of seats in a parliament. (We say a parliament and not a Congress and we hope that this distinction is clear to every reader. We know quite well that to discuss the question of proletarian revolution in connection with the Congress of the democratic United States is, if not downright subversive and illegal, very much frowned upon, and we are reluctant to be frowned upon. Hence, attention! We are speaking of any country you can find on the map but not, of course, of the United States.) Abstractly, we can even conceive of such a combination of circumstances under which the bourgeoisie is so demoralised and feels so hopeless and helpless that it decides not to offer resistance to the democratically-expressed will of the people and throws in the sponge. Good enough? We have granted the possibility!

Where does that leave us? Even the most optimistic social democrat, to say nothing of so intransigent a socialist as Erber, cannot guarantee that this is a certainty. It may turn out that way, but then again it may not. "It has not been disproved, but neither has it been proved." Now, if the fighting cadres of socialism, and the militant working class as a whole, are prepared, intellectually and organisationally, for nothing but the peaceful transition to socialism through the democratic parliamentary method, as it was in Germany and Austria[68] and so many other countries, what will happen when one fine day it is suddenly confronted with a damnably stubborn bourgeoisie, armed to the teeth, with its state apparatus ("bodies of armed men with prisons") and its unofficial fascist legions, determined to drown the peaceful, unprepared proletariat in its own blood? That might very well turn into an unpleasant situation for us.

On the other hand, however, if the socialist militants and the working class in general are prepared, intellectually and organisationally, to meet all conceivable forms of resistance to the inevitable social change which the tenacious

---

68. In Germany, the reformist Social Democrats gained hegemony after the revolution of 1918, and so it resulted in only a bourgeois republic with some precarious social reforms, which was then overthrown by using its own rules by Hitler in 1933. In Austria, too, there was a social upheaval after World War One, resulting in the fall of the monarchy and a Social Democratic government in 1919-20. The Social Democrats retained control of the city government in Vienna, but were then suppressed by the dictator Engelbert Dolfuss in 1934 after a few days of armed clashes.

bourgeoisie may offer, nothing is lost and everything is gained. We do not thereby cease or even impede our electoral and parliamentary efforts. If the bourgeoisie turns actively counter-revolutionary, the proletariat is prepared to meet it, blow for blow. But if we find, in the crucial days, that the bourgeoisie has decided to bow to the democratic will of the people, especially when this will is backed up by organised and unflinching power — why, so much the better? It will be an enormous relief to all of us and a happy augur for a fairly painless transition to socialist brotherhood and peace. Nothing can be lost by instilling the revolutionary concept into the minds of the working class! A lot can be lost, including seas of proletarian blood, by instilling the rosy reformist concept into the minds of the working class. A lot can be lost by forgetting the sanguinary tragedies of the working classes of Europe in the last thirty years of experience with reformism. Those who cannot remember the past, as George Santayana said, are condemned to repeat it.

Bourgeois democracy is an excellent thing for the working class. It is as superior to autocracy as it is inferior to workers' democracy. Universal suffrage is an excellent thing for the working class, which fought to get it and will fight to retain it. But like the prettiest girl in all of France, universal suffrage cannot give more than it has. Erber quotes Engels, but has only an ectoplasmic grasp of what he is quoting. We will cite for him another passage from our old Engels in which every word is a jewel in a modern setting, and which is so lucid, and simple that even Erber ought to perceive its true meaning:

"The possessing class rules directly through universal suffrage. For as long as the oppressed class, in this case the proletariat, is not ripe for its economic emancipation, just so long will its majority regard the existing order of society as the only one possible, and form the tail, the extreme left wing of the capitalist class. But the more the proletariat matures toward its self-emancipation, the more does it constitute itself as a separate class and elect its own representatives in place of the capitalists. Universal suffrage is the gauge of the maturity of the working class. It can and will never be anything else but that in the modern state. But that is sufficient. On the day when the thermometer of universal suffrage reaches its boiling point among the labourers, they as well as the capitalists will know what to do."

Engels is writing about the bourgeois democratic republic! Every word is a jewel.

Erber may not understand himself, but we understand him. Erber may think he has rejected the reformist road, but like Molière's doctor-in-spite-of-himself, Erber is a reformist-in-spite-of-himself. The passage from revolutionary Marxism to reformism is not hard in our times, especially in the United States. Its chauvinism eats into the working class and it eats into the revolutionary movement. Its aristocratism eats into the working class and it eats into the revolutionary movement.

The class struggle in this country is *still* in a primitive stage. It will not take as long here as it did in old Europe for the primitive stage to become a very advanced and sharp stage. But for the moment it is still primitive and moves,

Viktor Deni's Village 'Virgin' is a loose adaptation of the Mother and Child theme of traditional Russian religious painting. The poster substituted the Socialist Revolutionary Viktor Chernov for the Mother and the White Admiral Kolchak for the Child. Depicted above were the White leaders Yudenich (left) and Denikin (right).

or seems to move, slowly. The proletariat has not yet even constituted itself here as a separate class, although we are confident that once it starts it will move with startling speed. The proletariat here is still the tail, the extreme left wing, of the capitalist class, although we are confident that one day that tail will reach over and crack the head. It is these circumstances that make it possible for Erber and those like him to write such puerile, liberalistic, petty-bourgeois, fog-soaked, authentically "American" trash about the state, about democracy, about parliamentarism, about revolution.

Over forty years ago, Trotsky wrote that impatience is the hallmark of the opportunist. Even though time presses, we are patient. The American working class will not end as it is. When this giant really awakes, he can recreate America and perhaps the world. We will persevere and tell him the truth: about himself and about his leaders, about the bourgeoisie, about bourgeois democracy, about workers' democracy, about the state and revolution. On the day — and it will come — when the thermometer of universal suffrage reaches its boiling point, then, if he has learned the great lessons of our victory as well as the great lessons of our tragedies, he — and not only the capitalists — will know what to do.

State and revolution in the light of two experiences

Erber has also modified his views somewhat on the Russian Revolution.

He writes that "The October Revolution is an imperishable page in the history of the great movements of the masses to take their destiny into their own hands that began with the French Revolution." This can be found in the literary sobbings of any heartbroken liberal: "You know, the Russian Revolution was an interesting experiment. However, it didn't turn out so well. The laboratory blew up, the scientists blew up, the building blew up, everything around it blew up. No more experiments so far as I am concerned. No more totalitarianism for me. I am not enthusiastic about another war, a war against Russia. But if it has to be, it has to be. And if we have to use the atomic bomb, and Franco, and de Gaulle, and Peron[69] — well, you just have to be realistic in politics."

Yesterday, Erber was a champion of the Russian Revolution. Overnight, quick as a flash, he re-examined its history with the tender objectivity which distinguishes him. What he dreamed of rather set him back on his heels. By morning, it turned out that the world would be a better place to live in today if the imperishable page had perished before it was written. The Bolshevik Revolution was a mistake from start to finish — a grotesque and monstrous mistake. Everything we suffer from today, everything the Russian people suffers from, everything the rest of the world suffers from or has suffered from in the last thirty years — everything! Including Erber's readiness to join the great American crusade against Russia — is the result of the October Revolution.

Surely, this is an exaggeration? It may be possible to exaggerate Erber, but it is not necessary. We cite his own words.

"Rarely in history has a political leadership appeared that was so thoroughly motivated by a selfless idealism or so completely dedicated to the lofty mission of liberating mankind. But the course they chose had a terrible logic of its own. Once they embarked upon it, they became its prisoners and there was no turning back. This course could not be traversed without the suppression of the socialist opposition, without the Cheka[70] terror, without one-man management of the factories, without compulsory labour. They are all fatal links in a chain that began with Lenin's revision of the traditional Marxist concept of the relationship of democracy to socialism in favour of the anti-democratic view of

---

69. Francisco Franco, fascist dictator of Spain from the end of the Civil War in 1939 to his death in 1975; Charles de Gaulle, president of France 1944-46 and 1958-69; and Juan Peron, militarist-nationalist president of Argentina 1946-55 and 1973-4, are cited as examples of the right-wing politicians which the USA, despite its claims to represent the "free world", used as allies against Stalinism.

70. The Cheka was a "commission to combat counter-revolution", set up by the Bolsheviks and the Left SRs in December 1917 as the civil war began. As the civil war went on, it became increasingly reckless and difficult to control, and was able to mutate into the Stalinist GPU and KGB.

the party ruling on behalf of the masses, as was expounded by Lenin on the eve of taking power in his essay 'Will the Bolsheviks Retain State Power?'"

At the basis of the Bolshevik course was Lenin's revision of the Marxian concept of democracy and socialism (the one thing Erber finds unendurable is a revision of Marxian concepts), which led with a logic of its own to all the succeeding horrors. Including the Johnstown flood? Not yet, that comes later.

"Once the Bolsheviks had dispersed the Constituent Assembly and decided to rule alone, they had set foot on a course from which there was no turning back. The suppression of the socialist opposition, the terror, the secret police and the long, bloody, destructive Civil War, were now inevitable. As the latter developed, all intermediate solutions became impossible, and all forces that could help bring them about were ground to bits."

Coupled with the first error is the attitude toward the Constituent Assembly. Once dispersed there was no turning back, More horrors followed. So far, the "imperishable page" is not doing too well; it looks more like the most accursed page in history. But have patience and hold your nose. You are about to see how it is possible to catch up with and outstrip the rantings of an English aristocrat against the abominations of Jacobinism:

"The evidence is pretty conclusive that Lenin made a bad miscalculation in believing the German workers would make a successful revolution. Yet, he had staked his whole course in Russia on this gamble. If Lenin won, history would absolve the Bolsheviks of all the charges their socialist opponents made against then. But if he lost? The awesome consequences of Lenin's miscalculation are written in the last thirty years — the whole tragic history of Social Democratic sterility, Stalinist degeneration, fascist victory, a Second World War, and our world of Stalinist totalitarianism and capitalist decay."

Not less, and probably more! But in heaven's name, if that is true or even half true, how can you have one single kind word for people who, by revising Marx, dispersing the Constituent Assembly, and miscalculating the German revolution, inflicted upon mankind such an appalling series of calamities, disasters, and catastrophes as you describe? What place in history should be assigned to these people, from Lenin on down, whose "mistake" plunged a whole generation into our present abyss? Devil of a lot of good it does us to know that they were selfless idealists and completely dedicated to a lofty mission. They ruined the world, they did. A plague take them all and all like them, and thrice accursed be their names and the memory of them!

But there, we have become angry and lost our temper again! As everyone knows, this is a friendly discussion that should be conducted with courtliness, affability, dispassion, critical independence and a soft musical accompaniment. Right is right. In defence of our unseemly outburst, we have only this to say:

To us, the Bolshevik Revolution was the great dividing line between socialism in theory and socialism in practice — it was not yet socialism, but the end of socialism as mere theory and the beginning of socialism as living social practice. In the Paris Commune, the people ruled their own destinies for the first time in history, but only for ten weeks, without support from France, without support from Europe, without the strength, the time, the possibility of mustering such support; without clear consciousness, without clear leadership. It was the dawn, but a false dawn.

In the Russian Revolution, the people ruled for years, the socialist proletariat ruled with understanding of its status and its role, with a leadership such as no class in history ever equalled. They proved — in a backward, three-quarters ruined country! — and proved it once for all, that the socialist proletariat itself can take power in its own name, hold power, and proceed to put the inherited chaos into socialist order. Beset by every conceivable foe, handicapped by every conceivable difficulty, they proved it beyond anybody's dreams, beyond what they were required to prove, beyond what a working class encircled and isolated in one backward land could be expected to prove. They proved at long last that the proletariat does not have to have a master to exploit and oppress it, that there is no quality inherent in the proletariat that precludes its taking power for itself. They proved that in the dark mass for which all rulers and their retainers have such lordly contempt is hidden deep and powerful springs of resourcefulness, idealism, passion for liberty, capacity for brotherhood, enormous creative genius which awaits only revolutionary release to inundate and fructify the social soil corrupted by the rule of man over man until it blooms for a peaceful world.

They not only proved that this dark mass, once lighted up by revolutionary fires, can govern itself, but they found once more, to an infinitely higher degree than the Paris Commune or the Revolution of 1905 in Russia, that natural state form which the working class needs for its own rule until there is no rule by anyone over anyone — the Commune-type of state, the Soviet-type of state. It was there to be found, not because it had been invented and artificially imposed upon the people by some doctrinary, but because it developed naturally in their class struggle first as a fighting weapon and then also as the form of their rule. With this new and highest form of democratic representative government, they passed above and beyond bourgeois parliamentarism just as surely as parliamentarism had in its time passed above and beyond monarchical rule by divine right. And even after they succumbed to powers beyond their strength, the proof was not undone, for they succumbed not because the socialist proletariat had dared to take power but because the proletariat in other countries had not dared.

Practically every bourgeois in the world recognised whose revolution it was and who ruled Russia. Millions of workers and colonial slaves recognised it too. That is why revolutionary Russia was able to light the fires of freedom all over the world. That is why it aroused passions and hopes, combativity and confidence in millions and tens of millions who were never before inspired. History does not know of another event like it; this old world was never before shaken as it was by the triumph in Russia.

It was our revolution. It remains our revolution, our victory and vindication — even now. Even now when it has been killed, strangled by its imperialist encirclers, deserted by those who should have been its socialist comrades-in-arms, speared in the back by its Stalinist assassins, dragged in the mud by every backslider and faint heart, it is our revolution. How easy and contemptible it is to draw near the slain Achilles and kick his head now and spit on him. Every craven, every deserter, every dilettante can now track his dirty boots on to the imperishable page and relax his wretched bowels over it. That is cheap, it is popular in the most respectable quarters, it requires no courage.

The defence of the Russian Revolution moves along with the attack on it in the same way and with the same aim that the fight for socialism goes on by the side of the fight against socialism. Read all the "socialist" attacks on the Revolution that have been written in recent times, including Erber's. If their authors have the slightest awareness of what is really involved, they give no evidence of it. The attack upon the Russian Revolution conducted nowadays by the traditionally anti-socialist and anti-working class bourgeoisie is not different in a single essential from what it was beginning with 1917-1918. All that is new in it is the ammunition that Stalinism has provided it with. Otherwise, the attack remains the same. It has a political meaning determined by its class aim and the class interests that prompt it. What does it boil down to? We have indicated that before:

"You workers, whatever else you do, do not take state power, do not even think in such terms. We have been warning you against it since the days of Marx. In Russia, they didn't listen to us, and look what happened. Lenin carried through a Marxian revolution. we are even ready to admit that Lenin himself was a noble idealist, but that didn't mean very much. Once started on the road, the movement had an iron logic of its own. Its inevitable outcome is the Stalinist state they have today, which even the radical Trotskyists say is an inferno for labour. Once you abolish private property, once you put all economic power into the hands of the state, we are all done for, you as well as we. Socialism is a Utopia. Capitalism is not absolutely perfect, but so long as we have free enterprise and democracy, you can get as much out of it as we can. Go ahead with all the reforms you want to. We will disagree with you here and there. But learn from Russia. Do not think of revolutionary socialism!"

That is the *only political meaning* of the "theoretical struggle" to prove that Stalinist totalitarianism was the inevitable outcome of the Russian Revolution, that Stalinism flowed "logically" from Leninism which the bourgeoisie understands perfectly to have been nothing but revolutionary Marxism, That "theoretical struggle" is part and parcel — a very, very big part today — of the bourgeois struggle against socialism and against the working class. You have to be a permanent resident of one of the remoter planets not to see this demonstrated a hundred times over in the daily ideological and political life of our times.

When we defend the Russian Revolution, its great principles and its great achievements, it is not because we are hopeless stick-in-the-muds. We are not idol-worshippers or iconoclasts in principle. We are not traditionalists or innovators in principle. We do not believe that what is old is gold or what is new is true. Our defence of that Revolution, even more than our defence of its pioneer, the Paris Commune, is nothing but the continuation of our fight for the socialist emancipation of the people. And whoever does not know ought to, know that the whole line of the bourgeois attack on the Revolution is the continuation of the century-old fight against socialist liberty and part of the century-old fight against the working class.

"Come now, are you saying that any attack on the Russian Revolution or criticism of it is reactionary, bourgeois, a blow to socialism and the working class? Isn't that dangerously close to the method of argument used in the notorious Stalinist amalgams?" We anticipate the familiar question. Since that is not what we said, we are no closer to the Stalinist method than we ever were.

Lenin submitted the Russian Revolution and its course to criticism; Trotsky had his criticism, during the great days and later in his life; Rosa Luxemburg criticised it; we ourselves have a critical re-evaluation to make. Marx criticised the Paris Commune, but so also did the British bourgeoisie. It all depends on what you are criticising, what you are attacking, how you criticise or attack it, on what is your political point of departure, on what is your political conclusion. We do not even dream of denying anyone the right to criticise the Russian Revolution or the labour movement in general. We hope in turn that we shall not be denied the right to criticise the critics. There is the proletarian, the socialist or Marxist criticism of the Russian Revolution; there is the bourgeois criticism of it. And there is the "intermediate" criticism which renounces or rejects the struggle for socialism without yet adopting in full the position of our class enemy.

Erber's criticism, like those it is patterned on, does not belong in either the first or the second category. It is "intermediate" between the two... but not equidistant from them. Read Marx's memorial to the Paris Commune. It was not an uncritical eulogy of everything the Communards thought or did — far from it. Indeed it was written by a man who, a few weeks before the establishment of the Commune, regarded the idea as preposterous! But every line vibrated with a challenging defence of the revolution.

"Workingmen's Paris, with its Commune, will be forever celebrated as the glorious harbinger of a new society. Its martyrs are enshrined in the great heart of the working class. Its exterminators, history has already nailed to that eternal pillory from which all the prayers of their priests will not avail to redeem them."

Read Rosa Luxemburg's criticism of the Bolshevik Revolution which she set down in 1918 in her fragmentary prison notes. She did not draw back from what she felt she had to say about the regime of Lenin and Trotsky. But she was blood-kin of Marx, she was a revolutionist to her finger tips who never for a moment relaxed the struggle against the enemy for socialist freedom. What right — political, moral or any other — do the backsliders and tired and retired radicals have to pull into their camp the revolutionist who ended her critical notes with those clarion words:

"What is in order is to distinguish the essential from the non-essential, the kernel from the accidental excrescences in the policies of the Bolsheviks....It is not a matter of this or that secondary question of tactics, but of the capacity for action of the proletariat, the strength to act, the will to, power of socialism as such. In this, Lenin and Trotsky and their friends were the first, those who went ahead as an example to the proletariat of the world; they are still the only ones up to now who can cry with Hutten: 'I have dared!'[71]

"This is the essential and enduring in Bolshevik policy. In this sense theirs is the immortal historical service of having marched at the head of the international proletariat with the conquest of political power and the practical placing of the problem of the realisation of socialism, and of having advanced mightily the settlement of the score between capital and labour in the entire world. In Russia the problem could only be posed. It could not be solved in Russia. And in this sense, the future everywhere belongs to 'Bolshevism'."

Where can you find so much as a lingering trace of this spirit, this attitude, this intellectual axis, in all of Erber's document, which sniffles as though it had a perpetual cold, which moans and groans and squalls as though it ached in every joint? No use looking for it. It isn't there. His political fever has burned it out of him.

"Still, abuse him all you want, you ought to take up his views objectively, oughtn't you? Just because he rejects Leninism and Scheidemannism, does it follow that his views should not be given a hearing, or that they should not be answered — *if* you can answer them?"

Yes, you are right. He should be given a hearing and an answer, which is why we printed his document who needn't have, and why we are writing this commentary. You are right, dear reader, but do not be too surprised if Erber's rejection of Lenin "and" Scheidemann turns out to be of electrono-microscopic importance. As Morris Raphael Cohen[72] once said with a scepticism we frankly share: "The notion that we can dismiss the views of all previous thinkers surely leaves no basis for the hope that our own work will prove of any value to others." We shall soon see what value Erber's work has.

It is no great problem to attack the Bolshevik Revolution today. Buy any of a dozen such attacks and you find all the raw materials required by an enterprising person. For a modest investment, you got two or three standard blueprints plus a wide range of parts to choose from for the finished product. You get: what Trotsky said about Bolshevism before the revolution; what Luxemburg said about Bolshevism after the revolution; what Lenin said about the revolution before and after; a few loose facts and figures about the Constituent Assembly; a selection of stories about Kronstadt by any number of people who weren't there, authenticity guaranteed or your money back, plus a choice of figures on how many sailors were murdered by Lenin or Trotsky or Dzerzhin-

---

71. Ulrich von Hutten was a scholar and minor aristocrat who sided with Martin Luther in his fight in the 16th century against the power of the Catholic Church and the Pope. The phrase "I have dared" is the opening of his declaration of 1520 in support of Luther, *Aufwecker der teutschen Nation*.

72. Morris Raphael Cohen was a liberal academic, professor of philosophy at the City College of New York from 1912 to 1938, in a period when CCNY was a big pool of recruitment for revolutionaries and Trotskyists.

sky (one, one hundred, one thousand, ten thousand — whichever looks better); a selection of quotations from Lenin and Trotsky about (1) dictatorship and (2) democracy; a calendar showing that Stalin took power after Lenin died, proving with actual dates that Stalinism flowed from Leninism; differently coloured bits of gossip, all very spicy and revealing, about various Bolsheviks, certified by a number of political Peeping Toms; labels marked "Cheka Terror," "Secret Police," "Suppression of Socialists", all lithographed in scarlet to imitate bloodstains and scare children; an assortment of wiring, string, nails, screws, matchsticks, nuts and bolts, and a bottle of rubber cement. All the quotations are easily pasted together, for they come carefully chopped out of context and cut down to convenient size. No special skill or training is required; any child can follow the directions and assemble all sorts of articles from the kit, including a full series for the *New Leader*[73] which can be expanded, with the aid of a little more work and ingenuity into a full-length book for a venturesome publisher.

Once you have this handy little kit (more elaborate ones are available if you want to invest in more of these second-hand books), you can write a critique of the Bolshevik Revolution as good as Erber's. We urge the reader to believe that very little skill is required for this sort of job. They will all come out looking about the same.

In the beginning, there was the Error. At the other end are the woes of the world today. The shortest distance between two points is a straight line. You draw a straight line between the beginning and the end, and you get a clear and complete understanding not only of the development of the Russian Revolution but of all world politics for the past thirty years. It is a triumph for one of the elementary principles of plane geometry.

But where is the three-dimensional reality of the country known as Russia in all this, with its class stratifications and their reciprocal relations, with its economic and social situation and the urgent political and social problems it posed at a given time, with its relations to the rest of the capitalist world in which it actually lived? Doesn't exist. It is just a discrete point on the straight line.

Where are the classes in Russia at the time of the revolution and afterward, what was their position, what were they thinking, what were they doing, what did they want? Were there other political groupings in Russia, apart from the Bolsheviks, and did they play any role in the development of Bolshevik policy, in the development of the Revolution? Not important. Each class and each party gets no more than one discrete point on the straight line.

The same question with regard to the classes and their struggle outside of Russia, the political groupings, especially the Social-Democratic parties, and their policies, and what effect they all had on the Bolsheviks and the Revolution, gets the same answer. Not worthy of note. A few more discrete points on the straight line.

But at least the straight line is made up of all those discrete points? No sir! Got nothing to do with it! The line projected itself by a logic of its own right

---

73. For the *New Leader* see note 15.

out of the heart and substance of the Error itself, just as mundane wickedness emanates from original sin. Erber knows all there is to know about the inter- action between one force, the economic, and another, the political. What about the interaction of the multitude of economic and political forces which affected the development of Bolshevism and the Revolution? Not important. Waste of time. Therefore, not one word about it in his "reexamination" of the Revolu- tion. What about Trotsky's studies and analysis of the course and the causes of the degeneration of the Revolution, which Erber must surely have read while he was still alive? Of no value. Doesn't have to be refuted. Doesn't even have to be mentioned. Waste of time. I'm working with my kit.

"Wait a minute! Do you mean to tell me that Erber has written a re-evalua- tion of the Russian Revolution and Stalinism without dealing with Trotsky's analysis, which he himself shared for so many years?" Yes. "And he doesn't even mention it?" That's right. "But that's utterly impossible, I can't believe you!" Then read Erber's document for yourself, dear friend, and don't be so dogmatic about what is impossible nowadays. We, meanwhile, will proceed to Erber's re-evaluation in some detail.

We start with the three quotations in which Erber explains how Bolshevism, ruined the world.

The ruination began, in the theoretical field, "with Lenin's revision of the traditional Marxist concept of the relationship of democracy to socialism in favour of the anti-democratic view of the party ruling on behalf of the masses," and it gathered real momentum in the political field "once the Bolsheviks had dispersed the Constituent Assembly and decided to rule alone." That's plain enough and straightforward. There's nothing muddleheaded about that; or more accurately, nothing more than usual.

What is the traditional Marxist concept which Lenin, says Erber, held to firmly up to 1917 and revised in that year and afterward? We will repeat from Erber the quotations wherein it is set forth: "If there is anything that is certain, it is this, that our party and the working class can only come to power under the form of a democratic republic. This is, what's more, the specific form for the dictatorship of the proletariat, as the great French revolution has already shown."

That from Engels. And this from Rosa Luxemburg:

"Democratic institutions — and this is of the greatest significance — have completely exhausted their function as aids in the development of bourgeois society... We must conclude that the socialist movement is not bound to bour- geois democracy, but that, on the contrary, the fate of democracy is bound with the socialist movement".

And further from her *Reform or Revolution?*:

"We must conclude from this that democracy does not acquire greater chances of life in the measure that the working class renounces the struggle for its emancipation, but that, on the contrary, democracy acquires greater chances of survival as the socialist movement becomes sufficiently strong to struggle against the reactionary consequences of world politics and the bour-

The Paris Commune, 1871

geois desertion of democracy. He who would strengthen democracy should want to strengthen and not weaken the socialist movement. He who renounces the struggle for socialism renounces both the labour movement and democracy."

The quotation from Engels is taken from his long-concealed criticism of the draft of the Erfurt program of the German Social Democracy in 1891. It is not directed at some over-radical opponent of parliamentarism, but at the opportunists in the party. In a letter to Kautsky accompanying the criticism, Engels writes that he "found an opportunity to let fly at the conciliatory opportunism of *Vorwärts* (the German party organ) and at the cheerful, pious, merry and free 'growth' of the filthy old mess 'into socialist society.'" This gives us a hint of what Engels would let fly today at Erber. What occasioned Engels' reference to a democratic republic? Perhaps someone in the German party who wanted to disperse a Constituent Assembly and set up a Soviet government? Quotations from our teachers do not decide political questions for us; but if they are used, they should be used in context so that their real sense and purpose is conveyed. Engels complained bitterly about "the inroads which opportunism is making in a great section of the Social-Democratic press. For fear of a revival of the (Bismarckian Anti-) Socialist Law and from recollection of all manner of premature utterances which were let fall during the reign of that Law, the present legal position of the party in Germany is now all of a sudden to be treated as sufficient for the carrying out of all the demands of the party by peaceful means. People talk themselves and the party into the belief that the present society will grow into socialism without asking themselves if for this it is not equally necessary that society should grow out of its old social constitution and burst its old shell just as violently as the crab bursts its old shell — as if in Germany society had not in addition to smash the fetters of the still semi-absolutist and moreover indescribably confused political order..."

This already gives us quite different picture from the one our muddlehead wants to draw for us. We will not grow gradually into socialism, insists Engels. The old shell will have to be burst. And the opportunists are keeping quiet about the need to destroy in the very first place the semi-absolutist political order, the Hohenzollern monarchy. That is why he concludes that "our party and the working class can only come to power under the form of the democratic republic." Engels is simply posing the democratic republic in opposition to monarchical semi-absolutism! Not an inkling of this from Erber.

Was the democratic republic synonymous, for Engels (and Marx), with bourgeois democracy and parliamentarism? If that is the concept Erber wants to convey, it is his right; if he wants to make Engels responsible for it, it is not his right. Engels, in his *Origin of the Family*, calls the democratic republic the "highest form of the state", adding that "the last decisive struggle between proletariat and bourgeoisie can only be fought out under this state form. In such a state, wealth exerts its power indirectly, but all the more safely." In his letter to Bernstein on March 24, 1884[74], Engels writes that: "The proletariat too requires democratic forms for the seizure of political power, but, like all political forms, these serve it as means... Further, it was not be forgotten that the logical form of bourgeois domination is precisely the democratic republic, which has only become too dangerous owing to the development already attained by the proletariat, but which, as France and America show, is still possible as purely bourgeois rule... the democratic republic always remains the last form of bourgeois domination, that in which it is broken to pieces".

And that is precisely what the Paris Commune almost succeeded in demonstrating, and what the Russian Commune did succeed in demonstrating to the fullest. Not, as we shall see, if the Russian Commune had followed the free advice of the eminent Marxist Erber, but because it followed the leadership of Lenin. Does the shattering of bourgeois rule mean that the proletariat dispenses with a democratic republic? Not at all! That follows only for parliamentary cretins who cannot absorb the idea that there can be any democratic republic other than the bourgeois democratic republic and the bourgeois parliamentary system. The Paris Commune was not a bourgeois state. Engels called it a dictatorship of the proletariat, But the Paris Commune was a democratic republic nevertheless, and a thousand times more democratic than the finest bourgeois democracy! The democratic republic is "the specific form for the dictatorship of the proletariat, as the great French revolution [the Paris Commune] has already shown".

That is precisely what the great Russian Revolution also showed. The Russian Commune was not a bourgeois democracy, but a democratic republic. Neither in 1871 nor in 1917 did the revolutionary proletariat, in establishing its own democratic republic, set up a parliamentary state — but a Commune-type of state. Engels calls the Paris Commune a democratic republic in full knowl-

---

74. Eduard Bernstein, the editor of the then outlawed German Social Democratic Party's illegal paper *Sozialdemocrat*, was then in Zürich. He moved to London in 1887. bit.ly/eng-be.

edge of the fact that it was not a parliamentary regime. How does Erber explain that? He doesn't. He gives no sign of realising that there is something here that merits explanation. In the fog with which he has surrounded himself to match the state of his political mind, democratic republic and bourgeois democracy become synonymous and inseparable, democracy and representative government become synonymous with parliamentarism and inseparable from it. He sees the bourgeois republic and parliamentarism as a tremendous advance over autocracy and despotisms of all kinds, he sees the great advantages they offer the working class. But he cannot see beyond bourgeois democracy and parliamentarism to a workers' republic which is neither bourgeois nor parliamentary.

Lenin devoted page after page of his classic, *State and Revolution*, to showing — in our opinion irrefutably — what Marx and Engels saw with their critical eye in the Paris Commune, what they learned from it, and what they tried to teach to the working-class movement. It is simply inconceivable that Erber is unacquainted with what Lenin shows in these passages. That is precisely where Lenin should be grabbed by the throat and exposed for having revised the traditional Marxian concept. There is not a peep out of Erber on this score, not a hint, not even a wink of the eye. Why? Because he's an honest critic, a scholarly thinker and an objective one. Lenin quotes striking and illuminating sections of Marx's study of the Commune:

"The Commune was to be a working, not a parliamentary body, executive and legislative at the same time... instead of deciding once in three or six years which member of the ruling class was to misrepresent the people in Parliament, universal suffrage was to serve the people, constituted in Communes, as individual suffrage serves every other employer in the search for workmen and managers of his business."

A good deal more can be and has been written about parliamentarism by Marxists, but to save your life could you compress the revolutionary criticism of parliamentarism into so few words as succinctly and unambiguously as Marx has done here? You can argue for years on whether Marx was right or not, but no debate is possible on where Marx stood in this question. We will return to it later.

What Erber does not understand (as you see, we are very polite) is that Lenin opposed parliamentarism not because it was democratic and not because he was "for dictatorship," not in order to replace democratic by anti-democratic institutions, but for contrary reasons. "The way out of parliamentarism," wrote Lenin, "is to be found, of course, not in the abolition of the representative institutions and the elective principles, but in the conversion of the representative institutions from mere talking shops into working bodies." On what grounds did Lenin attack parliamentarism? Because of its inferiority to despotism or because of its inferiority — from the working-class point of view, of course — to the Commune-type of state?

"Take any parliamentary country from America to Switzerland, from France to England, Norway and so forth — the actual work of the 'state' there is done behind the scenes and is carried out by the departments, the offices and the staff. Parliament itself is given up to talk for the special purpose of fooling the 'common people'." The real government machine of the bourgeois-democratic state is the locust horde of bourgeois and bourgeois-minded bureaucrats, growing in number, power, arrogance and contempt for the masses every year. Even an Americanised "Marxist" ought to know this by now. If the American people as a whole do not know it better than the people of other countries, they are being forced to learn fast.

Even an Americanised "Marxist" ought to know — that is, ought still to know what he once knew and taught others — that of all the bourgeois democracies, the American is the most reactionary and the least responsive to the will of the masses. No other bourgeois democracy has a political system so cunningly calculated to thwart the will of the people: with its states' rights, its division into a bicameral legislative body, its enormously bureaucratised executive with unprecedented powers, its appointed judiciary with law-making and law-breaking powers, its outrageously undemocratic system for amending the Constitution, its broken-field system of electing Congressmen every two years, Presidents every four and Senators every six, with its boss-patronage political machine, which parallels and mocks the legal government machinery from top to bottom — to mention only a few of the traditional and fundamental characteristics of our bourgeois democracy. The Congress proposes; the bureaucracy disposes. The mass is allowed to vote once a year and to "petition" the government at all times. The rest of the time, it has nothing to do with running the government, with the adoption of the laws of the land, and even less to do with carrying them out. The parliament talks; it adopts the laws; the executive, the locust-horde of the bureaucracy, carries them out in its own fashion. That is in the very nature of parliamentarism. And that is why the Paris Commune and the Soviet system marked such an enormous advance in genuine democracy. In the Paris Commune, Lenin noted:

"Representative institutions remain, but parliamentarism as a special system, as a division of labour between the legislative and the executive functions, as a privileged position for the deputies, no longer exists. Without representative institutions, we cannot imagine democracy, not even proletarian democracy; but we can and must think of democracy without parliamentarism, if criticism of bourgeois society is not mere empty words for us, if the desire to overthrow the rule of the bourgeoisie is our serious and sincere desire, and not a mere election cry for catching workingmen's votes, as it is with the Mensheviks and S.R.s, the Scheidemanns, the Legiens, the Sembats and the Vanderveldes."

This was written by Lenin in the middle of 1917, before the Soviets took power, while the Bolsheviks were calling for the convocation of the Constituent Assembly which the bourgeoisie and the Mensheviks and the S.R.s and all the later champions of the Assembly were sabotaging with all the strength and tricks at their command. It was not written after the Bolsheviks dispersed the Assembly and in order to give a "theoretical cover" to their action. It was written in broad daylight, for everyone to see, and no political person had the right to misunderstand what the Bolsheviks stood for.

So far as Luxemburg is concerned, again Erber just doesn't understand what he reads and so imprudently or inappropriately quotes. Luxemburg is attacking the Bernsteinites, the revisionists, the opportunists, the very ones whose pathetic ideas Erber has already swallowed hook and line and is preparing to swallow sinker too. If any criticism is to be made of Luxemburg's formulation, it is that it tends to be a little absolute, but that does not concern us at this time. As a general statement of the Marxist view it is unassailable. It does not in the least speak against Lenin or the Russian Revolution; it speaks against the muddlehead! "We must conclude that the socialist movement is not bound to bourgeois democracy, but that, on the contrary, the fate of democracy is bound with the socialist movement."

Luxemburg is a Marxist. She distinguishes between bourgeois democracy and democracy. She is saying nothing more than this (it is a good deal): The victory of socialism does not depend upon the preservation of bourgeois democracy; genuine democracy depends upon the victory of socialism, upon strengthening the socialist movement, upon the independence and militancy of the proletariat, upon the unrelenting struggle for the socialist goal, on no compromise with bourgeois politics. "He who renounces the struggle for socialism renounces both the labour movement and democracy".

Does Erber understand whom Rosa Luxemburg is speaking of here? Of the man who, decades later, was to attack the Bolsheviks for establishing a workers' state instead of a bourgeois democracy, for expropriating the Russian bourgeoisie and taking socialist measures instead of maintaining capitalist economic relations. That man's name? Erber will find it on his birth certificate. "You mean Erber?" We do. "The same Erber who just quoted Luxemburg?" The same. "But Erber could not have attacked the Bolsheviks that way; it's impossible; I don't believe you." You will, as soon as we have quoted from Erber. Again, do not be so dogmatic about what is possible nowadays and what is impossible.

We are not finished with the question of Lenin's "revision" of the Marxist theory. We will not even attempt to finish with this theoretical and historical question in these pages. The reader is referred to the rewarding study, so neglected in our disorganised and superficial times, of at least three basic documents without which a serious discussion of the question is impossible: Lenin's *State and Revolution*, Karl Kautsky's reply to Lenin, and the indispensable sequel by Lenin, *The Proletarian Revolution and Kautsky the Renegade*[75]. If we return to the question, it will be in connection with the political reality of the Constituent Assembly and the struggle for the Soviet power.

"The Bolsheviks," writes Erber, "rose to power in the Russian Revolution on democratic slogans: 'Down with the Kerensky Dictatorship! Only the Soviet Power Will Convene the Constituent Assembly!' However, after the Bolsheviks dissolved the Constituent Assembly, democratic slogans became a weapon of

---

75. Kautsky published a pamphlet, *The Dictatorship of the Proletariat*, attacking the October Revolution, in 1918.

their socialist opponents, while they tried to give the relationship of democracy to socialism a new interpretation: not through political democracy, but through its overthrow would socialism be achieved, ran the new Bolshevik doctrine. Democracy was considered the fortress of the bourgeoisie, dictatorship the weapon of the working class. Democratic processes and institutions were described as bourgeois weapons to blind the masses."

It is doubtful if the editor of a liberal weekly would sink to such political vulgarity or such studied stupidity — it is hard to say which. There is no doubt that the editor of some cheap bourgeois rag could rise to it without any difficulty. It positively stinks with the odour of the unscrupulous and illiterate bourgeois journalist.

Lenin, Erber explained to us, first revised the Marxian concept on democracy and socialism in the early and middle parts of 1917. In its place, he adopted the "anti-democratic view of the party ruling on behalf of the masses." But, continues Erber's explanation, after adopting the anti-democratic view Lenin still put forward democratic and not anti-democratic slogans. Why? Was there a "cultural lag" in Lenin's mind? No, democratic slogans were the only ones by which the Bolsheviks could rise to power. A supremely clever trick! For, once in power by exploiting the democratic sentiments of the masses, the Bolsheviks dropped their mask and showed in practice what their revision of Marxism really meant. It meant the destruction of political democracy and the establishment of dictatorship. Democracy was denounced as bourgeois, so were democratic institutions and processes; and "democratic slogans became a weapon of their socialist opponents."

A clever trick which shows what the Bolsheviks really were, and, you will agree, a trick so despicable that it makes Erber's delicate nose twitch and curl.

You may have your own opinion on how clever the Bolshevik trick was, but only one opinion is possible about Erber's trick — it is not clever at all. Perhaps we are doing him an injustice in speaking of his "trick." We humbly and quickly apologise. Instead we will do him such justice as nobody can challenge: his head is drenched to super-saturation with the petty-bourgeois mode of thought and expression.

Because the Bolsheviks attacked bourgeois democracy as bourgeois and defended proletarian democracy as a thousand times more democratic, it follows, like apricots from acorns, that the Bolsheviks were against democracy. Because the Bolsheviks attacked bourgeois representative institutions and the bourgeois democratic process as bourgeois, it follows, like an oak tree from an apricot pit, that they were against democratic representative institutions and processes. Because the Bolsheviks attacked the twisted, subverted, formal political democracy that exists under bourgeois rule, under the ownership of the means of production and exchange by an exploiting minority which gives it the power of life and death ever the masses; because they supported the Soviet system of government as one which gives the masses real control, and power — it follows, like Erber follows Marx, that they were for destroying political liberty.

The poor fellow simply cannot think in any but bourgeois terms. His mind is tightly boxed in by them. Proletarian democracy, soviet democracy, is a blank space to him; he cannot see it. The minute you are opposed to bourgeois democracy, not from the standpoint of despotism or Fascism, but in the name of a Soviet democracy, you are in for it so far as Erber is concerned. He cannot forgive you. It is clear to him that opposition to bourgeois democracy is opposition to democracy — full stop. That's the only kind of democracy there is. No other democracy is possible, and don't try to fool him. The minute you are opposed to parliamentary representation, it doesn't matter what you are for — Erber knows you are against representative government, democratic institutions and processes, and political democracy in general. And don't try to confuse him with a lot of talk about classes and class antagonisms, because you'll be wasting your time too.

Let us see just what it was that the Bolsheviks did do, and what happened with the "democratic slogans which became the weapon of their socialist opponents."

The key to the first door of the mystification is given in the second part of the sentence above. How is it that "democratic slogans" became the weapon of the Mensheviks and Social-Revolutionists only after the Bolshevik Revolution? It was certainly a powerful weapon before November 1917. These slogans were certainly popular with the masses of the workers and peasants, in fact, they were so powerful and popular that the Bolsheviks were able to rise to power with their aid, as Erber notes. How is it that the "socialist opponents" didn't use this weapon against the Bolsheviks before they came to power, in order to prevent them from coming to power and bringing our whole world to its present dismay? Weren't they in an exceptionally favourable position to raise these slogans and to carry them out in political life? They were chiefs of Kerensky's Provisional Government and at the same time they were the chiefs of the Workers' and Soldiers' and Peasants' Soviets. They enjoyed, for months after the overturn of Tsarism, the undeviating and enthusiastic support of the masses. They had all the political power and support that anyone would need to do anything he wanted to, certainly so far as raising democratic slogans and carrying them out is concerned.

Now that our memory is refreshed by Erber, we too seem to recall that they didn't do anything of the kind. They refused to convene the Constituent Assembly. They refused to give the Finns, Ukrainians and other peoples yoked by Tsarism their national independence. They suppressed the peasants who tried to throw off the landlords and take the land. They persisted in carrying on the Tsar's imperialist war which had bled and sickened and tired the people. They did nothing of consequence against the bourgeoisie which was sabotaging and crippling the economy. They did nothing of consequence to crush the counterrevolutionary monarchist nests in the country. But they did what they could to crush the Bolsheviks, their press and their freedom of action.

Strange, isn't it? What inhibited these unterrified democrats? Were they lured away from the Marxist concept of the relation between democracy and socialism by Lenin's revision of it?

That was not quite the case, as we remember. Or did Erber fail to reach them in time with a copy of Engels' letter to Conrad Schmidt (if not in full, then in selected excerpts) plus his own theory on how the state can adapt itself to the economic movement and fulfil, not a class, but a "social" function? That explanation, too, while interesting, does not seem to be adequate. Or maybe the Bolsheviks, or the workers and peasants themselves, prohibited these "socialist opponents" from becoming champions of democracy? That, too is worth considering, but obviously not for long. The "socialist opponents" did not fight for democratic slogans and democracy when they were in the best position to fight for them. Why? Because to fight consistently and militantly, for democracy under conditions of the sharpest conflict between the classes, that is, under conditions of revolution, required a break with pure-and-simple parliamentary methods and modes of thought, a break with the bourgeois democrats and bourgeois democracy. Because such a fight required, for the realisation of its objectives, the installation of the working-class state power and led inexorably to this state power.

There is the hitherto well-kept secret of the failure of the "socialist-opponents" and the success of the Bolsheviks, which we are at last compelled to make public under Erber's ruthless pressure.

"Come now, you are joking. Erber knows this 'secret' and he even mentions it in one sentence." We are joking only a little bit. Erber "knows the secret" like a village journalist knows the Rosetta stone[76]. He can take a picture of it, he can describe its dimensions, he can even copy the writing, but he hasn't the remotest notion of what it means. Erber "knows" the secret, but he has no idea of what it means, although that requires none of the abstruse and esoteric skill of an Egyptologist. He writes:

"Against the Menshevik policy of subordinating the aims of the Revolution to the imperialist program of the bourgeoisie, Lenin advanced the policy of subordinating the Revolution to the full or maximum socialist program of the proletariat".

That he doesn't know what he's talking about in describing Lenin's policy is clear to anyone who has read what Lenin advocated and did in 1917. But his ignorance on this score has no special distinction since it is no greater than his general ignorance (the part is never greater than the whole, we were taught in school). In any case it belongs to another discussion. What he says about the Menshevik policy, however, has the distinguishing merit of being a fact, and shows that he has some notion of the "secret".

Now if we adopt the daring hypothesis that a policy of subordinating the Russian Revolution to the program of the bourgeoisie, its imperialist program, no less, was not quite the right thing to do, what, in the opinion of the Wise One, was the right policy for Marxists to pursue? On this question, we can call

---

76. The Rosetta Stone is a stone discovered in 1799 which has three versions of a decree issued in ancient Egypt — in Ancient Egyptian in two different scripts, and in Ancient Greek — and thus provided a key to deciphering Ancient Egyptian.

on Marx himself for a suggestion. Scared to death of being denounced as Marx-worshippers, we hasten to say that Marx's words do not "settle" the problems of the Russian Revolution. But. they do help to "settle" them and, at the very least, they show how Marx (not Lenin the revisionist, but Marx the Marxist) would have approached these problems.

Marx is writing about the bourgeois-democratic revolution in Germany, in his famous and all-too-little-known *Address to the Communist League* in 1850.

We do not ask Erber to read it, because it is much too simple, crude and static for him to understand, and we have no desire to overtax the facilities at his disposal. But the reader is asked to read our very long excerpt with the patient and rewarding attention it merits, bearing in mind, as he reads it, what actually happened in Russia, what the Bolsheviks actually said and did in Russia, and what the bourgeois democracy and the "socialist opponents" actually said and did in Russia.

"As heretofore, so in this struggle the mass of the petty bourgeoisie will maintain as long as possible an attitude of temporising, irresolution and inactivity, and then as soon as the victory is decided take it in charge, summon the workers to be peaceful and return to work in order to avert so-called excesses, and so cut off the proletariat from the fruits of the victory. It does not lie in the power of the workers to prevent the petty-bourgeois democrats from doing this, but it does lie in their power to render their ascendancy over the armed proletariat difficult, and to dictate to them such terms that the rule of the bourgeois democrats shall bear within it from the beginning the germ of its destruction, and its displacement later by the rule of the proletariat become considerably easier. Above all things, during the conflict and right after the battle, the workers must to the fullest extent possible work against the bourgeois measures of pacification, and compel the democrats to carry into action their present terroristic phrases. They must work to prevent the immediate revolutionary excitement from being promptly suppressed after the victory. They must keep it going as long as possible. Far from setting themselves against so-called excesses, examples of popular revenge against hated individuals or public buildings with only hateful memories attached to them, they must not only tolerate these examples but take in hand their very leadership. During the struggle and after the struggle the workers must at every opportunity put forth their own demands alongside those of the bourgeois democrats. They must demand guarantees for the workers the moment the democratic citizens set about taking over the government. They must if necessary extort these guarantees, and in general see to it that the new rulers pledge themselves to every conceivable concession and promise — the surest way to compromise them. In general they must restrain in every way to the extent of their power the jubilation and enthusiasm for the new order which follows every victorious street battle, by a calm and coldblooded conception of the situation and by an open distrust of the new government. Side by side with the new official governments, they must simultaneously set up their own revolutionary workers' governments, whether in the form of municipal committees, municipal councils, or workers' clubs or workers' committees, so that the bourgeois democratic governments not only immediately lose the support of the workers, but find themselves from the very beginning supervised and threatened by authorities behind which stand the whole mass of the workers. In a word: from the first moment of victory our distrust must no longer be directed against the vanquished reactionary party, but against our previous allies, against the party which seeks to exploit the common victory for itself alone.

"But in order to be able energetically and threateningly to oppose this party, whose betrayal of the workers will begin with the first hour of victory, the workers must be armed and organised. The arming of the whole proletariat with muskets, rifles, cannon and ammunition must be carried out at once, and the revival of the old bourgeois militia, directed against the workers, resisted. Where this cannot be effected, the workers must endeavour to organise themselves independently as a proletarian guard with chiefs and a general staff elected by themselves and put themselves under orders not of the state but of the revolutionary municipal councils established by the workers. Where workers are employed in state service, they must arm and organise in a separate corps or as a part of the proletarian guard with the chiefs elected by themselves. Arms and munitions must not be given up under any pretext; every attempt at disarmament must if necessary be thwarted by force. Destruction of the influence of the bourgeois democrats upon the workers, immediate independent and armed organisation of the workers, creation of the most difficult and compromising possible conditions for the momentarily unavoidable rule of the bourgeois democracy — these are the main points which the proletariat, and consequently the League, must have in mind during and after the coming uprising".

This utterly amazing document — amazing for the compactness and unequivocalness of its summary of Marx's views on the bourgeois democratic state and bourgeois democracy, on the role and tactics of the proletariat in the bourgeois-democratic revolution, and amazing for its almost uncanny word-for-word anticipation of the course of the Bolsheviks in the revolution — deserves reading in full, down to the last line in which the German workers are told that "Their battle cry must be: — the Permanent Revolution!"

(If we may be permitted a "personal note", we add the incidental information that it is to be found complete in a compilation of Marx's most important writings made by Max Eastman in 1932, which we helped to assemble, translate and edit. In his introduction, Eastman, referring to Marx's *Address* of 1850, says that "it will perhaps more than anything else written by Marx convey a full sense of the degree in which he was the author and creator of all the essential, outlines of what we call 'Bolshevism'." Right, a hundred times over and over again. Eastman's advantage over Erber lay in his knowledge, understanding and attempt at consistency. When, therefore, he repudiated the Russian Revolution and Bolshevism, he also repudiated Marx and the fight for socialism. It was a triumph for logic and what he called "the Anglo-Saxon mind".)

Let us jump from Marx in Germany in 1850 to Lenin in Russia in 1917. The Wise One writes:

"The Kerensky regime had done its utmost to block its further advance by frustrating the efforts of the masses to end the war and divide the land. The regime sought to stretch out its undemocratic authority as long as possible by repeatedly postponing the elections of a Constituent Assembly. If the revolution was to advance, Kerensky had to go. Only the Bolshevik party was able to show the way to the teeming, creative, democratic Soviets of 1917. The revolution broke through the impasse and opened a road toward a solution of the land and peace questions. Far from carrying out a coup d'etat, as their opponents charged, the Bolsheviks rode to power on the crest of an upsurge that sought to realise the long-promised objectives of land and peace."

We are beginning to get an idea of what the Marxist policy should have been, and it's not bad as a starter. "If the revolution was to advance, Kerensky had to go." Right is right. But Kerensky alone? Really, how would that have been fair? Should Kerensky have been made the scapegoat for the "Kerensky regime," that is, for the Kerensky government? What about the "socialist opponents" — the Mensheviks and S.R.s — who made the existence of the regime possible, who were part and parcel of it, who were fully co-responsible with Kerensky in trying to "stretch out" the "undemocratic authority" of the regime "as long as possible," in doing "its utmost to block" the advance of the revolution "by frustrating the efforts of the masses to end the war and divide the land"? What gives them immunity and not Kerensky? Whatever our opinion may be, we know the opinion of the Russian workers and peasants: the whole kit and caboodle had to go! Their place had to be taken by — write it down again! — "the teeming, creative, democratic Soviets of 1917." Led by whom? By Lenin and Trotsky, because write this down, too — "only the Bolshevik party was able to show the way" to the Soviets. Only the Bolsheviks.

That way was the seizure of power by the workers' and peasants' Soviets, which proceeded to give the land to the peasants, control of the factories to the workers, peace to the whole country, and to usher in the greatest victory for the socialist working class in all its history.

But what about the Constituent Assembly — didn't the Bolsheviks demand that it be convened and then, after tricking the workers into giving them power on the basis of this democratic slogan, didn't these same Bolsheviks disperse the Assembly when it did convene?[77] This brings us to Erber's second pontifical bull against the Bolsheviks, the second Error which brought about the subsequent thirty-years' horror. And for a second time, Erber is counting on the

---

77. The long-delayed Constituent Assembly elections went ahead two-and-a-half weeks after the Congress of Soviets declared itself the supreme power and the Provisional Government was dispersed. The elections were on lists of candidates drawn up a while before. They made the peasant-based S.R.s the biggest party, but in the meantime the S.R.s were splitting, and the anti-revolutionary Right S.R.s got the majority of the elected deputies while the pro-revolutionary Left S.R.s, who formed a coalition government with the Bolsheviks, got a minority. When the Constituent Assembly convened in January 1918, the Soviet government asked it to recognise the authority of the Soviets. When the Assembly instead demanded that the Soviets cede power to the Assembly, it had the Assembly building locked up and called a new Congress of Soviets.

possibility that his reader's ignorance is greater than his own.

The Bolsheviks along with the Left Social-Revolutionists did indeed disperse the Constituent Assembly. But this means that they refused to disperse or dissolve the revolutionary workers' government in favour of a counterrevolutionary and unrepresentative parliament. That's the first point and the main point!

What was the revolutionary Soviet power? It was "far from... a coup d'etat," it was the triumphant revolution of the "teeming, creative, democratic Soviets" which "broke through the impasse and opened a road toward a solution of the land and peace questions." This impasse was broken through against the opposition and resistance not only of Kornilov and Kerensky, but above all of the Mensheviks and S.R.s. The workers and the peasants, in their democratic Soviets, repudiated the two old parties and their leadership. They turned to the leadership of the left wing of the S.R.s and above all the leadership of Lenin's party because — we are still quoting from the Wise One — "only the Bolshevik party was able to show the way." That way was lined with the slogan, was it not, of "All power to the Soviets!"

What was the Constituent Assembly that finally convened in 1918, after the Soviet revolution? It was a faint and belated echo of an outlived and irrevocable political situation. It was less representative and less democratic than the Kerensky regime had been during most of its short life. During most of its existence, the Kerensky regime was supported by the bulk of the workers, soldiers, and peasants who were democratically organised in their Soviets. It was supported by the Menshevik and S.R. parties and party leaderships which, at that time, dominated the Soviets, had their confidence and support, and represented (more or less) the actual stage of political development and thinking of the masses at the time. Given the charge in the political development and thinking of the masses, this regime had to go, says the Wise and Stern One. But what did the Constituent Assembly represent when it finally came together, despite the months of Kerenskyite, Menshevik and S.R. sabotage? It was elected on the basis of outlived party lists. It was elected by a working class and peasantry that politically speaking no longer existed. The S.R. party, which held about half the seats, had already split in two. But while the official party, controlled by the right wing, held most of these seats, the new left-wing S.R. party which was collaborating with the Bolsheviks in the Soviet power and which already had or was rapidly gaining the support of the great majority of the peasants, held very few of the S.R. seats. The official S.R. list had been voted by the peasants before the tremendous revolutionary shift had taken place in their ranks. The official S.R. peasant supporters no longer existed in anything like the same number that had, earlier, cast their vote for the party list. Substantially the same thing held true for the Menshevik group in the Assembly, which represented the votes of voters who had since turned completely against the Mensheviks and given their allegiance to the parties of the Soviet Power, the Bolsheviks or the Left S.R.s. The composition of the Assembly, on the day it met, no longer corresponded even approximately to the political division in the country. The sentiments and aspirations of the masses had changed radically since the party lists for the Assembly were first drawn up and after the voting had taken place. By its composition, we repeat, the Assembly was less representative than the Kerensky government in its heyday.

It is not surprising, then, that the Constituent Assembly turned out to be a counterrevolutionary parliament. The Bolsheviks and the Left S.R.s called upon the parties of the Assembly to recognise the Soviet Power. The Mensheviks and right-wing S.R.s, to say nothing of the bourgeois Kadets, refused. Understandably! They had opposed the democratic slogans which brought about the revolution. They had brought the revolution against the monarchy to an impasse. They resisted tooth and nail the attempts to "open a road toward a solution of the land and peace questions." They had opposed the slogan of "All Power to the Soviets!" Their leadership had been repudiated and overturned by the "teeming, creative, democratic Soviets" which turned to the Bolsheviks as the "only" ones able to show the way. They had "subordinated the aims of the Revolution to the imperialist program of the bourgeoisie." They capped this not very glorious, not very socialist, not very democratic record by presenting a little amendment to the Soviet Power, namely, that it give up power and all claim to power, and take its orders henceforward from them! They asked the revolution to renounce itself, dig its own grave, jump into it and cover itself with earth hallowed by bourgeois democracy. From its very beginning, the Constituent Assembly declared war upon the Soviet Power.

Erber, the democrat, is merciless in his criticism of the Bolsheviks for dispersing the counterrevolutionary Assembly. But nowhere does he even indicate that what was involved was the demand by the Assembly to disperse and dissolve the revolutionary Soviet Government installed by the "teeming, creative, democratic Soviets of 1917". Erber is for the Soviets so long as they confine themselves to teeming, but not if they exercise their democratic rights end mission to create a proletarian, socialist power. What is the difference between the Russian Assembly, which he accepts, and the German Scheidemann whom, he says, he rejects? Only that Scheidemann succeeded in crushing the German Soviets and the Assembly failed to crush the Russian Soviets — that's all.

It may be asked: "Even if it is granted that this Assembly was unrepresentative, why didn't the Bolsheviks call for new elections which would have made possible the convocation of a parliament corresponding democratically to the political division in the country?"

The Bolsheviks preferred the Soviet (Commune-type) form of government to the parliamentary form from the standpoint of the working class and democracy and as the only state form under which the transition to socialism could be achieved. The Bolsheviks did not invent the Soviets, they did not create them. The Soviets developed spontaneously among the masses and, without asking anybody's approval, became organs for the defence of the demands of the masses and organs of power. The wisdom and superiority of the Bolsheviks consisted in understanding the full meaning and social potentiality of these democratic organs which they themselves did not fabricate artificially but which they found at hand as a natural product of the revolution. Among the Bolsheviks, it was Lenin who understood them best. His views were not concealed, hidden in his pocket to be brought out only after the masses had been tricked into giving the Bolsheviks state power. Immediately upon his return to Russia, Lenin saw that the Soviets were already a state power, a unique power, dual to the official state power and in immanent conflict with it. Almost the first words he wrote on the subject (*Pravda*, April 22, 1917) were these: "It is a power entirely different from that generally to be found in the parliamentary bourgeois-democratic republics of the usual type still prevailing in the advanced countries of Europe and America. This circumstance is often forgotten, often not reflected on, yet it is the crux of the matter. This power is of exactly the same type as the Paris Commune of 1871.

"The fundamental characteristics of this type are: (1) the source of power is not a law previously discussed and enacted by parliament, but the direct initiative of the masses from below, in their localities — outright 'usurpation', to use a current expression; (2) the direct arming of the whole people in place of the police and the army, which are institutions separated from the people and opposed to the people; order in the state under such a power is maintained by the armed workers and peasants themselves, by the armed people itself; (3) officials and bureaucrats are either replaced by the direct rule of the people itself or at least placed under special control; they not only become elected officials, but are also subject to recall at the first demand of the people; they are reduced to the position of simple agents, from a privileged stratum occupying posts remunerated on a high-bourgeois scale, they become workers of a special 'branch,' remunerated at a salary not exceeding the ordinary pay of a competent worker.

"This, and this alone, constitutes the essence-of the Paris Commune as a specific type of state."

Lenin prized the Soviet type of state, from the very beginning of the revolution, for its superiority from the standpoint of the workers and of genuine democracy. His view on the Constituent Assembly, furthermore, is most concisely and clearly set forth in the first two of his theses on the subject:

"1. The demand for the convocation of a Constituent Assembly was a perfectly legitimate part of the program of revolutionary Social-Democracy, because in a bourgeois republic a Constituent Assembly represents the highest form of democracy and because, in setting up a parliament, the imperialist republic which was headed by Kerensky was preparing to fake the elections and violate democracy in a number of ways.

"2. While demanding the convocation of a Constituent Assembly, revolutionary Social Democracy has ever since the beginning of the revolution of 1917 repeatedly emphasised that a republic of Soviets is a higher form of democracy than the usual bourgeois republic with a Constituent Assembly".

Lenin wrote his views about the Soviets, and repeatedly stated that: "Humanity has not yet evolved and we do not yet know of a type of government superior to and better than the Soviets of Workers', Agricultural Labourers',

Peasants' and Soldiers' Deputies," not after the Soviets had rallied to the support of his party, but from the very start — in April, when the Soviets were overwhelmingly under the leadership and control of the Mensheviks and S.R.s, with the Bolsheviks as a small minority among them. Lenin wrote his views on the Soviets and the Constituent Assembly, on the Commune type of state and the parliamentary type of state, for the entire political public to see and read. Anyone able to understand anything in politics was able to understand Lenin.

Once the Soviet power had been established with the decisive support of the masses of workers and peasants, the Constituent Assembly could not represent anything more than a throwback to bourgeois democracy, a throwback in the course of which the new Soviet power would have to be crushed, as it was crushed later on in Germany, Bavaria, Austria and Hungary. To have tried to bring into life a "good" bourgeois parliament when life had already made a reality of a far more democratic form of government established by the masses themselves and enjoying their support and confidence, would have meant a victory for reaction. That in the first place.

In the second place, we do not hesitate to say that, abstractly, a second and a third or fourth attempt to establish a more democratic parliament could not be ruled out as impossible, or unnecessary, or contrary to the interests of the working class. Similarly, you cannot rule out a decision by the revolutionists themselves, under certain circumstances, to dissolve Soviets that came into existence under different circumstances. The Soviets may be too weak to take supreme power in a country but strong enough to prevent the bourgeoisie and the petty-bourgeois parties from consolidating their power on a reactionary basis; the bourgeoisie may be too weak to crush the Soviets but strong enough to hold on to its rule. The revolutionists or the Soviets may not enjoy sufficient popular support; the bourgeoisie may hesitate before a civil war in which everything is at stake. Decisive sections of the people may believe insistently in the possibility of finding a solution in a more democratic parliamentary system and at the same time refuse to allow the new proletarian democracy to be destroyed. History knows all sorts of combinations of circumstances and is very fertile in creating new combinations. How long it would be possible for revolutionary Soviets (a semi-state) to exist side by side with an uncertain bourgeois parliament (another semi-state) under any and all conceivable circumstances, cannot be answered categorically or in advance. All we need to say is this; there are historical laws of revolution, we know these laws, and we also know that there have been and will probably continue to be exceptions to these laws.

However, it is not this abstract question that is being discussed, important though it is in its own right. We are not saying that in every socialist revolution, regardless of the country, the period, the economic and political conditions in which it develops, Soviets will arise; or if they do that they will develop just the way they did in Russia, that the workers' organs will come into existence in head-on conflict with the bourgeois parliamentary system, that these work-

Kornilov's troops surrender weapons

ers' organs will have to disperse or dissolve the parliament in the same way that we saw in Russia, that the bourgeoisie will have to be overturned by violence, that the ousted bourgeoisie is absolutely certain to resist with armed force, that a civil war is absolutely inevitable. It is conceivable that the rise of the socialist proletariat is so swift, mighty and irresistible; that the economy is in such a state of disorder and the bourgeoisie in such a demoralised, depressed and hopeless state, that it decides to throw in its hand without a real fight. It is conceivable that under such or similar circumstances the classical bourgeois parliament can be so drastically revised from within its own organs that it becomes transformed into something radically different. All laws, including historical laws, have their exceptions. But again, that is not what we are discussing here. We are discussing what actually happened in the Russian Revolution.

And what actually happened, that is, the way the social and political forces actually meshed and drew apart and clashed in Russia during the revolution, shows that the Bolsheviks acted as revolutionary socialists in the struggle around the Constituent Assembly and not like political science professors drawing diagrams on a high-school blackboard.

Which brings us to the third place — the political reality. Once the Soviets took power, the counterrevolution instantly adopted the slogan of the Constituent Assembly even before the Constituent actually convened. The true representatives of the classes regarded neither the Soviets nor the Constituent Assembly as abstractions. For the reaction as well as for the petty-bourgeois democracy (each from its own standpoint), the Constituent Assembly became the rallying cry, the banner, the instrument for the struggle to overthrow the Soviet Power of the workers and peasants, which also meant to overthrow all the achievements obtained by this power and expected from it. The conflict between "Soviet" and "Assembly" on the blackboard is one thing. In the Russia of 1917-1918, it was a violent and irreconcilable conflict between the classes.

In Erber's document, it need hardly be said, the class struggle does not exist. Or if it does, why, it can easily be straightened out by men of good will.

The Assembly demanded the capitulation of the Soviets; it could not exist without such a capitulation. Men of good will were of little use in this conflict. A civil war broke out, and as the German phrase has it, the weapon of criticism gave way to the criticism of weapons

The civil war that followed is clearly the fault of the Bolsheviks. Of that, there is no doubt in Erber's mind. It's notoriously true, too! If the Bolsheviks had not taken power, there would have been no need for a civil war to crush them! Even before the Bolsheviks took power, as a matter of fact, if the Soviets (we mean, of course, the teeming, democratic Soviets) had not existed at all, there might not even have been a Kornilovist-monarchist plot to drown them in a bloodbath. Indeed, we may even state it more generally: If workers were not so insistent and militant in trying to impose their modest demands on obstinate and reactionary employers, the latter would find no need of subsidising thugs and fascists to beat and shoot workers. You can hear that philosophy expounded in any high-school (third term), from a thousand pulpits and ten thousand newspaper pages: If labour gets unreasonable in its demands and doesn't know its proper place, well then, we don't like it, you know, but if that happens, Fascism just is inevitable. Yessirree! It's notoriously true. It is also true that if you stop breathing altogether, not even your worst enemy will dream of strangling you.

Oh, wait a minute! Erber is not defending the bourgeoisie and the reaction! He's really radical, and he doesn't care much about what is done to the bourgeoisie. What upsets him is that the Bolsheviks took power and dispersed the Assembly in opposition to the workers. Do you see now? Listen to this little sneer, lifted right out of the literature of the professional anti-Bolshevlk (and the professional anti-unionist, we might add): "As for the masses who constituted the Soviets, Lenin held that they would be won to the idea in time. It was for the vanguard to act and explain later. Those of the workers who refused to accept this concept of the dictatorship of the proletariat had to be handled firmly, for their own good".

Our little animal is a vicious one, isn't he? Lenin was for imposing his dictatorship upon the masses and explaining to them later. And if they didn't go along, why, shoot the rabble down for their own good! He turned out pretty bad, this Lenin. Fights for months with democratic slogans; fools everybody, including the democratic Soviets which brought him to power on the crest of their upsurge and without a coup d'état on his part, and then, a very few weeks later, the mask is off! He acts for their own good; he shoots them for their own good.

There's an authentic portrait of Lenin for you, an unretouched photograph of him. What is the proof for this insolent charge? The proof is the famous "demonstration" of January 18, 1918, organised by the reactionary City Duma of Petrograd against the Soviet Power and for the Constituent Assembly. The "demonstration" was dispersed by Red Guards. To show the magnitude of this Bolshevik atrocity, Erber quotes an article by Maxim Gorky, "whose honesty as a reporter of the events can be accepted." We hear Gorky burning with indignation at the charge that this was a bourgeois demonstration and denouncing the Bolsheviks for encouraging "the soldiers and Red Guards [to] snatch the revolutionary banners from the hands of the workers".

Gorky's honesty, guaranteed by Erber personally, makes him a good reporter of events! Gorky was, to be sure, an honest man and a socialist. But on revolutionary problems, he had no more qualification than the next man, except perhaps that he was warmly sentimental, almost always confused in the political conflicts of the Marxian movement, and a bitter enemy of the Bolshevik Revolution for a long time, above all at the time it occurred. If Erber picks him out as his reporter of events, it is a clear case of like calling unto like. Erber is attracted by Gorky's impressionism, and by his confusion, which he likes to think is no greater than his own muddleheadedness.

You read Erber's lurid quotation from Gorky, and your mind's eye conjures up the image of Scheidemann, Noske and Ebert mowing down the German workers with machine guns. Erber has his countries, parties and men mixed up a little. Who was involved in this huge demonstration which, if you follow Erber, you might think was terminated with workers dead and dying by the thousands? Three days before Gorky's anguished article, his own paper, *Novaya Zhizn*, reported the demonstration as follows: "About 11:30 some two hundred men bearing a flag with the words, 'All Power to the Constituent Assembly,' came across the Liteiny Bridge." There is the imposing number of the Petrograd population that followed the clarion call of the bourgeoisie, the Mensheviks and the S.R.s to proclaim the sovereign rights of the Constituent Assembly which they had so successfully sabotaged for six months. One hundred plus one hundred, making a grand total of two hundred men, all good and true.

The other proof is this: "Gorky is quite correct in asking what the bourgeoisie had to cheer about in the convocation of a Constituent Assembly in which the bourgeois party, the Kadets, held only fifteen seats out of 520, and in which the extreme right Social Revolutionaries, who had been identified with Kerensky, were thoroughly discredited".

We will even try to explain to this innocent what the bourgeoisie had to cheer about. A Constituent with only 15 Kadets out of 520 seats and a majority for the S.R.s, even right-wing S.R.s, would give the bourgeoisie very little to cheer about, if this Constituent were proclaiming its sovereignty against the Tsarist Duma. The same Constituent, however, in proclaiming its sovereignty against the revolutionary power of the democratic Soviets of the workers and peasants, would give the bourgeoisie, inside Russia and all over the world, plenty to cheer about. And it did cheer about it! How explain that mystery? And how explain a few other mysteries?

Between them, the right-wing S.R.s and the Mensheviks had the majority of the seats in the Constituent. Since it was an ever-so-democratic Constituent, this must have meant that the two parties were supported by the majority of the population. The Constituent is dispersed by the Bolsheviks, who do not have the masses but who act for them and explain later, and who shoot them down for their own good. So far, so good. The outraged S.R.s and Mensheviks return to the outraged masses, with the declaration, as one of them put it, that "The Constituent Assembly alone is capable of uniting all parts of Russia to put an end to the civil war which is speeding up the economic ruin of the country, and to solve all essential questions raised by the revolution". The masses want democracy and the solution of all these essential questions. The Mensheviks and S.R.s promise to solve them. In fact, Erber tells us, they are now really for peace and for land to the peasants. What is more, the roles are reversed on the matter of democracy. The Bolsheviks are for the despotic dictatorship over the masses and "democratic slogans became a weapon of their socialist opponents".

We are in 1918. The Bolshevik power is established in only a very tiny part of Russia and consolidated in none. The anti-Bolsheviks have political control in a multitude of localities — the great majority — and they even have considerable armed forces at their disposal. The Bolsheviks do not have what Stalin, for example, has today: a huge, tightly-knit political machine, hordes of privileged bureaucrats, a tremendous array, an all-pervading and terrifying G.P.U., and the like. They cannot simply dispose of their opponents by force or terror, as Stalin does. It is still a fair and square political fight, with the big odds still apparently in favour of the "socialist opponents" who now have democratic slogans as their weapons and the democratic Constituent Assembly, in the flesh, as their banner.

The unexplained mystery, hidden to Erber behind seven of his own fogs, is this: How account for the fact that the "socialist opponents" get nowhere with their "democratic slogans" and their Constituent Assembly? Aren't they the parties of the workers and peasants, as proved by the majority they registered at the opening of the Constituent? Aren't they now armed to the toes with "democratic slogans" which, only a day ago, were so vastly popular with the masses that the cunning Bolsheviks won power with their aid? Thorny questions, aren't they? But Erber is not going to get any thorns in his fingers if he can help it. Solution? He leaves the questions strictly alone.

That's a solution for him, but it does not answer the questions. The answer gives us the second key to the mystification: The bourgeoisie had everything to cheer about in the convocation of the Constituent Assembly — everything. It could not expect to restore its power in its own name in the Russia of 1917-1918. But it could hope to restore it behind the stalking horse of bourgeois democracy, the Constituent Assembly and its Menshevik-S.R. champions. Shall we look into this point for a minute?

Here, for example, we have the report of the U.S. Consul Dewitt Poole to the American Ambassador in Russia, written in Petrograd exactly one week after the final session of the Constituent. He is reporting on his visit five weeks earlier, to Rostow-on-Don "to investigate the question of the establishment of an American Consulate in that city." During his visit, Mr. Poole meets with notorious monarchist and Cossack counterrevolutionists like General Kaledin, General Alexeyev and others connected with General Kornilov. The anti-Bolshevik united front is being formed into a "Council" in the Southeast of Russia immediately after the Soviet Power is established and before the Constituent even assembles. Let us read, and with profit, every one of the lines that we have room to quote from Mr. Poole's report:

"Negotiations are in progress for the admission to the Council of three representative Social Democrats, namely, Chnikovsky, Kuskova and Plekhanov; and two Social Revolutionaries, namely, Argunov and Potresov.

"On the conservative side the Council, as now constitute, includes, besides the three generals (Alexeyev, Kornilov and Kaledin), Mr Milyukov; Prince Gregory Trubetskoy; Professor Struve; Mr Fedorov, representing the banking and other large commercial interests of Moscow; two other Kadets or nationalist patriots yet to be chosen; Mr. Bogayevsky, the vice ataman of the Don Cossacks; and Mr. Paramonov, a rich Cossack. The Council will undoubtedly undergo changes in personnel, but a framework of an equal number of conservatives and radicals, not counting the three generals, appears to have been adopted.

"In pursuance of the agreement with Mr. Savinkov, a proclamation to the Russian people has been drafted. It refers to the suppression of the Constituent Assembly and asks for the support of the people in defending that institution. It is sound on the subject of the continuance of the war. The proclamation will be issued in the name of the league, unsigned because it is frankly admitted that it has not yet been possible to obtain the names of persons who, it is thought, would be thoroughly acceptable to the people at large."

Isn't every line of our wonderful Mr. Poole covered with mother-of-pearl, even though he never, we suppose, read Engels' letter to Conrad Schmidt? What did the bourgeoisie have to cheer about in the convocation of the Constituent Assembly? Gorky didn't know. Erber doesn't know yet. False modesty prevents us from saying we know. But Generals Alexeyev, Kornilov and Kaledin — they know. Prince Trubetskoy — he knows. Gospadin Fedorov, "representing the banking and other commercial interests of Moscow" — he knows. Gospadin Paramonov, a Cossack who happens also to be rich — he knows. Alas, every one of them has passed from our midst to enjoy the reward of the pious; not one of them is alive today to tell Erber what he knows. And that's a double pity, because the proclamation of the Council was so "sound on the subject of the continuance of the war" — which is another subject that is of interest to Erber.

General Denikin issued a proclamation on January 9, 1918, before the hideous Bolsheviks dispersed the Assembly, proclaiming the aims of his "Volunteer Army."

"The new army will defend the civil liberties in order to enable the master of the Russian land — the Russian people — to express through their elected Constituent Assembly their sovereign will. All classes, parties, and groups of the population must accept that will. The army and those taking part in its formation will absolutely submit to the legal power appointed by the Constituent

Assembly".

This Tsarist general did not have much luck either. He was ready to "absolutely submit" to the Constituent, but he couldn't find anyone else who panted to follow his inspiring democratic lead. The volunteer movement, he wrote later in his memoirs, "did not become a national movement. At its very inception... the army acquired a distinct class character." Erber should be compelled by law — democratically enforced — to read this. There are classes in society and their interests are irreconcilable. Above all in revolutionary times, all groups, movements and institutions "acquire a distinct class character." So distinct that a Tsarist general finally sees it. But not Erber.

Here is another Tsarist general, Kornilov, and here are five instructive points from his program of February, 1918:

"(3) To reestablish freedom of industry and commerce and to abolish nationalisation of private financial enterprises.

"(4) To reestablish private property.

"(5) To reestablish the Russian Army on the basis of strict military discipline. The army should be formed on a volunteer basis, without committees, commissars, or elective officers...

"(8) The Constituent Assembly dissolved by the Bolsheviks should be restored.

"(9) The government established by General Kornilov is responsible only to the Constituent Assembly... The Constituent Assembly, as the only sovereign of the Russian land, will determine the fundamental laws of the Russian constitution and will give final form to the organisation of the state."

It's a double pity that Kornilov joined his ancestors in the unsuccessful attack on Yekaterinodar a few weeks later, so that he can't explain what the bourgeoisie had to cheer about, either.

Maybe we can find a hint from the other paladin of the Constituent, General Alexeyev, who is also armed to the teeth with "democratic slogans" (after the Bolsheviks take power but not, we regret to note, before), plus 100,000,000 rubles appropriated for his democratic efforts by the no less democratic government of France. In a perplexed and gloomy letter to the Chief of the French Mission in Kiev, the General writes in February, 1918:

"The Cossack regiments coming from the front are in a state of complete moral dissolution. Bolshevik ideas have found a great many followers among the Cossacks, with the result that they refuse to fight even in defence of their own territory. (Alexeyev means, of course, that these stupid Cossack regiments refuse to fight for the French banks.) They are firmly convinced that Bolshevism is directed solely against the wealthy classes, and not against the region as a whole, where there is order, bread, coal, iron and oil".

We have found the hint! In the eyes of the masses, even of the politically backward and privileged Cossacks, the Constituent Assembly, the fight for it, the men and groups leading that fight, represent not democracy but the wealthy classes, the restorationists, the reaction, and at best, the compromisers and confusionists. In the eyes of the masses, the Bolsheviks and the Soviets represent the fight for freedom and the assurance that it can be won. They represent the movement "directed solely against the wealthy classes".

That is why the Mensheviks and S.R.s, with all their votes, and with all their "democratic slogans" and their Constituent Assembly, never and nowhere inspired the masses, never and nowhere recruited them to the banner of struggle to overturn the Soviet Power and succeeded only in bringing the most shameful discredit upon themselves. That is why the "anti-democratic" Bolsheviks consolidated the Soviet Power among the democratic masses in spite of odds almost without historical parallel. The "theoretical dispute" was decided freely by the masses, decided in struggle.

So far as main lines are concerned, could the Bolsheviks have followed some other course? Erber has an alternative to suggest. He writes: "It is one of the unquestioned myths of our movement that the Bolsheviks, once they were in power, had no other alternative but the course they pursued."

What did that course they actually took lead to? Erber knows. They fooled the masses into putting them into power. They shot down workers for their own good. They shot down socialists... for their own good, They refused to dissolve the Soviet Power and, to maintain themselves, they terrorised everyone else. A terrible business, taken all in all. But there was once another myth about the Bolsheviks, and this is how it was once described:

"Far from the historical myth forged in the recent past by the anti-Bolshevik moralisers, the Bolsheviks did not take power with any wickedly conceived plans for iron dictatorship and terror. Witness their naive generosity toward armed counter-revolutionaries in the first months, the so-called honeymoon period, of the Soviet power. The morals of a revolutionary class are the morals of an army in combat. Its own code of conduct is bound to be conditioned by the sort of an enemy it faces and the conditions under which it fights. It is very naive — and most dangerous — to believe that the revolutionary forces will place adherence to a preconceived code of conduct above the need of survival in battle when confronted with those alternatives."

Not bad. Who wrote it? Naturally, Erber, in one of his long series of documents, just a few years ago. Against whom did he write it? Against those who are "very naive and most dangerous," that is, against muddleheads. Against whom, then? Against himself, in anticipation. But is it such a bad thing to dispel historical myths? No, it is an excellent thing. Any myth you dispel is a step in human progress. Only, it helps if you know how to distinguish myth from reality.

Now, what was the good, democratic, socialist, practical alternative course that the Bolsheviks could have pursued if they hadn't revised Marx and dispersed the Constituent Assembly? Before we quote further from Erber, we must halt for a solemn moment. The reader would be well advised to make his peace with his Maker, for he is about to die laughing. He can save himself from this fate only by deciding to read no further. We, however, who like a good joke as well as the next man, can take the chance.

"Instead of dispersing the Assembly, the Bolshevik course should have been"

— press down hard on your sides now — "a government that was responsible to the Constituent Assembly, either an S.R. government or a coalition of the worker and peasant parties (Bolshevik, Menshevik, Left S.R. and Right S.R. parties)... It would have experienced many internal crises and may have found it necessary to refer the disputes to the people in the form of new elections. However, such a government would have had a much wider base than the Bolshevik regime and the victory over the Tsarist and bourgeois counterrevolution would have been far easier, quicker and less costly".

By Jupiter, this is no commonplace genius we have here! What dazzling audacity and sweep of thought, hammerlock grip of logic, boulder-crushing simplicity, graceful persuasiveness of argument, blade-edge keenness of concept, anxiously-concealed modesty!

This is no small triumph. For thirty years now, and longer, men and women of all faiths and estates have ripped their brains to shreds trying to find the answer to the question of how the bloody conflict of the parties of the Russian Revolution, and the consequences of this conflict, could have been averted. It is the story of "The Purloined Letter"[78] all over again, but on what a world-shaping scale. The answer was there all the time, right in front of us. But like the purloined letter, to perceive it took the combination of genius and simpleness which distinguishes the Auguste Dupin from the gendarme on the street.

How avert the conflict of the parties and its consequences? By stopping the conflict! By uniting the parties! By forming a coalition! Where? In the Constituent Assembly, the Temple of the Faith and Fountain of All Blessings. How simple that would have been, had anyone been gifted enough to think of it in 1917-1918. Just take the first two ingredients for the coalition — the Mensheviks and the Bolsheviks — and mix them together thoroughly. The Mensheviks, Erber told us, had the "policy of subordinating the aims of the revolution to the imperialist program of the bourgeoisie." Lenin "advanced the policy of subordinating the revolution to the full or maximum socialist program of the proletariat." After all, is that so much of a conflict in policy as to prevent men of good taste and good will from getting together? After all, don't you have to be practical in politics? After all, when there is a difference of opinion you can't go biting a man's head off. You compromise on a little point here and a little one there. You take a little, you give a little. All you needed, then, was for the Mensheviks to give a little to the "socialist program of the proletariat," and for the Bolsheviks to give a little to the "imperialist program of the bourgeoisie." That would be the least any reasonable person could do in all fairness to both classes.

Naturally, such a coalition "would have experienced many internal crises." If not many, then at least a few. Erber is a statesman of the new day, and such things do not throw him into unseemly panic. A way can always be found by men of good will. If, for example, the first question on the agenda of the won-

---

78. *The Purloined Letter* is a short story by Edgar Allen Poe, reckoned to be one of the first detective stories ever written. In the story, the searched-for letter of the story's title is hidden where no-one thinks to look for it: in plain sight. Auguste Dupin is the detective in the story.

derful coalition with its "much wider base," was what to do with the Soviet Power, that could easily be disposed of. The Mensheviks and S.R.s make a motion to dissolve the Soviet Power. The Bolsheviks and, let us say, the Left S.R.s oppose the motion. There is a fair and square discussion, two speakers on either side, with equal time for all, and then — the democratic vote. It is carefully counted by two impartial tellers. The majority of the coalition cabinet has voted for the motion. The Soviet Power is dissolved, so is the Red Guard. The revolution is buried in a simple but dignified casket. The Bolsheviks, democrats to the marrow of their bones, shrug their shoulders. "Guess we lost that one," they say with perfect good humour. "What's the next point on the agenda?" The first little internal crisis of the coalition, and just see how easily it was overcome! All we needed was men of good will, a sense of humour, and a thorough knowledge of Engels' letter to Conrad Schmidt with glossary notes by Erber.

Well and good, the Grand Coalition is formed. The Mensheviks do not demand as the price for collaborating with the Bolsheviks that Lenin and Trotsky be kicked out summarily (as they did demand). The Left S.R.s promise that if there is a disagreement over signing a peace treaty with the Germans, they will not rush into the streets with rifles to overthrow the Bolsheviks or the coalition as a whole (as they did do). The Right S.R.s also swallow themselves whole and agree to the coalition if the Soviets are suppressed. At last, the coalition is here. Now, "what," asks Erber, "would have been the nature of the state that would have emerged under such a regime, and what would have been its social basis?"

To this he answers: — "In its essentials it would have been what Lenin had in mind for Russia until February, 1917, under the formula of a democratic dictatorship of the proletariat and peasantry: a state which would have cleansed Russia of the vestiges of feudalism and curbed the power of the big bourgeoisie through the nationalisation of monopolies and trusts, while leaving private enterprise and the market undisturbed. The participation of the workers in economic life through collective bargaining and measures of workers' control of production would have been far more extensive and democratic under such a regime, despite capitalist economic relations, than was the case after a year of Bolshevik rule..."

It is hard to bring ourselves to discuss the merits of this mouldy piece of *Scheidemannistic counter-revolution* at any length. It shows such a wilful ignorance of the social structure and the class relations in Russia (we insist, *wilful ignorance*, because we *know* that Erber *once knew* something about them), that it has no claim at all on a detailed reply.

One of the unique features of Russian social and political life lay precisely in the inseparability of the landlords and capitalists. It is precisely this feature of Russia that brought Marxists like Lenin, Trotsky, Luxemburg, Mehring and (the pre-war) Kautsky on one common side, against the Mensheviks on the other side. The real Marxists all agreed on at least this much: The coming bourgeois-democratic revolution in Russia cannot be carried out under the leadership of the bourgeoisie because the real class and economic relations in Russia prevent it from conducting a struggle against the feudalist landlords with whom they are either identical or inseparably identified in a hundred social and economic ways. The democratic revolution can be carried out in Russia only under the leadership of the proletariat. Up to that point — agreement.

The disagreements began only after that point, and it is after that point that Trotsky developed his now familiar theory of the permanent revolution. Trotsky has demonstrated — which is of course the reason for Erber's prudent silence on this score — that whatever Lenin's formula may have been before the revolution, the actual course of events forced the workers, supported by the peasants, to take power. He has demonstrated in numerous writings that nobody has successfully challenged that once in power, the working class and the Bolsheviks, who had not advanced the policy of the "full or maximum socialist program of the proletariat" (as Erber ignorantly or maliciously states), found themselves compelled to pass over to the "maximum program" in order to carry out the program of the democratic revolution. What compelled them? Among other things, and not least of all, the uncontrollable class reaction of the bourgeoisie. It understood what neither the Mensheviks nor their dim feeble echo understood: that the assault upon "the vestiges of feudalism" and the "curb" upon their own power, if carried through seriously, could not but be an assault upon its own class position, could not but mean the end of all its economic and social power. The only way to maintain "capitalist economic relations," now so dear to Erber, was to abstain from a real cleansing of the stables of feudalism in Russia, Why, at bottom, didn't the Mensheviks use the "democratic slogans" as their weapons before 1918? Because they were wicked men without good will? No, they loved the workers and loved democracy. But to carry out the democratic slogans against the "vestiges of feudalism" and despotism would have brought them into violent conflict with the bourgeoisie. It would have forced them to break with the bourgeoisie. It would have forced them to lead in the establishment of a democratic republic without the bourgeoisie and against it, but that is exactly what they could not do and did not do because they were tied to their dogma that, since the Russian Revolution is a bourgeois-democratic revolution, it must remain under the leadership of the bourgeoisie and within the framework of capitalist society. They remained captives of this dogma throughout the revolutionary storm. Result: they were paralysed in the struggle for freedom, they lost the support of the workers, and came out of the revolution eternally and shamefully compromised.

All our muddlehead proposes, with that insight you get only from hindsight, is that the Bolsheviks should have followed exactly the same course as the Mensheviks. The high revolutionary role which he asks that the revolutionary party should have assigned itself in the Russian Revolution, boils down to this! Maintain capitalist economic relations (in plain English, capitalism) which the workers themselves were smashing; and smash the state power of the "teeming, creative, democratic Soviets" which the workers themselves were maintaining. And what, we ask in a conspiratorial whisper, what if some of the workers resisted the execution of this modest program? No problem. It would be, to quote the muddlehead, "for the vanguard to act and explain later." The workers would have "to be handled firmly, for their own good." The only trouble is that when the Mensheviks and S.R.s, with a little help from Kaledin, Alexeyev, Kornilov, Churchill, Wilson and Poincaré[79], did try to execute this program, for the good of the workers and peasants — the rabble didn't know what was good for them.

"Lenin subjected to merciless ridicule Trotsky's theory of the permanent revolution right up to the outbreak of the February Revolution."

Thus Erber, from whom nothing can be kept secret. So what? This: if Lenin had kept up this merciless ridicule, and stuck to the theory of the democratic dictatorship of the proletariat and peasantry," the Constituent Assembly would be going full blast to this day and "despite capitalist economic relations," the world would be a distinctly pleasant place to live in, free from Fascism and above all from Stalinism. But what happened? We don't know exactly why it happened or how, we don't know what got into Lenin's democratic head, but that which did happen was the worst thing imaginable: Lenin took over Trotsky's mercilessly ridiculed theory. From then on Russia, along with the rest of the world, was a gone goose. The Bolsheviks "set foot on a course from which there was no turning back."

Now you know everything, for just as no secret can be kept from Erber, so Erber keeps no secrets from you. We have Stalinism today (and Lord alone knows what else) because of the theory of the permanent revolution. All you Social Democrats, liberals, ex-Marxists, ex-Trotskyists and other professional anti-Bolsheviks, take note. Do you still think Stalinism flowed from Leninism? You are wrong. From Leninism, the authentic, the genuine, the unrevised and unretouched, would have flowed milk, honey, democracy and a world of other blessings. The real truth has finally been discovered by the Wise One Who Drips Water after thirty years of world history and thirty weeks of concentrated meditation. It deserves to be set off in a separate paragraph:

Stalinism flows from Trotskyism!

We don't even dream of questioning a conclusion of such epoch-making proportions. We can only say, once our breath has come back to us, that sweeping though it is, it is not sweeping enough.

Why from Trotskyism alone? The course of the Bolsheviks, seduced by Trotsky, followed very strictly (more strictly than even they knew at the time) the prescriptions of Marx and Engels on the course that revolutionary socialists should follow in the bourgeois revolution, the course they will have to adopt toward "democracy." In this connection, a few highly revealing words from Engels will suffice to show where the responsibility for the Error really lies, where it all began. We quote from his letter to August Bebel on December 11, 1884[80]:

---

79. Wilson was the president of the USA, and Poincaré president of France. Winston Churchill was minister of munitions, and then secretary of state for war, in the British government at the time of the October Revolution and the Civil War. Kaledin, Alexeyev, and Kornilov were, as above, Tsarist generals.

80. Bebel was a leader of the German Social Democratic Party. For the letter see bit.ly/eng-beb.

"As to pure democracy and its role in the future I do not share your opinion. Obviously it plays a far more subordinate part in Germany than in countries with an older industrial development. But that does not prevent the possibility, when the moment of revolution comes, of its acquiring a temporary importance as the most radical bourgeois party... and as the final sheet-anchor of the whole bourgeois and even feudal regime. At such a moment the whole reactionary mass falls in behind it and strengthens it; everything which used to be reactionary behaves as democratic... This has happened in every revolution: the tamest party still remaining in any way capable of government comes to power with the others just because it is only in this party that the defeated see their last possibility of salvation.

"Now it cannot be expected that at the moment of crisis we shall already have the majority of the electorate and therefore of the nation behind us. The whole bourgeois class and the remnants of the feudal landowning class, a large section of the petty bourgeoisie and also of the rural population will then mass themselves around the most radical bourgeois party, which will then make the most extreme revolutionary gestures, and I consider it very possible that it will be represented in the provisional government and even temporarily form its majority. How, as a minority, one should not act in that case, was demonstrated by the social-democratic minority in the Paris revolution of February 1848...[81]

"In any case our sole adversary on the day of the crisis and on the day after the crisis will be the whole collective reaction which will group itself around pure democracy, and this, I think, should not be lost sight of".

An absolutely remarkable letter! An absolutely remarkable anticipation of what happened in Russia in 1917-1918 and of just how it happened! No wonder we revolutionists have such tremendous esteem for our two great teachers, Marx and Engels. Such intellectual titans come but once a century, and not always that often.

And what did Engels mean by how "one should not act in that case?" What was his reference to the February 1848 revolution in Paris? He explained in a letter to the Italian socialist Turati, on January 26 1894[82]:

"After the common victory we might perhaps be offered some seats in the new Government — but always in a minority. Here lies the greatest danger. After the February Revolution in 1848 the French socialistic Democrats (the Réforme people, Ledru-Rollin, Louis Blanc, Flocon, etc.) were incautious enough to accept such positions. As a minority in the Government they involuntarily bore the responsibility for all the infamy and treachery which the majority, composed of pure Republicans, committed against the working class, while at the same time their participation in the government completely paralysed the revolutionary action of the working class they were supposed to represent".

---

81. Engels in a passage cited by Shachtman just a few lines further on indicts the error that "the social-democratic minority" made in accepting posts in the bourgeois Provisional Government set up after the revolutionary overthrow of the Orleanist monarchy in France in February 1848.

82. Letter at bit.ly/eng-tur.

Now perhaps Erber will know, from Engels, who was not present in the Russian Revolution, but who, in exchange, had over Erber the advantage of revolutionary understanding and revolutionary spirit, what the bourgeoisie and the "whole collective reaction" had to cheer about when it "grouped itself around pure democracy" in the form of the Constituent Assembly and its struggle against the Soviet Power. Now perhaps he will know why the Bolsheviks refused, in advance, his kind proposal that they enter as a minority in the Grand Coalition he dreamed up. Perhaps he will also understand why Eastman and others like him, in rejecting the Bolshevik Revolution, also and necessarily reject Marx, why they do not come to a ridiculous halt after condemning Lenin and Trotsky but go right back to the "source" of the Error. Once they have rejected the Russian Revolution and Marx, they have the elementary decency not to pretend that they are still interested in the fight for socialism.

We will dwell even loss upon Erber's third "charge" against Lenin — that he gambled everything on a victory of the revolution in Germany. Not only because the charge is false and was specifically repudiated more than once by Lenin himself, as everyone who has read Lenin's writings knows and as everyone who takes the right to talk about the question ought to know, but because it does not merit more than a few contemptuous words.

To Lenin, you see, the whole Bolshevik revolution, the whole course he pursued in it, was a gamble. "If Lenin won, history would absolve the Bolsheviks of all the charges their socialist opponents made against them." A fine democrat's morality this is! A fine piece of gross and typically "American" success-philistinism this is. If the revolution had triumphed in Germany, all Lenin's crimes against Marxism, against socialism, against democracy, all his anti-democratism and contempt for the masses who have to be shot for their own good, all of this and more would have been pardonable. Scheidemann, who saved Germany from the socialist revolution, therewith saves our stern judge from the unpleasant task of pardoning the criminal.

Contrast this low-quality philistinism with Marx, whose revolutionary spirit Erber will never see in a mirror — with Marx writing about the Paris Commune which, we may recall, he was at first opposed to establishing. "World history would indeed be very easy to make, if the struggle were taken up only on condition of infallibly favourable chances," he wrote to Kugelmann. Give Erber infallibly favourable chances for victory and he'll plunge into the revolution with only the least bit of hesitation. And not he alone...

Contrast Erber and every word he writes with the critical appraisal of the Bolsheviks written in prison by Rosa Luxemburg, who is invoked against revolutionary socialism nowadays by every turncoat and backslider who wouldn't reach up to her soles if he stood on tiptoes: "That the Bolsheviks have based their policy entirely upon the world proletarian revolution is the clearest proof of their political farsightedness and firmness of principle and of the bold scope of their policies."

You will never see that quoted by the turncoats who have drafted Luxemburg into the crusade against Bolshevism against her will. Nor will you see this quoted:

"The party of Lenin was the only one which grasped the mandate and duty of a truly revolutionary party and which, by the slogan — 'All power in the hands of the proletariat and peasantry' — insured the continued development of the revolution... Moreover, the Bolsheviks immediately set as the aim of this seizure of power a complete, far-reaching revolutionary program: not the safe-guarding of bourgeois democracy, but a dictatorship of the proletariat for the purpose of realising socialism. Thereby they won for themselves the imper-ishable historic distinction of having for the first time proclaimed the final aim of socialism as the direct program of practical politics".

We can see now how much right Erber has to drag Rosa Luxemburg into court as a fellow-detractor of the Bolsheviks, how much right he has to mention her views in the same breath with his own. Fortunately, Luxemburg is not a defenceless corpse. She left a rich political testament to assure her name from being bandied about by soiled lips. Read this:

"The real situation in which the Russian Revolution found itself narrowed down in a few months to the alternative: victory of the counter-revolution or dictatorship of the proletariat — Kaledin or Lenin. Such was the objective sit-uation, just as it quickly presents itself in every revolution after the first intox-ication is over, and as it presented itself in Russia as a result of the concrete, burning questions of peace and land, for which there was no solution within the framework of bourgeois revolution".

This is a revolutionist writing — not an idol-worshipper of Lenin and the Bolsheviks, but still a revolutionist, a tireless, defiant, unflinching champion of the proletariat in the class struggle.

"In this, the Russian Revolution has but confirmed the basic lesson of every great revolution, the law of its being, which decrees: either the revolution must advance at a rapid, stormy and resolute tempo, break down all barriers with an iron hand and place its goals ever farther ahead, or it is quite soon thrown backward behind its feeble point of departure and suppressed by counter-rev-olution. To stand still, to mark time on one spot, to be contented with the first goal it happens to reach, is never possible in revolution. And he who tries to apply the home-made wisdom derived from parliamentary battles between frogs and mice to the field of revolutionary tactics only shows thereby that the very psychology and laws of existence of revolution are alien to him and that all historical experience is to him a book sealed with seven seals".

Read it over again, especially that wonderfully priceless last sentence!

"Still, didn't Rosa criticise the Bolsheviks for dispersing the Constituent As-sembly?" No, she did not. She criticised them for not calling for elections to a new Constituent; she criticised them for the arguments they made to justify the dispersal. But in the first place, her criticism has next to nothing in common with that of the latter-day anti-Bolsheviks (or, for that matter, of the anti-Bol-sheviks of the time). And in the second place, she was wrong, just as she was wrong in her criticism of the Bolshevik position on the "national question" and of the Bolshevik course in the "agrarian question". And in the third place, what she wrote in prison, on the basis of "fragmentary information" (as the editor of the American edition of her prison notes admits), was not her last word on the question. Before her cruel death, she altered her position on the basis of her own experiences, on the basis of the living realities of the German revolution. Lenin's *State and Revolution* was checked twice — first in the Russian Revolution and then in the German revolution! We will give the reader an idea of what she wrote before her death so that he may see why our present "champions" of Luxemburg never find time, space or inclination to quote her to the end.

The German workers, a year after the Bolshevik Revolution, overturned the Hohenzollern monarchy and, just as spontaneously as did the Russians before them, they formed their Workers' and Soldiers' Councils (Räte, Soviets). The German Mensheviks — Scheidemann, Noske and Ebert — feared and hated the Councils just as much as did their Russian counterparts. They championed the National Assembly (German counterpart of the Russian Constituent) instead, calculating thereby to smash the Councils and the struggle for socialism. Haase and Kautsky, the centrists of the Independent Socialists, oscillated between the Councils and the Assembly. What position did Rosa Luxemburg take, what position did the Spartacus League and its organ, *Die Rote Fahne*, take? Here once more was the problem of workers' democracy versus bourgeois democracy, the democratic republic of the Councils versus the bourgeois republic, dictatorship of the proletariat organised in the Councils versus the National Assembly — not in Russia but in Germany, not in 1917 but a year later, not while Rosa was in Breslau prison but after her release.

Here is Rosa Luxemburg in *Die Rote Fahne* of November 29, 1918, writing on the leaders of the Independents:

"Their actual mission as partner in the firm of Scheidemann-Ebert is: to mystify its clear and unambiguous character as defence guard of bourgeois class domination by means of a system of equivocation and cowardliness. This role of Haase and colleagues finds its most classical expression in their attitude toward the most important slogan of the day: toward the National Assembly.

"Only two standpoints are possible in this question, as in all others. Either you want the National Assembly as a means of swindling the proletariat out of its power, to paralyse its class energy, to dissolve its socialist goal into thin air. Or else you want to place all the power into the hands of the proletariat, to unfold the revolution that has begun into a tremendous class struggle for the socialist social order, and toward this end, to establish the political rule of the great mass of the toilers, the dictatorship of the Workers' and Soldiers' Councils. For or against socialism, against or for the National Assembly; there is no third way".

On December 1st, Luxemburg spoke on the situation at a meeting of the Spartacus League in the hall of the Teachers' Union. At the end of the meeting, a resolution was adopted setting forth her views and giving approval to them:

"The public people's meeting held on December 1st in the Hall of the Teachers' Union on Alexander Street declares its agreement with the exposition of

Comrade Luxemburg. It considers the convocation of the National Assembly to be a means of strengthening the counter-revolution and to cheat the proletarian revolution of its socialist aims. It demands the transfer of all power to the Workers' and Soldiers' Councils, whose first duty is to drive out of the government the traitors to the working class and to socialism, Scheidemann-Ebert and colleagues, to arm the toiling people for the protection of the revolution, and to take the most energetic and thorough-going measures for the socialisation of society".

In her first editorial in *Die Rote Fahne* of November 18, she writes under the title, "The Beginning": "The Revolution has begun... From the goal of the revolution follows clearly its path, from its task follows the method. All power into the hands of the masses, into the hands of the Workers' and Soldiers' Councils, protection of the work of the revolution from its lurking foes: this is the guiding line for all the measures of the revolutionary government... [But] What is the present revolutionary government [i.e. Scheidemann and Co.] doing? It calmly continues to leave the state as an administrative organism from top to bottom in the hands of yesterday's guards of Hohenzollern absolutism and tomorrow's tools of the counter-revolution. It is convoking the Constituent Assembly, and therewith it is creating a bourgeois counterweight against the Workers' and Peasants' representation, therewith switching the revolution on to the rails of the bourgeois revolution, conjuring away the socialist goals of the revolution.

"From the *Deutsche Tageszeitung*, the *Vossische*, and the *Vorwärts* to the *Freiheit* of the Independents, from Reventlow, Erzberger, Scheidemann to Haase and Kautsky, there sounds the unanimous call for the National Assembly and an equally unanimous outcry of fear of the idea: Power into the hands of the working class. The 'people' as a whole, the 'nation' as a whole, should be summoned to decide on the further fate of the revolution by majority decision. With the open and concealed agents of the ruling class, this slogan is natural. With keepers of the capitalist class barriers, we discuss neither the National Assembly nor about the National Assembly... Without the conscious will and the conscious act of the majority of the proletariat — no socialism. To sharpen this consciousness, to steel this will, to organise this act, a class organ is necessary, the national parliament of the proletarians of town and country. The convocation of such a workers' representation in place of the traditional National Assembly of the bourgeois revolutions is already, by itself, an act of the class struggle, a break with the historical past of bourgeois society, a powerful means of arousing the proletarian popular masses, a first open, blunt declaration of war against capitalism. No evasions, no ambiguities — the die must be cast. Parliamentary cretinism was yesterday a weakness, is today an equivocation, will tomorrow be a betrayal of socialism".

It is a pity that there is not space in which to quote far more extensively from the highly remarkable articles she wrote in the last few weeks of her life, before she was murdered by those whose "parliamentary cretinism" became the direct betrayal of socialism — by those for whom Erber has now become a

The Petrograd Soviet

shameful apologist by "showing" that the defeat of the revolution in Germany was as much the responsibility of the masses as it was of the Scheidemanns and Noskes! The articles as a whole show the veritable strides that Luxemburg took away from her prison criticism and toward a policy which was in no important respect different from the one pursued by the Bolsheviks toward the bourgeois and petty-bourgeois democrats, toward the Mensheviks and other "socialist opponents", toward the Constituent Assembly and the Soviets. With these criticisms of hers in print, to mention her today as an enemy of the Bolsheviks, as a critic of their attitude toward bourgeois democracy and the Constituent, is excusable only on the grounds of inexcusable ignorance.

The course of the German Revolution, life, the lessons of the struggle — these left us the heritage of a Rosa Luxemburg who was, in every essential, the inseparable comrade-in-arms of the leaders of the Russian Revolution. To claim that this firm solidarity did not exist, is simply an outrage to her memory. What is worse, it shows that nothing has been learned of the lessons of the Russian Revolution and nothing of the lessons of the German Revolution — the two great efforts of the proletariat to test in practice what is, in the long run, the question of life and death for us: the state and revolution. And on this question, with Lenin and with Luxemburg, the real Luxemburg — we remain under the banner of Marxism.

Revolution and violence

Erber is still a little restless over the traces of what he learned when he was a Marxist. So he comes to the noteworthy conclusion that: "Even the most perfect organisation of the workers in industry, transport, communications, etc., will not guarantee a non-violent accession to power. Since the working class may be challenged by force on the democratic road to socialism, let it be prepared to take up arms not to overthrow a democratic state but to 'win the battle of democracy'. Standing as the defenders of the best traditions of American democracy, its cause will be immeasurably strengthened. A Marxist in the United States can commit no greater folly than to view the workers' road to power as culminating in an armed insurrection against a state that rests on political democracy".

Against whom is all this pomposity directed? Against the Bolsheviks, perhaps? What democratic state did they overthrow with arms in hand? The state of Kerensky and Company ("Company" meaning the Mensheviks and S.R.s). And what was this state? We turn to the most authoritative reference work on the subject, Erber's document, and read that the Kerensky "régime sought to stretch out its undemocratic authority as long as possible by repeatedly postponing the elections of a Constituent Assembly. If the revolution was to advance, Kerensky had to go". Very well, go he did, along with everyone whose teeth were sunk in his coat tails.

Is it directed, perhaps, against Erber's newly-acquired chums and exemplars, the Mensheviks and the S.R.s? They did indeed take up arms to overthrow a democratic state, the state of the "teeming, creative, democratic Soviets". Is it directed against the Left S.R.s, that minority which sought to impose its will on the Soviet government with arms in hand because it disagreed with the decision to sign the Brest-Litovsk peace treaty? Or against the still tinier minority of anti-Bolshevik anarchists who likewise sought to overthrow the régime?

Or maybe it is directed against us here? Maybe he wants us to give a solemn pledge not to resort to armed insurrection against "a state that rests on political democracy"? Very well, we do not hesitate to give our pledge to Citizen Pompous Muddlehead, and therewith to state once more our credo.

We do not and will not call for armed insurrection to overthrow a "democratic state", a "state that rests on political democracy". It is an oath. The political infants who led the early communist movement in this country and who had little in common with Marx or Lenin — they issued such calls and advocated such a course. We — never! Not yesterday, not today, not tomorrow. We are not for violence in principle any more than we are for parliamentarism in principle. If anything, our principles call for an abhorrence of violence, as a capitalistic and uncivilised means of settling disputes among people. We are compelled every day to defend ourselves, with whatever organised strength we can muster, from the violence, open or concealed, with which the ruling classes impose their exploitation upon the masses. We are not putschists because we are not bureaucrats — and in every putschist, who has no confidence in the people, is concealed the bureaucrat, who has contempt for the people. Overthrow the bourgeois state by armed insurrection! Who, we? Not today and not tomorrow, and not if we had a hundred times as many members and followers as we have now!

The bourgeois state, bourgeois democracy, still has the confidence and support of the overwhelming majority of the people, including the working-class people. They think bourgeois democracy can solve all their basic problems. We Marxists do not. We believe that such a solution requires a working-class democracy, the rule of the proletariat, which develops into a more extensive and even more genuine democracy, the rule of all the people, which in turn develops into the end of all rule (democracy is a form of rulership) by dissolving into the classless society of socialism and communism. The workers do not share our belief. Can we even dream of imposing our views, the views of a tiny minority, by merely wishing, or by decree, let alone by armed insurrection?

Without abandoning our views for a moment, we say to the workers: Unite into your own economic and political organisations, free from the control and influence of your sworn class enemy. You have confidence in bourgeois democracy? Then organise your own political party. Challenge your enemy not only on the economic field but also on the political. Send your own representatives into the legislative bodies to work and fight for your interests. We say, with Engels, that "universal suffrage is the best lever for a proletarian movement at the present time". We say, with Engels, that "universal suffrage is the gauge of the maturity of the working class". We will therefore do everything we can to raise the red line in that thermometer which measures the maturity of the working class. "On the day when the thermometer of universal suffrage reaches its boiling point among the labourers, they as well as the capitalists will know what to do".

But what Marxist would try to displace the bourgeois state with a workers' state before that boiling point has been reached? Certainly no intelligent and educated Marxist would think of it, any more than he would think of walking out into the street with only his shirt on in winter time. A "state that rests on political democracy" is a state which, deservedly or not, still enjoys the confidence of the masses. To think of overturning such a state by armed insurrection under such conditions is putschist madness and adventurism, not revolutionary Marxism.

When the masses no longer have confidence in the bourgeoisie and the bourgeois state, when they have reached the point where they are ready to take the state into their own hands[83], ready to undertake a radical solution of the social problem, ready to take control of their own destinies — the situation changes! Once the masses have expressed their decision to take power and expressed it clearly and democratically — be their will expressed in the organs of parliament or in organs of their own which they find at hand or in organs of their

83. Evidently Shachtman doesn't mean to propose the workers taking the existing, bourgeois, state into their own hands, as if such a thing were possible. He must mean: the workers are ready to take power into their own hands.

own which they create spontaneously in the course of the struggle — the situation changes! If the bourgeoisie and the bourgeois state bow to the democratic will of the people, so much the better! As we wrote before, nobody would be more delighted than we and with us the whole working class. Up to now, however, history has been very frugal with examples of such bowing to the democratic will of the revolutionary people. But if there is nevertheless one chance in a thousand of that happening, then it is possible only if the working class confronts the bourgeoisie not with votes alone (the "boiling thermometer") but with serried class strength, with organised power. Such power is nothing else than potential violence, that is, violence that can be summoned the minute the democratically rejected bourgeoisie tries to perpetuate its domination over the people by the use of armed force.

But what if the bourgeoisie and its state do not bow to the will of the people? What if the régime seeks "to stretch out its undemocratic authority as long as possible"? With all due respect to the fanatical democratism of the bourgeoisie and its state, we think such an alternative is... quite possible. If it occurs, we will manage to say what Erber says about the Kerensky régime: It has to go! And if it does not go quietly, it will have to be pushed a little.

We, followers of Marx and Lenin, want to make sure that on the day the thermometer boils over — not today, not tomorrow, but on that day — "the labourers... as well as the capitalists will know what to do". We want to make sure that on that day the labourers not only have enough votes in their hands, but enough power to enforce their will. Is that the folly of viewing "the workers' road to power as culminating in an armed insurrection against a state that rests on political democracy"? Not at all! Erber is just repeating the drivel of the social democrats. What we Marxists call for is the good common sense of the workers' road to power culminating in the armed dispersal — if the stubborn bourgeoisie insists on it — of the state which no longer enjoys authority among the workers, which no longer has the confidence or support of the people, which, therefore, no longer "rests on political democracy" in the real sense of the term. It is with this view that we want to imbue every socialist militant, every vanguard fighter, every worker whom we can reach with our voice and pen.

That, stated for the hundredth time, is our credo. Take note of it, O Pompous Muddlehead! Take note of it, all workers! Take note of it, too, Mr Public Prosecutor!

Why are we so impassioned and tenacious in our defence of Marx, of Lenin, of Luxemburg, of Trotsky, of the Paris Commune, of the Russian Revolution, from all their falsifiers and detractors? Out of academic considerations? Because we are mere historians concerned with an accurate record of our past? Because we are Talmudists with our noses buried in the ancient books of wisdom? We are revolutionary socialists, and the fight to keep our heritage clean is an indispensable part of our fight for socialism. Does anyone think we would consume all this paper just to prove that this one is a liar, that one a deserter, that one a muddlehead, the other one a traitor? There are more pleasant and important things to do in life.

We are fighting for socialist freedom, and in this fight we are now on the defensive. The working class is on the defensive all over the world. It is lacerated by defeats, it is confused and disoriented, it has lost a lot of confidence in its power. It has been backed into a corner like the quarry in a hunt. It is surrounded by baying hounds.

The hunters challenge it: "Give up! Surrender!" The Stalinist dogs bark: "You cannot free your own self. Wherever you tried it yourself, you failed. You are too weak, too stupid, too undisciplined if left to your own democratic devices. You need an iron hand over you, an iron hand with a whip. Under Lenin's Soviets, the state was feeble. Under Stalin's GPU, the state has been consolidated and enormously strengthened. We have extended it over Europe and Asia. You must have a bureaucracy to lead you out of the wilderness!"

The bourgeois dogs bark: "You cannot free your own self. Wherever you tried it, you failed. You failed because there have always been the rulers and the ruled, and that is how it will always be. Look what happened in Russia! You tried, and you failed. Socialism is an ideal, but a Utopian ideal. Marx brought you Lenin. Lenin brought you Stalin. Under our rule you will at least have a bed to sleep in and unemployment insurance".

The reformist dogs bark: "Learn the lessons of the Russian Revolution! Revolution brought you only misery. Lenin was a gambler and he lost. Don't take power into your own hands. All evil flows from that. Just send enough of us to parliament and we will patch up the bourgeois state without you having to do a thing. Collaborate with the decent bourgeois elements, in plant and government and war, and don't lose your temper and get violent. It will do you no good. Revolution only brings Stalinism".

Erber whimpers: "Woe is me and woe is the world. Lenin ruined us all. I really don't know what to say or do about it. I really don't know what road you should take to get out of this universal ruin. Wait here, don't budge. I'll figure out something presently. But whatever you do, don't follow Lenin, don't take the road of revolution, don't take power into your own hands".

Hook nods philosophically: "Lord Acton was right, dear pupils. Power corrupts and absolute power corrupts absolutely. I want no part of it".[84]

Eastman and Louis Waldman conclude: "Vote for Dewey".[85]

Social interests shape ideas. Ideas serve social interests. The ideological campaign against Marxism is still what it always was: an integral part of the fight against socialism and the interests of the working class. The campaign against Lenin, the Bolsheviks and the Russian Revolution is still what it always was: an integral part of the fight to turn the working class away from the idea of taking their fate into their own hands. The attempt to bind and gag Rosa Lux-

84. For Hook see note 8. Lord Acton was a conservative Catholic historian in the late 19th century, famous for his reported remark that "power corrupts...", etc.

85. For Eastman see note 15. Louis Waldman was in the 1930s a leader of the Social Democratic Federation, a right-wing split from the Socialist Party of America. After about 1940 he was politically inactive, but prominent as a lawyer. Thomas Dewey was the Republican candidate in the 1948 presidential election.

emburg and kidnap her to an alien camp, so that we cannot hear what she really said and wanted, is still what it always was: an integral part of the turncoats' campaign to dirty the great Russian Revolution so that their capitulation to democratic imperialism will look clean.

The defence of the Russian Revolution is the defence of Marxism. The defence of Marxism is the fight for socialism, the fight to drive away the baying hounds, to enable the working class to leap forward again with renewed confidence in its own strength, in its great emancipating mission, in its eventual triumph.

We are not idol-worshippers. We are not uncritical eulogists of Marx or Lenin or Trotsky or Luxemburg or the Russian Revolution itself. From its grandeur, we have learned what to do. From its decay, what not to do. We think we understand now more than we ever did before why Lenin, for all his disagreements with Rosa, called her an eagle. Even in her prison notes she wrote these words which are so timely thirty years later:

"Everything that happens in Russia is comprehensible and represents an inevitable chain of causes and effects, the starting point and end term of which are: the failure of the German proletariat and the occupation of Russia by German imperialism. It would be demanding something superhuman from Lenin and his comrades if we should expect of them that under such circumstances they should conjure forth the finest democracy, the most exemplary dictatorship of the proletariat and a flourishing socialist economy. By their determined revolutionary stand, their exemplary strength in action, and their unbreakable loyalty to international socialism, they have contributed whatever could possibly be contributed under such devilishly hard conditions. The danger begins only when they make a virtue of necessity and want to freeze into a complete theoretical system all the tactics forced upon them by these fatal circumstances, and want to recommend them to the international proletariat as a model of socialist tactics. When they get in their own light in this way, and hide their genuine, unquestionable historical service under the bushel of false steps forced upon them by necessity, they render a poor service to international socialism for the sake of which they have fought and suffered; for they want to place in its storehouse as new discoveries all the distortions prescribed in Russia by necessity and compulsion — in the last analysis only by-products of the bankruptcy of international socialism in the present world war.

"Let the German Government Socialists cry that the rule of the Bolsheviks in Russia is a distorted expression of the dictatorship of the proletariat. If it was or is such, that is only because it is a product of the behaviour of the German proletariat, in itself a distorted expression of the socialist class struggle. All of us are subject to the laws of history, and it is only internationally that the socialist order of society can be realised. The Bolsheviks have shown that they are capable of everything that a genuine revolutionary party can contribute within the limits of the historical possibilities. They are not supposed to perform miracles. For a model and faultless proletarian revolution in an isolated land, exhausted by world war, strangled by imperialism, betrayed by the international proletariat, would be a miracle...

"In Russia the problem could only be posed. It could not be solved in Russia. And in this sense, the future everywhere belongs to 'Bolshevism'."

"The danger begins when they make a virtue of necessity". That danger is inherent in every great revolution and every great revolutionary party suffers from it. The Bolsheviks were no exception. They could not be, especially given a socialist revolution, for which there was no blueprint worked out in advance and could not be. Improvisation was imperative. What is remarkable is that in all this convulsive turbulence, so much order prevailed, so much was done according to plan, so much was done to turn the helm when the ship of state hit uncharted and unexpected reefs. In that respect, no revolution, no social transformation in history, even equalled the Russian Revolution.

"Es schwindelt", said Lenin to Trotsky when the storm lifted the Bolsheviks to the first socialist power in history — it makes you dizzy. Everybody was made dizzy. The Bolsheviks alone kept their heads. The others lost them completely. The bourgeoisie, the landlords, international capital, plunged into their mad and sanguinary adventure to crush the revolution. The Mensheviks and S.R.s joined with what Engels predicted would be "the whole collective reaction which will group itself around pure democracy". In the wild civil war that followed, in which millions of ordinary workers and peasants proved to be far fiercer and more intolerant toward the opponents of the Soviet power than the Bolsheviks themselves, there was no room for the "home-made wisdom derived from parliamentary battles between frogs and mice". One by one, the parties and groups that took up arms against the Soviet Power were outlawed. No one has found another solution in civil war.

The Bolsheviks performed no miracles; they promised none. They were summoned to hold the first revolutionary citadel against frenzied and maddened besiegers until the relief columns of the Western proletariat could be brought forward. They held the citadel, better and longer than anyone expected, even they themselves. Julius Martov, the Menshevik leader, wrote in October 1921 that "The political tactic of our party in 1918 and 1920 was determined primarily by the fact that history had made the Bolshevik Party the defender of the foundations of the revolution against the armed forces of the domestic and foreign counter-revolution". Alas, what he says about the Menshevik party is not true in its entirety, but only for a tiny part of it. What he says about the Bolshevik party is true in its entirety. But the Bolsheviks were not gods. In seeking to master necessity, they also had to bend to it. War, especially civil war, especially when your enemies on a world scale outnumber you a hundred to one, is not the ideal culture medium for democracy to flourish in. "The Bolsheviks have shown that they are capable of everything that a genuine revolutionary party can contribute within the limits of the historical possibilities. They are not supposed to perform miracles. For a model and faultless proletarian revolution in an isolated land, exhausted by world war, strangled by imperialism, betrayed by the international proletariat, would be a miracle".

Once the civil war came to a triumphant end for the Soviet Power, necessity became more and more a virtue. What was imposed on the Bolsheviks by the exigencies of war was gradually transformed into an article of faith for the period of peace as well. One-party government, which is anything but abnormal in all countries at all times, and was just as normal and unexceptionable in Russia, was transformed to mean: Only one party can enjoy legal existence in the country. To this, Stalinism succeeded in adding: Only one faction can enjoy legal existence in the party. The extension of full democratic rights — not the right to armed putsches but full democratic rights — to all parties, without exception, would have strengthened the country and reinvigorated the Soviets themselves. It should now be clear that without the presence of other political organisations capable of freely debating (debating, not shooting at) the proposals presented to the Soviets by the Bolshevik party, the Soviets would rapidly and inevitably deteriorate to the position of a superfluous duplicate of the ruling party, at first only consulted by the latter, then disregarded by it, and finally discarded altogether for the direct rule of the party alone (the bureaucracy of the party at that!) In this process, the decay of democracy within the Bolshevik party and the decay of Soviet democracy went hand in hand, each having the same deleterious effect upon the other until both were suppressed completely and, along with them, all the achievements of the revolution itself. Deprived of the saving oxygen of revolution in the West, the democratic organism was suffocated. Poisons accumulated throughout the whole system which could not be thrown off, new poisons were added (necessity becoming virtue), and the Revolution moved tragically toward its death.

Of the period of decay, much has been written — in advance by Luxemburg, in his time by Lenin, later by Trotsky. Much more can be written and much more will be written as the distance of years sets off the Revolution in clearer perspective. We close no doors. We file away nothing as an absolutely closed case. But it does not follow that no conclusions at all can be drawn from the great revolution. It does not follow that anybody's conclusions, no matter how superficial or trivial or reactionary, are as valid as any others. Our struggle has been hurled back — that is now a commonplace. But it does not follow that we start with tabula rasa — knowing nothing, learning nothing, believing nothing. From the grandeur of the Russian Revolution, we have learned something: the superiority of proletarian democracy to bourgeois democracy. From the decay of the Russian Revolution we have learned something: that proletarian democracy cannot exist for long if it is confined to one faction or one party, even if it be the revolutionary party, that it must be shared equally by all other working-class and even — under favourable circumstances — bourgeois parties and groups, for without it the proletarian party and the proletarian democracy both die and with them die the prospects of socialism.

These are not the happiest days for socialism. We know that. We know that the grotesque outcome of the Russian Revolution, the failure of the proletariat anywhere else to come to power, has raised more than one gloomy doubt about the social ability of the working class to reorganise society rationally, about the very possibility of a socialist future.

It is precisely in this regard that the Russian Revolution is our lasting triumph! It is precisely in this regard that the Russian Revolution continues to fortify our convictions!

What the Marxists claimed for decades, the Russian Revolution proved. What did it prove? That the rule of the capitalist class is not eternal. That the power of the capitalist class is not invincible. That the working class can overthrow the rule of capital and the bourgeois state, not in the books of Karl Marx, but in the living struggle of organised workers. To this day and hour we say: If the Russian working class could take power, the working class can take power in any other country under similar circumstances. This we consider proved. To disprove it, it is only necessary to show that the Russian proletariat had national or racial characteristics which determined its victory and which are not to be found in any other proletariat. Or, it is only necessary to prove that no other country can ever reproduce a combination of circumstances similar to those which made possible the triumph of the Russian Revolution. Nobody has done this up to now. Until it is done, we regard as proved the ability of the working class to take power in its own name and for its own self. That's a tremendous acquisition for the Marxian and working-class movements. Why should any socialist or even a non-socialist worker be fool enough to bagatellise this acquisition, let alone relinquish it?

"But surely you cannot claim that the Russian Revolution also proved that the working class can hold power and use it to usher in a socialist society!" We make no such claim. That, the Russian Revolution did not prove and, by itself, could not prove. But to prove that the proletariat has some basic and inherent social incapacity to hold power and establish socialism, you must be concrete, and not confine yourself to going from concept to concept to concept without once touching material ground. To be concrete, dear skeptic, to be scientific, you have to show why the Russian proletariat lost power. Merely to point to the fact that the proletariat, one hundred years after the Communist Manifesto, has not yet liberated itself, is not one whit more serious an argument to prove that it cannot liberate itself than pointing to the fact that the failure of generations of scientists to find a cure for cancer proves that cancer cannot be cured.

This brings us back to what is, after all, the very essence of the dispute over the Russian Revolution: Why did the proletariat lose power and, therewith, lose the indispensable instrument for constructing socialism?

Exactly ninety-nine per cent of the critics of Bolshevism answer the question in this way, at bottom: The Russian workers lost power because they took power. Stalinism (the destruction of the Russian workers' power) followed ineluctably from the seizure of power by the proletariat and Lenin's refusal to surrender this power to the bourgeois democracy. Exactly ninety-nine per cent of the revolutionary Marxists answer the question in this way, at bottom: The Russian workers lost power because the workers of the other countries failed to take power.

There is the difference. It is fundamental and yields to no compromise.

We also know that our "proof" is not final. There is no way of making it final in the realm of concepts. It can be made final only in struggle. By ourselves, we cannot provide this final proof. We will not even attempt it. The proletariat alone can provide this proof, in the course of the struggle which it must carry on in order to survive, and in which it must triumph if it carries it on to the end. The Russian proletariat, the Russian Revolution, proved all it could prove. The rest will come. And one of the not unimportant reasons why it will come is that we remain loyal to the fight for socialism. We remain defenders of the imperishable Bolshevik Revolution. We remain under the banner of Marxism.

Around 250 satirical journals appeared during the period of the first Russian revolution of 1905-7. This image, entitled 'Invasion', by Boris Kustodiev, depicts the brutal repression of the Moscow workers' uprising of December 1905.

Poster for May Day 1920. By S I Ivanov.

Left: 'The Spider and the Flies' by Viktor Deni (1919).

Below: 'Retreating before the Red Army' by A Apsit (1919).

ПАУК И МУХИ.

Дили - бом !.. Дили - бом !..
Стоит церковь -божий дом,
А в том доме - паучек,
Паучек - крестовичек,
Паучек семи пудов-
„Все от праведных трудов" :
Нахлобучивши клобук,
Ухмыляется паук
И трезвонит целый день:
Делень - день !.. Делень - день !..

Паук весело живет,
Паутиночку плетет,
В паутину ловит мух :
Молодаек и старух,
Молодцов и стариков-
Богомольных мужиков.·
Мухи жалобно жужжат,
Пауку несут деньжат,
Все, что нажили горбом,
Дили - бом !.. Дили - бом !..

Ой вы, братцы - мужики,
Горемыки - бедняки,
А давно уж паука
Взять пора вам за бока:
Ваши души он „спасал"-
Вашу кровушку сосал
Да кормил жену и чад-
Паучиху, паучат,
И пыхтел, осклабив рот:
- Ай дурак - же наш народ !

№45

Демьян Бедный·

ОТСТУПАЯ ПЕРЕД КРАС= НОЙ АРМИЕЙ БЕЛОГВАРДЕЙЦЫ ЖГУТ ХЛЕБ.

'The Tsar, the Priest and the Kulak' by Pet (1918).

Right: '1st May' by A Apsit (1919).

Left: 'Comrade Muslims' by D S Moor (1919)

Below left: 'To the Deceived Brothers' by A Apsit (1918)

Below right: 'Year of Proletarian Dictatorship' by A Apsit (1918).

'The Entente under the Mask of Peace' by Viktor Deni (1920).

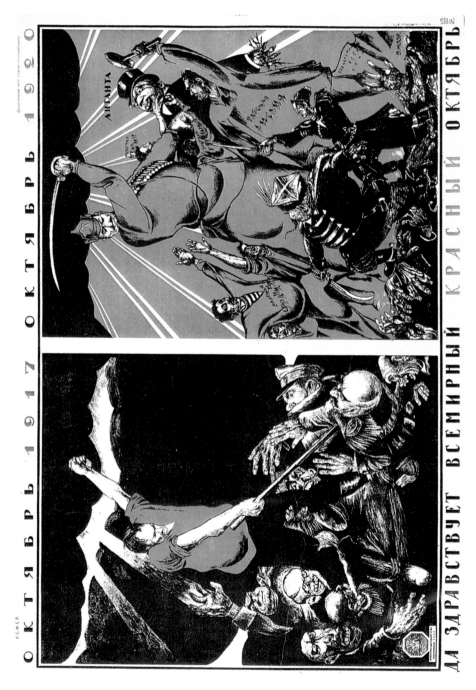

'October 1917 — October 1920' by DS Moor (1920).

'Have you enrolled as a volunteer?' by D S Moor (1920).

Вот **Капитал и Компания**—
Сволочи всякого звания
Это — буржуев „икона"
Тут все столпы их „закона",
Все, кто им служит опорою.
Шайка злодеев, с которою
Сбросивший цепи народ—
Бьемся мы третий уж год.

Как **Капитал** ни бахвалится.
Трон его скоро развалится:
С туши, оплывшей от жира,
Сорвана будет порфира.

В°° Типо-литография М.Т.С.Н.Х.

№ 63.

Взмах пролетарского молота—
Рухнет твердыня из золота,
Похоронив под собой
Международный разбой».

Рухнет все зло многоликое.
Братья, за дело великое
Духом не падая, стойте.
Жизнь нашу новую стройте
Вольную, светлую, мирную.
Стройте коммуну всемирную,
Тесный союз трудовой
Всей бедноты мировой!

Демьян Бедный.

'Capital and Co.' by Nikolai Kochergin (1920).

Part 2: The party that led the revolution

Trotsky as St George, killing the dragon of reaction. By Viktor Deni.

The party that won the victory

The rise and fall of the Russian Revolution are both linked to the Bolshevik Party. Since 1917, revolutionary situations have developed in a dozen countries, with all the elements required for a working-class victory present to at least the same degree as in Kerensky's Russia — all but one: a revolutionary party prepared for just such a situation and capable of utilising it to the utmost. This difference provides the decisive reason why the revolution triumphed in Russia and was defeated everywhere else. It also provides the basis for explaining the subsequent victory of the counter-revolution in Russia itself. The more generally this fact is acknowledged, the less trouble is usually taken to analyse it.

Political parties as we know them today are a comparatively recent development. They were quite unknown under feudalism. There were partisans of this or that group, of this or that idea, but there were no parties in the modern sense. That is understandable. Even though the young bourgeoisie created rudimentary political organisations in its struggle against feudalism in some countries, these were not an indispensable condition for the victory of the new society and its consolidation.

Two basically different revolutions

The revolution against feudalism and the socialist revolution against capitalism are alike only in that both bring a new class to power and organise a new social system. In every other respect, they are fundamentally different.

The bourgeois revolution takes place with the elements of capitalist economy already developed within, coexisting with and constantly transforming feudalism itself. The revolution consists fundamentally in undoing the feudal shackles on the existing and growing capitalist organisms. Its task is not so much to "establish" capitalist relations as to liberate them for their freest unfoldment.

The proletarian revolution does not find the socialist economic forms or relations at hand. All that dying capitalism provides it with — no trifle, to be sure! — is a tremendous economic machine, the socialisation of production, and a modern working class capable of reorganising society. Socialism itself does not exist; the revolution must first create it, establish it. The bourgeois revolution need not necessarily be carried out by the bourgeoisie itself, that is, by the bourgeoisie as a class. The bourgeois revolution need not necessarily bring the bourgeoisie to political power. The basic requirements of this revolution are fulfilled when the main feudal shackles upon capitalist economic relations are broken. This can be accomplished by the bourgeoisie. But it can also be accomplished without the bourgeoisie and even against it. It can be carried out by the plebeian masses, with the bourgeoisie taking over power

Lenin in 1917

only later on by means of a counter-revolution; or it can be carried out "from above," in the Bismarckian manner, by the aristocracy, by feudal or semi-feudal lords themselves. The bourgeoisie can maintain and consolidate the social system peculiar to it and nevertheless share political power with the outdated classes; it can even be cheated of political power by the latter. What is more, it can maintain itself to its dying day without necessarily destroying all "residues" of feudalism; in fact, in vast territories of the world, its continued power is based precisely upon the preservation of pre-capitalist economy. For the bourgeoisie it suffices that its economic system predominates.

The proletarian revolution, on the contrary, cannot be made by any other class but the proletariat itself, inasmuch as only the proletariat is capable of establishing the socialist society which is the only aim of this revolution. The first and absolutely indispensable condition of this revolution is "to make the proletariat the ruling class, to establish democracy." The bourgeoisie, on the basis of already existing capitalist economy, strives for political power. The proletariat, on the other hand, must first conquer "its political supremacy in order, by degrees, to wrest all capital from the bourgeoisie," and then organise socialist production. Capitalism, the capitalist state — these are conceivable without the political power of the capitalists. The very beginnings of the transition to socialism, however, are inconceivable without a workers' state, "this meaning the proletariat organised as ruling class."

Consciousness and revolution

The bourgeois revolution is not (not necessarily) the conscious revolution of a class. It is carried out with a false ideology (or to use the term in its original sense, simply ideology). Its victory over feudalism is assured by its fundamental nature, that is, the predominance of capitalist over feudal property is assured to the former by the "superiority of its productive methods." Capitalist production takes place, grows, goes through crises, declines, as a natural economic movement, regardless of will and in defiance of plan. The economy is automatically renewed (be it on a higher or lower level).

The proletarian revolution, on the contrary, cannot but be a conscious revolution, purposeful, planned, prepared, organised, timed. It does not have the automatic character of the bourgeois revolution. The transitional economy through which the revolution moves to socialism (above all if the revolution is surrounded by a predominantly capitalist world economy) is not automatically assured of a unilateral development to a classless society. Until the "administration of things" can replace the "administration of men," the socialistic character of the new economic relations depends entirely on the proletarian character of the state. Whereas capitalist production, based on "private property and competition, have been working out their own destiny," the development of the productive forces in a socialist direction, following the proletarian revolution, is "indivisibly bound up with the new state" as repository of the new property relations. "The character of the economy as a whole thus depends upon the character of the state power." The movement toward a socialist society can, therefore, take place only as a result of conscious planning. And inasmuch as a socialist society is based on production for use, planning can only mean plans elaborated by the "users," that is, democratic, socialist planning. Without consciousness and plan, the proletarian revolution is impossible; lacking them, a working class that seizes power will never hold it. Without consciousness and plan, the establishment of socialism is impossible; if socialism is not consciously planned, it will never come. Consciousness and plan imply a self-active, aware, participating, deciding proletariat, which implies in turn a dying-out of coercion and bureaucratism.

Consciousness (socialist consciousness, that is) does not, however, come unfailingly to every worker at a given age, like hair on the head on a growing baby. Some acquire it early; some acquire it late; others go to their graves without it. The acquisition of a socialist consciousness equals the acquisition of an understanding of the indispensability of joint, deliberate and planned action for the fundamental task of reorganising society. The ingenuity of man has not produced a vehicle or an instrument for this action that equals the organised political party.

The revolutionary proletarian party is the repository of the socialist consciousness of the working class. Composed of the conscious workers, the party is a means by which the working class is saved from existing permanently in a bourgeois stupor, from living intellectually from hand to mouth. It is the organised memory of the working class. It not only connects up yesterday with today, but today with tomorrow. In every activity of the working class it keeps before it its historic goal, thus helping to unify these activities, to rid them of distortions, to give them a progressive meaning and a basic purpose.

Lenin's most important contribution
Of all the great contributions made by Lenin, none was as vitally important as the theory and practice of the revolutionary working-class political party

which he evolved. It is true that the elements of Bolshevism-as-a-party (Bolshevism without a party means nothing) are to be found in Marx. But Marx did not, and could not work up these elements into the rounded, systematized, theoretically-motivated and practically-tested whole which they became under Lenin's leadership.

Lenin's whole conception of the party began and ended with the idea of an organisation composed, trained and activated in such a way that it could be depended upon to lead the working class to power at the right time as the first step in the socialist reorganisation of society. All critics and improvers of Bolshevism, of Lenin's party, who ignore this, are guaranteed to miss the mark.

This conception meant, first of all, a party composed of politically-educated fighters, capable of subordinating all other interests and considerations to the cause of the socialist victory. If the party is to be the repository of the socialist consciousness of the working class, it must be made up of men and women whose action is based upon understanding. They had to understand the nature of the capitalist society whose overthrow they proclaimed; they had to understand the nature of the class that was to overthrow it; they had to understand the means, the strategy and tactics, by which it was to be overthrown.

Lenin's party was the best-educated political organisation in the world. The Bolsheviks were intolerant of theoretical sloppiness; toward inattentiveness or neglect of theory, they were absolutely merciless. Lenin's "Without revolutionary theory, no revolutionary practice" was an organic concept with them. The sniggering at "theory" which became current in most other socialist parties of his time was never stylish in Lenin's party.

Lenin was an alert and ubiquitous polemist, and not a mild one. His polemically harsh and even violent language against adversaries used to shock (and still does) the delicate sensibilities of bourgeois and petty bourgeois politicians who considered it perfectly normal, however, to have the ruling class answer their "critics" with police clubs and prison sentences, to say nothing of disposing of "arguments" by slaughtering millions of "opponents" in a war. Lenin's violence in polemic was due to his uncompromising fidelity to the socialist revolution and the policy best calculated to achieve it. He was deadly serious about the revolution. Those whose theories and policies led the workers off the track, reconciled them with their class enemy, frustrated their efforts, had to be challenged with a vigour that matched the peril they represented. He helped train a party which, like himself, was sufficiently confident of the superiority of its program and views to engage anyone in debate without fear of coming off second best. He understood that you often teach more by polemical presentation and criticism than by "straight" exposition — the correctness of your own views standing out more clearly when counterposed to the views of others. He understood that mere reiteration of your own views is not enough to build a firm party. These views must be constantly defended in public (or revised when they cannot be defended!) against all critics — and defended successfully — otherwise your followers either begin to lose faith in your views or else continue to support them out of blind "party patriotism." Lenin, who was a party patriot if there ever was one, had no use at all for this kind of "patriot," any more than he cared for dopes in general. His own words were even blunter: "Whoever takes anything on faith is an idiot who can be disposed of with a wave of the hand." (The epigones of Leninism everywhere do far more, alas, to raise idiots than to raise Bolsheviks.)

Lenin's polemics, like all his writings, were meant to educate the party and the working class, to clarify, enhance and steel their consciousness. He did not substitute harsh words for logical substance. (The epigones believe they have destroyed an opponent's argument completely, and revealed themselves as living incarnations of Leninism, when they bark: "You are a prostitute! You are an agent of the bourgeoisie!" and then sit down, exhausted but content and triumphant.) The monger of platitudes, however orotund or shiny, bored Lenin to death; the demagogue, he detested as "the worst enemy of the working class."

A revolutionary party of action

The Bolsheviks built up a revolutionary party of action, not a pleasant company of salon habitues, dilettante socialists, or hair-splitting debaters. Their party was not a debating society, but a fighting army which had bloody battles to engage in and a world-renovating victory to win against the most powerful and deadliest enemy a class ever faced. Add to this the special circumstances of existence under Czarist autocracy and it is easy to understand why the Bolshevik Party was and had to be strictly centralised and disciplined. The right-wing socialists, especially of western Europe, who never envisaged battles or revolution, who looked forward to capitalist society gradually filling up with socialism by painless osmosis, shrank from Lenin's conception of centralism and discipline. The only discipline they wanted enforced was against the "ultra-leftist madmen." But Lenin, who understood to perfection the class enemy, its power, its savage capacity for self-preservation, its desperate unscrupulousness, knew that the revolutionary party challenging the enemy for nothing less than all-power itself would have to be a party of steel, disciplined, tested and re-tested, its ranks and program constantly checked for weakness, its fighting capacity kept at a high pitch.

What other conception of a party can you have if it is the socialist revolution you really aim for — a revolution that has not proved to be as easy as rolling off a log? Take, for example, our own Socialist Party in this country. Examine it from this standpoint and see why we cannot take it seriously (assuming that anyone else does). Pretty near anybody can be a member (except a real revolutionist — these are expelled). Anybody can put forward pretty near any view he wants to in public (except again, a real revolutionist). The party can adopt one position on a vital question, but the party leader, not caring particularly for this position, can put forward one of his own, even if it is the diametric opposite of his party's official stand. Pretty near any member can act as he pleases in the labour movement, follow whatever policy his heart and mind dictate, whether or not it conflicts violently with the policy followed by his fellow party-member in the same union or even with that officially advocated by his party. His party obligations are microscopic — he can attend meetings or not; if he drops into party headquarters once a month to exchange the time o' day with the other boys, he is not frowned on particularly; if his "party life" takes up two hours a month and the rest of his life is devoted to the bourgeois world, that is not a very black mark against him.

A fine, democratic and ever-so-non-fanatical a party! Of course, it never can and never will carry out the fight for socialism. It can never lead the workers in any serious struggle. In exchange, however, it offers these advantages over Bolshevism: A worker who wants to know, "Where do you, the Socialist Party, which wants to lead me, to have my support, stand on this or that vital question on the class struggle?" can always get his choice of half a dozen answers, each enjoying equal standing and validity, i.e., zero. A worker who wants to join the party need commit himself to nothing more serious than paying his dues, yawning over the pages of the party paper, and voting quadrennially for the party "standard bearer."

Joining the Bolshevik Party meant becoming a soldier in a revolutionary army. It meant discipline and centralisation of efforts. It meant the ability to say: My party has this clear-cut policy, that clear-cut program, this answer to this problem; this is what it proposed to do about this situation; this is what it calls upon the people to do in that situation; if you agree with my party, support it, join it. My party means business; it is serious; it doesn't fool around with the interests and struggles of the working class; it calls upon labour to act as one man and it sets an example of how to act like one man.

In the last twenty years, there has been so much intellectual devastation in the revolutionary movement that Lenin's views on this point have been twisted and deformed beyond recognition. His insistence on discipline in action has been made to read discipline in thinking. His abhorrence of a "debating society," which he contrasted to a party capable of discussing policy thoroughly, coming to a decision by majority vote, and then unitedly executing the policy, has been made to read "no debates" in the party. The rich, even tumultuous, intellectual life of the Bolshevik party, for which there is no parallel anywhere; the continuous, passionate — and passionately interesting — and fruitful discussions of basic as well as topical questions which characterised it; the wide freedom of viewpoint which always prevailed in it as a matter of course, and not as a magnanimous bureaucratic dispensation once every two-three-four years, and even the freedom of political groupings and factions — all this has been wiped out by the not very sedulous apes of Leninism and its very opposite consecrated. Leninism, it now seems, boils down to this: We are rough and tough. We are hard people. We spit bullets. Shut up. Stop thinking. End debate — don't even start it. We know best. Our program is finished, amendments not admitted. Etc., etc. A party built on these "principles of Leninism" will do no more to bring about the socialist revolution than Norman Thomas' laissez-faire, laissez-aller party.

With the breaching of the world capitalist front in Russia, we have had, as Trotsky often noted, no lack of revolutionary situations. There has likewise been no lack of revolutionary initiative by the working class, resourcefulness, epic heroism, and repeated demonstrations that it is ready to extirpate the plague consuming civilisation. This has showed that capitalism is doomed, inasmuch as it can no longer maintain peace, order and social equilibrium, and that the force called upon to dispatch it is irrepressible. All that has been and still is lacking is… a party of the Leninist type, not an artificial copy of the Bolshevist party, but a party of that type, built and schooled in the same way.

Having one, the Russian proletariat was able to accomplish more than anyone had a right to expect of the working class of any one country, and of a backward country, to boot. Having lost it, the Russian proletariat lost all its revolutionary achievements. That it lost its party is not due to that mysterious "fundamental defect" in Bolshevism which its critics have yet to explain to us, but to the fact that the working class of the advanced countries failed in time to build parties like it and remained under the domination of the anti-Bolshevik parties of labour.

With this loss, the centre of revolutionary gravity has shifted further and further to the West. From Moscow there no longer come the liberating legions of the socialist revolution — as is unbelievably claimed by the self-patented "Trotskyists"[1] — or the liberating ideas of Lenin, but the rolling waves of black reaction. Once, Leninist Russia almost freed the West. Now only the West can free the Russia of Stalin, not the West of today but of tomorrow. Success depends entirely upon how well and how soon a party of Bolshevism is built in countries like the United States. We have, it would seem, more time than many others. Every hour of it must be utilised to prepare for the inevitable revolutionary crisis.

The urgent task of the day

If we do not succeed in having, at the crucial moment, the kind of party the Bolsheviks had in Russia in 1917, the absolutely inevitable catastrophe that would befall us all would have long-lasting effects. There is good reason, however, to believe that we shall not fail. The American working class has shown the most encouraging ability to move forward, not at a snail's pace but with leaps and bounds. It has not spoken its last word — only its first. Our bourgeoisie, "the most powerful in the world," has so little confidence in itself that it squealed with terror for months just at the sight of so limited and contradictory a step as the organisation of labour into an independent political force in … bourgeois politics! How will it feel when labour really declares its political independence as a class?

The difference between how it feels and what it really gets, depends primarily and decisively upon the building of the revolutionary party. We have not been hurled back to the starting point. We have learned what is important to learn from Lenin in the period of the rise of Bolshevism; we have learned what is important from Trotsky, in the period of Bolshevism's crushing by the counterrevolution. The vanguard now knows more and knows it better. It must now clothe the skeleton of its program with the flesh and sinews of tens of thousands of workers who are breaking intellectually from capitalism. That is the task of tasks of the Fourth International today.

---

1. The Orthodox Trotskyists in 1944 argued that the advance of the Russian Army into Europe would spur on socialist revolutions.

'The Last Hour' by Viktor Deni (1920)

Did Bolshevism produce Stalinism?

It is hard to say who has written more absurdities about Lenin's "organisational principles": the Stalinists who seek to prove that their totalitarian party regime conforms identically with the views set forth by Lenin or the modern anti-Bolsheviks who argue that if the two are not quite identical it is nevertheless Lenin's views and practises that led directly to the present Stalinist regime. They represent complementary and mutually parasitic parts of a division of labour which has successfully devastated the thinking of millions of people, with one saying that the totalitarian tyranny leads to (or is!) socialism and the other that socialism can lead to nothing but this totalitarian tyranny.

Either as perpetrators or victims of falsification, both are so thoroughly and extensively wrong that it would require volumes just to exhume and properly correlate the facts. It is not merely a matter of setting the historical record right — that is of secondary importance. It is above all a matter of resuming the lagging fight for socialism, which a Stalin abandoned so completely to pursue one reactionary course and a Wolfe[1] has abandoned just as completely to pursue a different reactionary course.

In Lenin's conception of the "party machine," of its role in relationship to the working class, Wolfe finds (as what popular writer nowadays does not?) "the germ of a party dictatorship over the proletariat itself, exercised in its name," that is, the germ of Stalinism. It is out of this feature of Bolshevism that Wolfe erects the third pillar of his analysis. He reminds us that at the beginning Trotsky warned against the inevitable outcome of Lenin's conception:

"The organisation of the party will take the place of the party; the Central Committee will take the place of the organisation; and finally the dictator will take the place of the Central Committee".

"Was ever prophecy more fatefully fulfilled by history?" exclaims Wolfe. The truth is that if prophets had no better example than this of how they are confirmed by history, the profession would be in sorry shape. With due respect to Trotsky, it can be said that to find in Stalinism a fulfilment of Trotsky's "Cassandra-like prevision" (Wolfe's phrase) of Lenin's conception requires a well-trained capacity for superficiality assisted by an elaborate ignoring — we will not say manipulation — of the historical facts. The "prevision" was not fulfilled at all; and Trotsky himself was not the last to understand this.

But before this is established, let us see what it is that makes Lenin's views so reprehensible in Wolfe's eyes. Rather, let us try to see, for on this score Wolfe is either ambiguous or obscure, or just plain silent. He makes the task of the reviewer almost baffling. Attentive reading of page after page of Wolfe fails to disclose exactly what it was in Lenin's ideas about the "party machine" that

---

1. Bertram Wolfe, in his book *Three Who Made a Revolution*. This article is an abridgement from Schachtman's review of Wolfe's book, *New International*, July 1950. The quotation from Trotsky is from *Our Political Tasks*, written in 1904 when Trotsky briefly sided with the Mensheviks.

Bertram D Wolfe was a founding member of the US Communist Party, then a leader of the "Lovestoneite" splinter. After 1940 he drifted towards liberal right-wing politics.

led to Stalinism.

Was it Lenin's conception of who is entitled to party membership? Wolfe describes the dispute at the party congress in 1903 on the famous Article I of the party constitution. Lenin's draft defined a party member as one "who recognises the party's program and supports it by material means *and by personal participation in one of the party organisations*." Martov, leader of the Mensheviks-to-be, proposed that the phrase in italics be replaced entirely by the following: "and by regular personal assistance under the direction of one of the party organisations." Martov's formula was supported by the majority of the delegates.

Wolfe describes Lenin's view unsympathetically, which is his God-given right. But what was wrong with it? Wolfe's answer is a significant wink and a knowing nod of the head, as if to say, "Now you can see where Lenin was heading from the very start, can't you? Now you know what Bolshevism was at its very origin. If you really want to trace Stalinism to its historical roots, there indeed is one of the sturdiest and most malignant of them."

But wink and nod notwithstanding, all that Lenin proposed was a provision that had been and was then and has ever since been a commonplace in every socialist party we ever heard of, namely, that to be considered a party member, with the right of determining the policy and leadership that the membership as a whole is to follow, you have to belong to one of the units of the party. That would seem to be, would it not, an eminently democratic procedure, to say nothing of other merits.

By Martov's formulation, the policy and leadership of the party to one of whose branches you belong are determined for you by persons who are given the title of party members in exchange for "assisting" it without the obligation of belonging to any of its established branches. It is the thoroughly bureaucratised bourgeois political machines that are characterised by the kind of party "membership" that Martov's draft proposed, and it is one of the ways in which leadership and party policy are divorced from control by the ranks. But what socialist party, regardless of political tendency, does Wolfe know that has ever adopted a party statute such as Martov defended? The Social Democratic Federation of August Claessens and Algernon Lee is not entirely corroded by Bolshevism, it is said. But suppose someone were to advocate that membership in the SDF be extended to persons who assist the Federation under the direction of one of its branches without actually joining a branch. These nonagenarians would immediately summon every remnant of their remaining muscularity to crush the hardy advocate as a madman who threatens the integrity of the SDF and the "Leninist organisational principle" which they take even more for granted than they do the atrocity stories about the history of Bolshevism.

Or suppose the roles had been reversed, and it was Lenin who had advocated the Martov formulation in 1903. Just imagine the speed with which heads would bob knowingly and eyes blink significantly, and how profound would be the conclusions drawn about the sinister character of Bolshevism as far back as the date of its birth! And the whole joke is that there was a reversal, at least on the part of Martov! Wolfe is oblivious to it; but in his history of the Russian Social Democracy Martov reminds us that under the influence of the 1905 revolution, the Mensheviks, at their Petersburg conference in December of that year, "abandoned Paragraph I of the old party statutes [that is, the Martov formula of 1903] which weakened the strict party-character of the organisation in so far as it did not obligate all the members of the party to join definite party organisations." So, about two years after the London debate, the Mensheviks themselves adopted Lenin's definition of party membership and there is no evidence that they ever altered it subsequently. From then on, at least, Lenin's view was never really in dispute. It is only in our time that it is splattered across the pages of anti-Bolshevik literature, with all sorts of dark but always undefined references to its ominous overtones, undertones and implications.

Was it Lenin's intolerance toward difference of opinion within the party, his conception of a party monolithism that allowed only for obedience to a highly-centralised, self-appointed and self-perpetuating leadership, his autocratic determination to have his own way regardless of the consequences, with a penchant for splitting the movement when he did not get his way? These are familiar charges against Bolshevism, and against Lenin in particular. Wolfe might have made an original contribution to these charges by providing some facts to sustain them. Instead he preferred to repeat them, and more than once.

We feel neither the desire nor the need to canonise Lenin as a saint, or to regard his works as sacred texts. He was the greatest revolutionary leader in history, and that is more than enough to assure his place against both detractors and iconographers. If we knew nothing at all about him, it would be safe to assume that he had his faults, personal and political; learning about him only confirms this innocent and not very instructive assumption. He was devoted to the cause of socialist freedom and his devotion was durable and passionate. As an adversary, Paul Axelrod, said, "there is not another man who for twenty-four hours of the day is taken up with the revolution, who has no other thoughts but thoughts of the revolution, and who, even in his sleep, dreams of nothing but revolution." This made him, in the eyes of dilettantes and philistines, let alone defenders of the old order, a fanatic. It was his strength. He was, in consequence, a passionate partisan of the instrument he regarded as indispensable for the revolution, the party, of the sharpness and clarity of its thought. This necessarily brought him into conflict with others, and not only with dabblers but with revolutionists no less devoted to socialism than he. In polemic and in factional struggle generally (neither of which was really invented by Lenin, and which can be avoided only by eschewing politics altogether), he was resolute, self-confident and uncompromising. It is easy to think of worse qualities. But they were qualities that made him incomprehensible or insufferable in the eyes of tergiversators and cobwebheads. If, as was often the case, he exaggerated or overreached himself, it was generally because nobody helped him by inventing a method of carrying on polemical and factional struggle without risk of exaggeration. (Reading Wolfe, for example, shows that such a method has still to be invented. Only, for his exaggerations there is not even that excuse.) But all this about Lenin, and a good deal more, does not begin to prove the "standard" charges against him.

Take splits. Wolfe says that "in the matter of splitting, Lenin was invariably the aggressor". It is a categorical statement — one of the few made by Wolfe who generally prefers indirection. To illustrate how much dehydrated bunk there is in the statement, we can take the famous 1903 party congress which split the Russian Social Democratic Party. There was a furious fight over the above-mentioned Paragraph 1 of the party statutes. Lenin was defeated after a two-day debate. But he did not bolt the congress or the party. Earlier in the sessions, however, the delegates led by Lenin and Martov, Axelrod, Trotsky and Plekhanov, overwhelmingly defeated the position of the Jewish Bund on the question of autonomy. The Bund, refusing to bow to the majority, split from the congress. No sermon from Wolfe on the virtue of unity and the vice of splitting.

Then the congress, Lenin and Martov included, voted against the separate organisation around the "Economist" journal, *Rabocheye Dyelo*. Whereupon, two Economist delegates split from the congress. Still no sermon from Wolfe. Then the congress, by a slender majority but nonetheless a majority, adopted Lenin's motion for an *Iskra* editorial board of Plekhanov, Lenin and Martov, as against the outgoing board which had included old-timers like Axelrod and Zasulich. Whereupon Martov announced his refusal to abide by the decision — to serve on the board — and the split between the now-named Mensheviks (Minority) and Bolsheviks (Majority) became a fact. Conclusion? "In the matter of splitting, Lenin was invariably the aggressor."

Of course Lenin was responsible for a split here and a split there! To deny it would be absurd; to feel apologetic about it, likewise. But it is interesting to see how Wolfe applies different standards in different cases — so sternly moralistic toward the Bolsheviks and so maternally tender toward their opponents. He quotes Lenin as writing that he could not understand why the Bund split from the congress since "it showed itself master of the situation and could have put through many things"; and then observes with haughty severity:

"Since, all his life, Lenin attached a feeling of moral baseness to 'opportunism,' he found it hard to understand that these men of the Bund and *Rabocheye Dyelo* could have firm convictions and principles of their own, and, defeated on them, would not content themselves with 'putting through' what he regarded as opportunistic measures."

Happy Bundists to have so sympathetic an advocate! Lenin found it hard to understand, but he, Wolfe, he understands. After all, if people have firm convictions and principles, they will not, if defeated in their own organisation, consent to forego them just for the sake of unity. They will not and they should not. Better a split than that! All this applies to Bundists, Economists, Mensheviks and other opponents of the Bolsheviks. But not to the Bolsheviks themselves. Even though their principles and convictions were no less firm, they deserve no such affectionate consideration. Why not? Because... because... well, because in the matter of splitting Lenin was invariably the aggressor.

The tale of Lenin's "intolerance" toward opponents inside the party has been told in a dozen languages. In the best of cases (they are rare enough), the record is seen through the completely distorting glasses of the present-day Stalinist regime; in the worst of cases (that is, as a rule), the record is falsified in whole or in part. At least nine times out of ten, Lenin's "intolerance" consisted, for the opponents, in the fact that he refused to accept their point of view on a question.

The phenomenon is familiar to anyone who has been active for any length of time in politics, especially in those working-class movements where politics is not an intellectual pastime but is taken most seriously. A man who puts forward a point of view on some question, but adds that his opponent's view is probably just as good if not better — there is a tolerant man for you. If he says that it really doesn't matter much whether the organisation adopts his view or not — there's a tolerant man. If he is not so impolite as to try vigorously to win supporters for his view and to plan, with his initial supporters, on how to win a majority for it — he is tolerant too. Or if his point of view miraculously wins the support of, let us say, the organisation's convention, and he then announces that he is ready to concede the leadership to his opponents who are against his position and who, with the best will in the world, could not carry out the adopted policy with enthusiasm or understanding — there is a most tolerant man. He is not at all like Lenin, granted. He differs from him in that he does not take his views or his organisation — or himself — very seriously. He is in politics for a week-end, warmed by the sunny thought that after he has returned to his normal pursuits he will have left behind a memory unmarred by the tiniest Leninist stain.

The references generally made to Lenin's "intolerance" are actually calculated to convey the impression that he imposed upon the Bolsheviks a uniquely dictatorial regime in which his word, or at best, the word of his Central Committee, was law that could be questioned only under penalty of the severest punishment. The unforearmed reader tends to think of Lenin's organisation in terms of Stalin's — not quite the same, to be sure, but as an only slightly modified version.

The comparison is utterly monstrous. Up to 1917, the Russian revolutionary movement was an illegal, underground movement, working under the onerous conditions of czarist autocracy. In spite of that, the Bolshevik movement had, on the whole, more genuine democracy in its organisation, more freedom of opinion and expression, a freer and healthier internal life, than at least ninetenths of the other socialist or trade-union organisations of Europe, most of which enjoyed legality and other facilities beyond the dreams of the Russians.

This was true not only of the relations between the Bolsheviks and Mensheviks when they represented only contending factions within a more-or-less united party, but likewise true among the Bolsheviks themselves, first as a faction and, after 1912, as an independent party. The hideous monolithism of Stalin's regime was entirely unknown — it was not even dreamed of — among the Bolsheviks. Political tendencies were formed without let or hindrance, and if they dissolved it was not under compulsion of any kind. The official leading committee always had its central organ — the spokesman of the faction or the party — but time and again periodicals would be issued on their own responsibility by political groupings or tendencies inside the party and even (or rather particularly!) inside the Bolshevik faction (later inside the Bolshevik Party) itself. Even after the Bolsheviks took power, this tradition was so strong and normal and deeply rooted that, in the most perilous period for the new Soviet regime, it was possible for groups of dissident Bolsheviks not only to publish newspapers and reviews of their own independently of the Central Committee but to attack that committee (and of course Lenin!) with the utmost freedom and... impunity.

These separate organs of tendencies or groups or factions discussed all questions of party theory, party policy, party organisation, and party leadership with a fullness, a freedom and an openness that was known to no other working-class organisation of the time and has certainly had no equal since the rise of Stalinism. The idea of "secret" or "internal" discussion of political or theoretical questions of the movement, introduced by Zinoviev and Stalin in the period of the revolution's decline and now considered perfectly good "Bolshevik" practice, alas, even by self- styled Marxist organisations, was simply not known among the Bolsheviks — mind you, among the Bolsheviks even while they were an illegal, police-hounded and police-infiltrated movement! Lenin's collected works, which are composed largely of open "inner-party" polemics and the files of a dozen different factional papers and pamphlets, provide inundating evidence of this rich, free and open party life. In this respect, no other socialist organisation of those days could even equal the Bolsheviks.

Even in its best days, the German Social Democracy did not have anything like so free and democratic an organisational-political life, while it was an outlawed party or afterward in the period of legality. Why, even Marx and Engels sometimes had to fight to get their views published in the German party press and their fight was not uniformly successful. Among the Bolsheviks, such a thing was unheard of, and not just with respect to a Marx or Engels or Lenin, but also to the spokesman of some unpopular grouping in the party or faction.

Read, or re-read, all the anti-Bolshevik histories or commentaries with the closest care, and see what facts are related about how Lenin's "organisational principles" worked out in party practice. You will find all sorts of hints, suggestions, innuendo, clouded allusions, grunts, grimaces, pursed lips, winks and nods; you will find gossip, chit-chat about factional excesses which are "normal" in heated factional fights, titillating tales about the "dubious" sources of Bolshevik funds calculated to shock the sensibilities of our pious business and trade-union circles and of course a lot of plain kiln-dried falsification without filler, shellac or varnish. But it would be astounding if you found even one fact about the regime in the Bolshevik party or fraction that contradicts the record cited here about what the regime actually was. And it is this regime, as it really existed, that is supposed to have led to Stalinism! This is the tradition that is said to have helped Stalinism appear and triumph! Stalinism rests upon it exactly the same way a stiletto rests on the heart it has stabbed.

Or just suppose that, in the search for facts about Lenin and the old Bolshevik movement, Wolfe or any other anti-Bolshevik writer had discovered about them the things that are known about other leaders and other political groupings. For example, in the early *Iskra* days, Plekhanov, in order to assure his domination of the editorial board that was evenly divided between the "old" and the "young," was given two votes as against one for all the other members! If that had happened with Lenin — then or at any other time in his life — can you imagine the pages — no, the chapters — filled with outrage in every line, that would be written to argue that this was the very essence of Bolshevism, the core itself of Leninism, the proof positive and irrefutable of how it was pregnant with Stalinism from the day it was born?

Or take the party of Rosa Luxemburg, who was, writes Wolfe generously and rightly, "the outstanding advocate of revolutionary policy and the outstanding defender of democracy within the labour movement." Yet, she shared the theory of the permanent revolution which, says Wolfe, led to Stalinism; her party was opposed, and not on very democratic grounds, to the Soviets of Workers' Deputies in the revolution of 1905; she and her party were opposed to the democratic slogan of the right of self-determination and on grounds that were, objectively, reactionary; her party (we refer to the Social Democratic Party of Poland and Lithuania) was opposed to the idea of mass, formally non-party trade unions and insisted that the unions must declare their allegiance to the revolutionary party; and in spite of her criticisms of Lenin's "organisational principles," the regime in her own party in Poland was exceptionally factional, narrow, super-centralistically disciplined and far more "bureaucratic" than anything the Bolsheviks were ever guilty of.

The anti-Bolsheviks, who have exactly nothing in common with Luxemburg, ghoulishly drag her into court against Lenin, but if that record were to be found in the history of the Bolsheviks, can you imagine the uproar in twelve languages?

Or take the Narodniks (Populists) for whom Wolfe has such an extravagant reverence. In their early days, these spiritual (and political) ancestors or the Social Revolutionists, convinced but primitive revolutionists, exploited — with the best intentions in the world — the anti-Semitic pogrom feelings of the Russian peasants and even issued leaflets spurring them on. Can you imagine what the anti-Bolshevik professionals would make of such a thing if it could be found in the record of the Bolsheviks or their forebears? Or what they would say if some Bolshevik argued that Kerensky's role in 1917 "flowed from" the anti-Semitic aberrations of the Narodniks four decades earlier?

Such examples could be cited almost indefinitely — but not with reference to Lenin and the Bolsheviks. If they and they alone are the targets today, it is not as a result of objective historical re-examination but because of the frenetic campaign against socialism by a desperate and dying bourgeoisie and by disoriented and disillusioned ex-revolutionists. And by the same token, if we defend the Bolsheviks today it is in the interest of historical objectivity but also because we remain loyal to the emancipating fight for socialism.

Wolfe does deal with two aspects of Lenin's "conception of the party machine" that are indeed of decisive importance. He separates them when they should be connected. Properly connected and focussed, they would throw a most revealing light on Bolshevism, the Russian Revolution, its decline and on the rise and meaning of Stalinism. Right here, perhaps, is Wolfe's most glaring failure. He fumbles the problem helplessly and hopelessly, where he is not utterly oblivious to its significance. You cannot help asking yourself what in heaven's name this man learned about Marxism during his long years in the communist movement — or since.

First, Wolfe finds in Lenin's views on the interrelations between the revolutionary movement, socialist consciousness and the spontaneous struggles of the workers, as he expressed them early in the century, the

"... dogma, obscure as yet in its implications, [that] was at the very core of 'Leninism'. From it flowed an attitude toward the working class, toward its ability to think for itself, to learn from experience, toward its capacities and potentialities for self-rule, toward its 'spontaneous' movements such as might take place without orders and control from the party of socialist theoreticians and professional revolutionaries. From it would spring a special attitude toward trade unions, toward the impromptu strikers' councils or Soviets, even toward two revolutions — in 1905 and the spring of 1917 — that would come not on order but by surprise."

Elsewhere, Wolfe finds something else that makes up "the real core of 'Leninism,' separating him by an abyss from the Mensheviks, and blurring to the vanishing point the dogmatic line which divided him from Trotsky." The "core" is this:

"In short, Lenin's real answer to the question: what happens after we get power? is: Let's take power and then we'll see."

This "core" separated Lenin not only from the Mensheviks but from Marx as well, and Wolfe argues the point with a brevity, if not erudition, which merits full quotation:

"To Marx it might have seemed that 'the forms of the state are rooted in the material conditions of life', that 'the economic structure of society. .. independent of men's will... determines the general character of the social, political and spiritual processes', and that 'no social order ever disappears before all the productive forces for which there are room in it have been developed'. But to Lenin's political-power-centred mind, for all his Marxist orthodoxy, such formulae were intolerable fetters unless subject to the proper exegesis. And the exegesis literally turned Marx on his head until the Marxist view that 'in the last analysis economics determines politics' became the Leninist view that, with enough determination, power itself, naked political power, might succeed wholly in determining economics."

Wolfe has more to say about these two points, but very little more.

Lenin's ideas about socialist consciousness and the struggle of the working class were not invented by him nor were they uniquely his own. They are nothing less than the intellectual underpinnings of any genuinely socialist party, and it is inconceivable without them. In an even deeper sense they underlie the very conception of a rationally-ordered socialist society. No one developed these ideas more sharply and profoundly, even if with polemical vehemence, than Lenin, and that was his special contribution. But the ideas themselves go back to the beginnings of the scientific socialist movement, back to Marx and Engels. A serious examination of Lenin could not have failed to establish this fact and draw conclusions that it indicates Wolfe cannot help but know that Lenin's views were an almost literal copy of those expressed earlier, just as the century turned, by Karl Kautsky. And his present-day venerators would be horrified to hear that, by virtue of what he wrote at that time, he was the fountainhead of what was inevitably to become Stalinism! Kautsky, before Lenin, wrote:

"Many of our revisionist critics believe that Marx asserted that economic development and the class struggle create, not only the conditions for socialist production, but also, and directly, the *consciousness* of its necessity... In this connection socialist consciousness is presented as a necessary and direct result of the proletarian class struggle. But this is absolutely untrue.

"Of course, socialism, as a theory, has its roots in a modern economic relationship in the same way as the class struggle of the proletariat has, and in the same way as the latter emerges from the struggle against the capitalist-created poverty and misery of the masses. But socialism and the class struggle arise side by side and not one out of the other; each arises out of different premises.

"Modern socialist consciousness can arise only on the basis of profound scientific knowledge. Indeed, modern economic science is as much a condition for socialist production as, say, modern technology, and the proletariat can create neither the one nor the other, no matter how much it may desire to do so; both arise out of the modern social process. The vehicles of science are not the proletariat, but the *bourgeois intelligentsia.* It was out of the heads of members of this stratum that modern socialism originated, and it was they who communicated it to the more intellectually-developed proletarians who, in their turn, introduce it into the proletarian class struggle where conditions allow that to be done.

"Thus, socialist consciousness is something introduced into the proletarian class struggle from without, and not something that arose within it spontaneously. Accordingly, the old [Austrian] Hainfeld program quite rightly stated that the task of Social Democracy is to imbue the proletariat with the *consciousness* of its position and the consciousness of its tasks. There would be no need for this if consciousness emerged from the class struggle." (Kautsky's emphasis.)

To this should be added: neither would there then be any need for a distinct, separate political movement of socialism — a socialist party — except, perhaps, to fulfil the not very useful function of passive reflector of the welter of ideological and political confusion that, to one extent or another, will always exist in the working class, at least so long as it is a class deprived of social power and therewith of the means of wiping out its own inferior position in society. It is kept in this inferior position under capitalism by force but only in the last analysis, only at times of crisis. As a rule, be it under democratic or even under fascist capitalism, the ruling class maintains or seeks to maintain itself by ideological means.

The whole of capitalism's "headfixing industry", as one Marxist wittily called it, is directed toward keeping the working class in ignorance or confusion about its social position, or rather about the purely capitalist reasons for its position, toward concealing from the working class the emancipating historical mission it has and the road it must travel to perform it.

So long as the workers do not acquire an understanding of their social position and their social task, their battles against the ruling class, be they ever so militant or massive, can only modify the conditions of their economic subjugation but not abolish them. Indispensable to their abolition is the socialist consciousness (an exact mathematical formulation of which is neither possible nor necessary) of the working class, which means nothing more and nothing less than its realisation of its position in society today, of its power, and of its obligation and its ability to reconstruct society socialistically.

Now the dispute over the ideas of Kautsky-Lenin on the subject boils down to this: either the working class, organised in its elementary trade union organisations or not, acquires this consciousness by spontaneous generation in the course of repeated struggles for the improvement of its conditions — or in its decisive section, it acquires it, in the course of these struggles, to be sure, with the aid of those who already possess this socialist consciousness and who are banded together (in a group, a league, a movement, a party — call it what you will) in order more effectively to transmit it, by word of mouth and by the printed page, to those whose minds are still cluttered up with bourgeois rubbish, that is, the products of the "head-fixing industry."

Between these two, there is not a single person today who calls himself a socialist of any kind who would venture to defend, flatly and frontally, the former conception. All you get from the anti-Bolsheviks is, as in Wolfe's case, murky reference to the "special attitude" that flowed from Lenin's formulation of the position, in which the only thing definite is a sneer at the very conception of a socialist party — the "socialist theoreticians and professional revolutionists." The reformists who distinguish themselves from Lenin by saying that while they too are for a socialist party, they look upon it as a "servant" of the working class and not as its "master" or "dictator"; as a means of the "socialist education" of the working class in whose "ability to think for itself" they devoutly believe and not for the purpose of "ordering and controlling" it from above — are either hypocritical or inane. Their daily practice, inside the labour movement and in politics generally, would indicate that it is less the latter than the former.

The question of socialist consciousness which Lenin developed has wider implications. Wolfe sees in it only the source for establishing a new slavery for the working class, the Stalinist tyranny in the name of the "dictatorship of the proletariat". The truth is not merely different, but in this case it is the exact opposite!

Workers' democracy and, indeed, that complete realisation of democracy which inaugurates the socialist society, are not only inseparable from Lenin's ideas on socialist consciousness but, without them, become empty words, un-attainable hopes, illusions at worst.

What was the obvious meaning of Lenin's insistence that the specific role of the socialist movement was to "introduce" a socialist consciousness into the working class? What, for example, was the clear implication of Lenin's "Aside from the influence of the Social Democracy, there is no conscious activity of the workers," which Wolfe quotes as a sample of the "dogma [which] was at the very core of 'Leninism'"... and from which "flowed an attitude toward the working class"? It should be obvious.

The "party of socialist theoreticians and professional revolutionaries" was not assigned thereby to trick the incurably blind and incurably stupid workers into lifting it to power so that it might establish a new kind of dictatorship over them. That makes no sense whatsoever. It was assigned the job of making the workers aware of the fundamental reasons for their exploited and subjected position under capitalism; of making the workers aware of their own class strength and having them rely only upon their class strength and independence; of assembling them in a revolutionary party of their own; of making them aware of their ability to free themselves from all class rule by setting up their own government as the bridge to socialist freedom.

Without a socialist consciousness, there would be working-class activity but the workers would continue to remain the ruled and never become the free. For the workers to rule themselves required conscious activity toward socialism.

What is Wolfe trying to convey with his suggestive prose? That Lenin dwelled so emphatically upon the need for the party to instil socialist consciousness or stimulate it in the working class because he did not believe in "its ability to think for itself, to learn from experience"? Or because he was sceptical about "its capacities and potentialities for self-rule"'? Did Lenin expect to imbue the unable-to-think-and-lead proletariat with socialist conceptions by intravenous hypodermic injections? Or is Wolfe just a little... careless with his innuendoes?

Let us go further. Lenin knew — he referred to it often enough and nowadays it is especially necessary to emphasise and elaborate it — one of the most basic and decisive differences between the bourgeois revolution and the socialist revolution. One of the outstanding characteristics of the former was that it could be carried through without a clear ideology, without an unequivocally-formulated consciousness on the part of the bourgeoisie whose social system it was to establish. In fact, not only could it be carried out in this way, but generally speaking that is how it was carried out.

The greatest bourgeois revolution, the French, was carried out by plebeians, without the bourgeoisie and in part against it; and it was consolidated by Napoleon, in part without the bourgeoisie and in part against it. In Germany it was carried out, that is, the supremacy of capitalism over feudalism was assured, in the Bismarckian or Junker-landlord way — again, in part without the bourgeoisie and in very large part against it. The passage from feudalism to capitalism in Japan is only another example of the same phenomenon. Yet, in all these and other cases, including those where the bourgeoisie was not raised to political power, the bourgeois revolution was nevertheless effected, consolidated, guaranteed. Why? As Lenin once wrote, in 1918:

"One of the main differences between the bourgeois and the socialist revolution consists in this, that for the bourgeois revolution which grows up out of feudalism the new economic organisations, which continually transform feudal society on all sides, gradually take form within the womb of the old society The bourgeois revolution faced only one task: to throw off and destroy all the fetters of the former society. Every bourgeois revolution that fulfils this task, fulfils everything that is demanded of it: it strengthens the growth of capitalism."

But if the bourgeois fetters upon production are thrown off and destroyed, that alone does not and cannot assure the growth of socialist production. Under capitalism, production is assured by the irrepressible tendency toward accumulation of capital which is dictated primarily, not by the will of the capitalist, but by the blindly-operating market as the automatic regulator of capitalist production. Socialist production is incompatible with market relations.

It is production for use and therefore planned production, not automatically regulated by a blind force. Given a certain level of development of the productive force available, everything then depends upon planning, that is, upon the conscious organisation of production and distribution by human beings.

Now, under capitalism, what and how much is produced is determined by the market, and the distribution of what is produced is determined basically by the relations between the class that owns the means of production and exchange and the class that is divorced from them. Overturn capitalism, and it is found that there is no market to determine what is produced and in what quantities, and there is no class that owns private property.

Until the distant day when all classes are completely abolished and socialism fully established, the conditions of production and distribution must necessarily be determined by politically-associated human beings — no longer by the blind market but by the state.

In other words, where the state becomes the repository of all the means of production and is in complete control of them, economy is for the first time subject to planned and conscious control by those who have the state in their hands. In this sense, politics determines economics! This may sound startling to Wolfe, as well as to all sorts of half-baked half-Marxists But if this simple and irrefutable fact is not understood, then the whole idea of the working class taking power in order to organise a socialist society becomes absurd and even meaningless. In revolution, but above all and most decisively in the socialist revolution, the relationship between economics and politics is not only reversed, turned upside-down, but it must be reversed!

But if politics now determines economics (again, within the limits of the given productive forces), or to put it differently, if the conditions of production and distribution are now determined by politically conscious individuals or groups, the question of the nature of the determining politics is immediately thrown open. What assurance is there that the politics will be socialist in nature, so that production relations are socialist or socialistic (by which is meant socialist in tendency or direction) and that distribution corresponds to them, so that what is produced is for the use of the people and not of a small privileged group?

To rely for that on the good will, the honourable intentions or the socialist past or professions of faith of a group of planners who hold the state power to the exclusion of the rest of the people, is naive, where it is not reactionary. In any case, it is not a socialist idea and certainly not Lenin's. A socialist development of the economy can be assured only by those who are to be its principal beneficiaries, the working class, and only if it has the power to make the decisions on production and distribution and to carry them out, hence only if it holds the power of the state. For politics now determines economics! And it cannot acquire this power or wield it unless it is permeated by a socialist consciousness, which means, among other things, an understanding of the decisive role it has to play in the new state, and therefore and only by that means, the role it has to play in assuring a socialist direction to the operation of the economy.

That is why Lenin, in distinguishing between bourgeois and socialist revolutions, underlined the fact that the Bolshevik revolution "found at hand" not socialist economic relations that had developed under capitalism as capitalist economic relations had developed under feudalism, but rather a democratic political factor: "victory depended solely upon whether already finished organisational forms of the movement were at hand that embraced millions. This finished form was the Soviets."

The same thought was in his mind when he urged that every cook should become an administrator, so that with everyone exercising the power of "bureaucrat" no one would be a bureaucrat. And the thought was even more pregnantly expressed in his famous saying that "Soviets plus electrification equal socialism." (It is impossible even to imagine Lenin saying that a totalitarian prison for the workers plus nationalised property equals a degenerated workers' state!)

The Soviets, before the Bolsheviks took power, were acclaimed by every Menshevik and Social Revolutionist as the "revolutionary democracy." That was right. What is more, the Soviets were a magnificent example of a spontaneous movement of the workers and peasants themselves, not set up by order of any party or according to its plan.

Wolfe finds that from Lenin's "dogma" about socialist consciousness "flowed" an attitude toward the working class which was uncommendable because, it would seem, it was most undemocratic and even contemptuous toward the working class, including "its 'spontaneous' movements such as might take place without orders and control from the party..." Like the Soviets of 1917, for example? Then how explain that every party in Russia, except the Bolsheviks, fought to keep the Soviets (the "revolutionary democracy") from taking over all power, and worked to keep them as a more or less decorative appendage to the never-elected but self-constituted Kerensky regime?

True to Lenin's "dogma", the Bolsheviks alone strove to imbue the Soviets with a genuinely socialist consciousness, which meant concretely that the workers (and even the peasants), more democratically and representatively organised in the Soviets than ever before or ever since in any other movement in any country of the world, should take command of the nation and therewith of their own destiny.

This example of what really was the "attitude" of Lenin and his party toward the "spontaneous" movements of the workers, their ability to think and learn for themselves, and their capacities and potentialities for self rule — not in some thesis or polemical article or speech, but in one of the most crucial periods of history is so outstanding, so overshadowing, so illuminating about Lenin's "conceptions" that Wolfe passes it by. We will not ask what this historian would have said about Lenin's "dogma" if the Bolshevik attitude toward the "revolutionary democracy" in 1917 had been the same as, let us say, that of Kerensky. But we wonder what he will say in succeeding volumes about the Menshevik and SR "attitude" toward the Soviets and the "dogma" from which it "flowed".

The revolution of 1917 was the decisive test for all political parties and groups. In spite of conservative trends in the ranks (all parties tend toward conservatism about some of their "dogmas"), Lenin showed that he had been able to build and hold together a party which proved, in this most critical hour, to be the only consistent champion of revolutionary democracy and revolutionary socialism, and the only "political machine" ready and able to lead both to victory. This is what brought Trotsky to the side of the Bolsheviks and caused him to "forget" his "Cassandra-like prevision" about how "the dictator will take the place of the Central Committee" and the party itself.

If Wolfe finds that Trotsky's prediction was "fatefully fulfilled by history", it is primarily because of his method of separating the history of the conflict of social forces from specific political events, or worse, of simply ignoring the former. The fact is that whatever grounds there may have been or seemed to have been in 1903-04 for Trotsky to utter his warning, the main tendency of the development of Lenin's group or party, particularly from 1905 onwards, was in an entirely opposite one from that feared by Trotsky.

The apparatus did not replace the party, nor the Central Committee the apparatus, nor the dictator (Lenin!) the Central Committee. The inner-party democracy and freedom of opinion and discussion of the Bolsheviks as an illegal movement, it is worth repeating, can be matched, without apology, against the regime of virtually every other working-class organisation, legal or illegal, that ever existed.

Here, too, the decisive test was 1917 itself. At least, you would think so, on the basis of almost universal experience in such matters. A working-class movement which is suffering from a fatal disease — opportunism, let us say, or bureaucratism — does not usually reveal it, not clearly, at any rate, in normal periods, in periods of social calm or political decay. It shows it, and most disastrously for itself and its followers, in the most critical and troubled periods of society, above all in the crisis of war and the crisis of revolution. But precisely in the critical period of 1917, the Bolshevik party passed the test, and so well that Trotsky found it possible to abandon his early apprehensions about it.

Now, why didn't Lenin's conception of organisation, which was one of the "roots of Stalinism ", manifest itself in 1917 in a way that would cause the Bolshevik party to play a conservative or reactionary role in the revolution, to be a brake upon the workers and peasants? The question is of first-rate interest. Therefore, Wolfe passes it by.

Did the Bolshevik party measure up to its task early in 1917? Of course not! But that was not because Trotsky's prophecy about Lenin's conception of organisation had been fulfilled, fatefully or otherwise. It was an entirely different prophecy of Trotsky's that was fulfilled or almost. Years earlier, Trotsky had written that the Bolshevik formula of "democratic dictatorship of the proletariat and peasantry" had its revolutionary side, as opposed to the Menshevik conception of a revolution in which it would he the role of the proletariat to bring the bourgeoisie to power. But, he added, if the Bolsheviks persisted in this formula, the coming revolution would reveal its reactionary side, that is, that which inhibited the proletariat from carrying the democratic revolution through to proletarian power and the inauguration of socialist measures.

Steeped in Lenin's old formula, most of the party leaders in 1917 adopted a position which paralysed the revolutionary possibilities of the party. It took a further fight by Lenin, after his arrival in Russia in April, to effect that "rearmament" of the party which finally assured the victory in October. But, this "prophecy" of Trotsky's — or rather, Lenin's rearming of the party in the direction of Trotsky's theory — is regarded by Wolfe as one of the sources of Stalinism!

Important is the fact that Lenin did not replace the Central Committee by a dictator in any sense indicated by Trotsky. He enjoyed, justly, immense authority among the Bolsheviks, but he had won it and kept it to the end of his life by his intellectual ability and character as a leader and not by any dirty manipulation or usurpation.

In 1917, most of the party leadership opposed his famous "April Theses." He was not only unable to dictate to the others, but did not dream of it. He won them over, one by one, partly by the pressure of the party ranks whom he convinced and partly by convincing the leaders as well. In 1917, or before, when his point of view won, it was not because the dictator had replaced the Central Committee; and when his point of view lost, as was more than once the case, it was not because the apparatus had replaced the party.

Yet, the Bolshevik party did degenerate; Soviet democracy was replaced by a unique Bonapartist dictatorship. But the process did not conform with Trotsky's prediction, which Wolfe transforms into an abstraction raised to the nth power. Reading Wolfe, you would think that the Bolshevik party was a sort of supra-mundane evolving out of some purely internal mechanism, unaffected by the strains and influences exerted by terrestrial forces.

It is only necessary to read what the Bolsheviks said and wrote in the period of the revolutionary upsurge to see what their real attitude was toward Soviet and socialist democracy, what ideas of working-class self rule they sought with all their strength to instil into the Russian people. The bureaucracy rose not because of these ideas, but in spite of them.

The revolution was soon plunged into a fierce civil war, and if it had not been for the Bolsheviks, including their "machine", the Soviet power would not have lasted 48 hours, to be replaced, in all likelihood, not by bourgeois democrats but by the czarist reaction which Anglo-French imperialism was sponsoring.

Civil war, unfortunately, is not the ideal culture for the growth of the democratic bacillus. The days of War Communism were harsh and stringent. At the front and at home, command inevitably took the place of free discussion and voting. The tendency to bureaucratic command gripped and held not only Bolshevik leaders, but rank-and-file militants, Bolshevik and non-party, as well.

Even so, Soviet democracy could have been restored after the civil war if the accursed backwardness of Russia had been overcome rapidly by the aid which a successful revolution in the advanced West could have contributed on a grand scale. It could have been maintained if, to start with, the Menshevik and SR parties had allied themselves with the "revolutionary democracy" in the civil war and not with the monarchist reaction.

Russian Populists of the old days once exclaimed: "Never will history forgive the autocracy for making terrorists out of us." With far more justice the Bolsheviks might have declared: "Never will history forgive the Mensheviks and Social Revolutionists for joining the war against the Soviets and forcing us to substitute our party for the Soviets."

Soviet democracy might have been restored by another road, the re-democratisation of the Bolshevik Party itself. And here it is interesting to note that the big fight for party democracy was launched by an outstanding section of the Old Bolsheviks who rallied to Trotsky's position; in fact, by the time Zinoviev broke with Stalin and joined the Trotskyists, it can be said that the bulk of the militants who had been most thoroughly trained in the old school of Bolshevism and in Lenin's "conception of organisation" lined up against the Stalinist bureaucracy, which was represented primarily by comparatively recent members or by obscure personages who had never played an important part in the life of the party.

Well or badly, consistently or not, the old Bolshevik cadres resisted the rise of the new Stalinist bureaucracy. If they failed, it was not due to the overpowering force of Lenin's organisational principles, but to an overpowering force of a radically different kind.

In passing, Wolfe writes:

"Nineteen five and nineteen seventeen, the heroic years when the machine was unable to contain the flood of overflowing life, would bring Trotsky to the fore as the flaming tribune of the people, would show Lenin's ability to rise above the confining structure of his dogmas, and would relegate Stalin, the machine-man, to the background. But no people can live forever at fever heat and when that day was over and Lenin was dead the devoted machine-man's day would come."[2]

Just in passing! But these two sentences contain more insight than can be found in any two chapters of Wolfe's book. Revolutions are periods of turbulence precisely because the people are so free to choose their course and their

2. Shachtman's note: Then why the title *Three Who Made a Revolution*? Up to now, only Stalinist forgers have presented Stalin as one of those who outstandingly led the revolution. The facts presented by Wolfe show this to be a falsification and the above quotation confirms it. The title he gives his book is therefore utterly misleading. It would of course be very awkward to load a book with a title like Two Who Made a Revolution and One who Made a Counter-Revolution, but one merit it would have: it would be accurate.

Leon Trotsky during the period of the post-Revolution civil war

leaders for themselves and so hard to control by a machine. Wolfe merely sets down the two deeply significant sentences and then goes on as though they were no more than a chance collection of words.

He seems to shy away from matters and statements of social importance spontaneously, without special effort, as if by instinct. But the sentences are important regardless of Wolfe. When the masses were free to choose democratically in the revolutions of 1905 and 1917, Trotsky and Lenin were lifted to power. (Their names can be used here as symbolic of Bolshevism as it really was.) And it is only when the masses were exhausted or apathetic or prostrate, that is, when revolution was succeeded by reaction, that the Stalinist counterrevolution could triumph over the masses and over the Bolshevik party.

There is the "core of Stalinism" indeed! The Stalinist bureaucracy did not grow out of an organic evolution of the Bolshevik party, as was implied by Trotsky's "prophecy". Its growth paralleled and required the destruction of that party. And its destruction, root and trunk and leaves and branch, until absolutely nothing is left of it today except the plagiarised name. This fact, too, is of such capital importance that the anti Bolshevik writers pass it by. Destroyed: the principles of Bolshevism, its program, its tradition, its history, its personnel down almost to the last man, including (how significant this is!) even those Bolsheviks who tried to capitulate to Stalinism, and yes, including even the big bulk of the original Stalinist faction of the old party! Preserved: the name of the party and a few renegades from the second and tenth ranks of the old Bolshevik party — that and nothing more.

The destruction of the Bolshevik party meant the destruction of socialist consciousness. The measure of the growth of the Bolshevik party was the growth of this consciousness among the workers it influenced; and in turn it grew among the workers to the extent that the party remained attached to the ideas which Lenin most conspicuously advocated. It is of tremendous interest that for the Stalinist faction to extend its initial victory inside the party apparatus (that's where its first victory occurred) to a victory inside the party generally, it had to flood the party.

The first big public step, so to speak, taken by the Stalinist bureaucracy was the notorious Lenin Levy organised right after Lenin's death. Hundreds of thousands of workers were almost liberally poured into the party. Who were they? Generally speaking, the more conservative workers and employees, people who had not shown any interest in joining the party in the tough days of the revolution and civil war but who could, in 1924-25, be persuaded to join it now that its power seemed consolidated, now that membership seemed to guarantee employment, privileges, a career.

Almost to a man they could be counted on by the bureaucracy in the fight against the Opposition, against the Bolsheviks, their principles, their revolutionary and socialist and democratic traditions. It was Stalin's first and least important step in literally dissolving Lenin's "machine" in order to substitute a despotic police regime that was utterly alien to it. This first step was typical of those that followed.

There is as much justification, then, for the theory that Stalinism was rooted in the Bolshevism which it extirpated, as there is, for example, in the kindred theory that the socialist movement, its methods and its theories in general form the roots of the fascist movement and its methods and theories.

The anti-Bolshevik democrat would feel outraged at seeing the latter argument put forward. He would declare indignantly that to explode such nonsense, nothing more is needed than the fact that Hitlerism crushed the socialist organisations, imprisoned or killed their leaders, outlawed their ideas, and so on and so forth. Yet the argument that Hitlerism had its authentic roots in the German Social Democratic Party is advanced in all coolness by so eminent an anti-socialist as Frederick von Hayek, and with the same reasoning, with the same analogies, with the same cavalier attitude toward decisive facts as is displayed by those who argue that Stalinism is rooted in Bolshevism.

Hayek is a defender of the capitalist status-quo-ante-state intervention and a sworn foe of socialism, and he has his means of discrediting its good name. The aim of the democratic or reformist anti-Bolsheviks is somewhat loftier, as it were, but the means they employ to discredit Bolshevism are in no essential different from Hayek's.

On the flyleaf of his book, Wolfe quotes, for his motto, the noble words of Albert Mathiez:

"The historian has a duty both to himself and to his readers. He has to a certain extent the cure of souls. He is accountable for the reputation of the mighty dead whom he conjures up and portrays. If he makes a mistakes, if he repeats slanders on those who are blameless or holds up profligates or schemers to admiration, he not only commits an evil action; he poisons and misleads the public mind."

Mathiez devoted much of his great work to defending the great French Revolution and its Jacobins from detractors. The socialist today has the duty to defend the great Russian Revolution and its Jacobins in much the same spirit. As to how faithfully Wolfe has heeded the injunction of Mathiez, the reader of his book will judge for himself.

'Stand up for the defence of Petrograd' by A Apsit (1919).

The working class and socialism

We consider ourselves as heirs of the Trotskyist movement when it was a living movement in the full sense of the word, when it represented the imperishable tradition of revolutionary Marxism. And today [1953], 25 years after the founding [in 1928] of that movement, looking backward with a minimum of maudlin sentimentality and a maximum of calm, objective and reasoned analysis — what do we celebrate on this 25th anniversary?

What do we seek to represent in the working class movement as a whole, of which we are an inseparable part? What fundamentally justifies our independent and separate existence, our stiff-necked obduracy in maintaining that existence, in refusing to give up in insisting not only that we will hold on to what we have but get more and more until our ideas infuse the bloodstream of the whole working class movement?

It is the essence of revolutionary Marxism — that respect in which it always differed, as it differs today, from every other social and political tendency, from every other movement, from every other mode of thought in society.

And that essence can be summed up in these four words: Marxism is proletarian socialism.

They say — by "they" I mean professors former professors, aspirant professors — that there are as many schools of socialism as there are socialists. Every Princeton student bursts his seams when he hears this: "There are other socialisms, and which of the 57 varieties are you referring to?"

I, who like a joke as well as the next man, would be the last man in the world to dream of depraving these poor, intellectually poverty-stricken apologists for a decaying capitalist social order of their little joke. And you will admit it is little.

So I will say: Yes, historically and actually — if it will make you happy, and after all we socialists are for the extension of happiness — there are 57 and even a greater number of socialisms.

When Marx came on the intellectual scene, in Germany, in France, in Belgium and in England, there were any number of socialisms; and there were socialisms before Marx was born; and there were socialisms promulgated after he died. Marx mentioned a few in his deathless Communist Manifesto. There were the "True Socialists", the Christian socialists, the reformer socialists, co-operative socialists, the reformer socialists, cooperative socialists, bourgeois socialists, feudal socialists, agrarian socialists, royal and imperial Prussian socialists. They existed and continue to exist. In our time we had "National-Socialists"; we have had if I may say so "Stalinist socialism". Stalinist socialism — I don't like to say that, but we do have all sorts of "socialists".

But even if it gives the professors and the Vassar students another burst seam, I say there is one socialism that we adhere to. Even if we will not say that this is the "true" socialism, that it is the "right" socialism, that it is the "genuine" socialism — we will say that it is our socialism.

If you don't find it "true" you can become a royal and imperial Prussian socialist, you can become a Stalinist "socialist", you can become (every man is entitled to his joke) a "Sidney Hook socialist". For we believe in everybody having the right to be any kind of socialist, or anti-socialist, he wants. We claim no more for our socialism, than the fact that it is ours.

Marxian socialism is distinguished from all the others, not in the fact that it holds to the so-called labour theory of value, and not even in the fact that it developed the ideas of dialectical materialism, and not even in the fact that it participates in and prosecutes the class struggle. Its fundamental and irreconcilable difference with all the others is this: Marxism is proletarian socialism.

The great discovery of Marxism — what distinguished it as a new socialism in its day, what distinguished the great discovery of Karl Marx in his search for a "bearer of philosophy" as he used to say in his early days, in his search for a "carrier" out of the contradictions of capitalism — the great discovery of Marxism was the revolutionary character of the modern proletariat.

That is the essence, that is the durable characteristic, of Marxian socialism. Proletarian socialism, scientific socialism as distinct from all other socialist schools, from utopian socialism, dates from that great discovery — the social revolutionary character of the modern proletariat.

When speaking of socialism and socialist revolution we seek "no condescending saviours" as our great battle hymn, the International, so ably says. We do not believe that well-wishing reforms — and there are well-wishing reformers — will solve the problems of society, let alone bring socialism.

We are distinguished from them all in this one respect above all others — we believe that task belongs to the proletariat, only the proletariat itself. That is a world-shattering idea. It overshadows all social thought.

The most profound, important and lasting thought in Marxism, the most pregnant thought in Marxism, is contained in Marx's phrase that the emancipation of the proletariat is the task of the proletariat itself. It is clearly the most revolutionary idea ever conceived, if you understand it in all of its great implications.

That is why we are in the tradition of the Paris Commune, for example, the first great attempt of the proletariat to emancipate itself. That is why we are in the tradition of the great revolution in Russia — the Bolshevik revolution — the second great attempt of the proletariat to emancipate itself. That's why we defend it from its detractors. That's why we are so passionate about it. That's why we are, if you will, so "dogmatic". We know what we are defending even if they do not always know what they are attacking.

And that is what we learn all over again from Trotskyists what we have begun to forget, what we have begun to ignore, what we have begun to take for granted.

If I may speak for myself, I can tell you I will never forget the explosion in my Communist smugness when for the first time I read Trotsky's criticism of the draft programme of the Comintern, written when he had already been banished to Alma-Ata in 1928, written for the Sixth World Congress of the Communist International. What a commentary it is on the Communist movement in 1928 that, so far back, that precious Marxian document, which is so fresh to this hour, had to be written in exile in Russian in 1928 — in exile! It had to be transmitted by theft; Cannon had to steal his copy in Moscow from the Comintern secretariat and smuggle it into the US. It had to be disseminated here in the Communist Party illicitly, to three or four people who would read it behind locked doors — because if the leaders of the CP found out that we had it (let alone that we were reading it, let alone that we were favourably influenced by it) they would put us on trial and expel us, and they did.

To read that work and to know what was really going on in that fight of Trotskyism, that it was always a question of international socialism versus national socialism, the coordinative efforts to bring about socialism of the entire working class of the world as against the messianic, nationalistic utopian idea that it could be established in one country alone by the efforts of a benevolent bureaucracy of the working class — that had a shattering effect upon our thinking.

We learned then from Trotskyism what we hold so firmly to now: There can be no socialism without the working class of the world, no socialism without the working class of Russia. Twenty-five years later we see the results of building socialism without the international working class — without the Russian working class and against the Russian working class. No matter how many books you leaf through, no matter how old they are, where will you find the story of such an unendurable tyranny as has been established in the Stalinist countries, where "socialism" has been built without the working class and against the working class?

We are the living carriers and embodiment of the ideas to be learned from these events. We are its living teachers, for those whom we can get to listen in these days of darkness, confusion and cowardice.

In this country we have learned far more about the meaning of the idea of an American labour party, a labour party based on the trade unions, than we ever dreamed was represented by that idea when we first put it forward in 1922 in the American Communist movement, than when we put it forward again and again later in the Trotskyist movement. To us it represents a declaration of independence of the working class, its first great step in the country toward self-emancipation, and also to us it represents the remedy for that series of tragedies, calamities, misunderstandings and frustrations represented by New Dealism — that is, collaboration of the working class with a benevolent liberal bourgeoisie.

And what it represents runs through everything we say and everything we do and everything we want others to do in the United States and elsewhere: Not with them — not under them — you yourselves are the masters not only of your own fate but the masters of the fate of all society if you but take control of society into your own hands! That is your destiny! That is the hope of us all.

We are optimistic because that will remain our hope in the greatest hours of adversity, while everywhere else lies pessimism. Our role is to teach Marxism, that Marxism which is proletarian socialism, Marxist politics, socialist politics.

Marx Shachtman (left) and James P Cannon in the early years of US Trotskyism.

Our idea of politics boil down to this revolutionary idea — to teach the working class to rely upon itself, upon its own organisation, upon its own programme, upon its own leadership. Upon its own ideas and need for democracy, and to subordinate itself at any time to the interest, the needs, the leadership, the programme, the movement, the organisation, or the ideas of any other class.

We regret that in other branches of the socialist movement or what is called the socialist movement, that idea does not dominate every thought. We are proud that in our section of the socialist movement it does dominate every thought and every deed. That's why we are Marxists; that's what we learned all over again in many intellectual and political battles under that peerless teacher and peerless revolutionary Trotsky.

And we start by teaching socialists to rely upon themselves.

When we read for the first time *The New Course* by Trotsky, his work directed against the first big and dangerous manifestations of bureaucratism in the Russian Soviet state, another explosion took place in our smugness. I venture to call it — it's an awkward phrase and I hope it's not too badly misunderstood — a bible of working-class democracy. This was Trotsky's brilliant simple overwhelming pamphlet on how a socialist movement should act inside and outside, how a socialist state should act, how socialist leaders and socialist ranks, the socialist elders and the socialist youth, should act toward themselves and one another...

What we have learned more sturdily than every before, what is more completely a part of our Marxian idea of proletarian socialism, is that there is no socialism and no progress to socialism without the working class, without the working class revolution, without the working class in power, without the working class having been lifted to "political supremacy" (as Marx called it) to their "victory of democracy" (as Marx also calls it). No socialism and no advance to socialism without it! That is our rock. That is what we build the fight for the socialist future on. That is what we're unshakably committed to.

Look at what has happened — I hold them up as horrible examples — to all and singly who have renounced this struggle after having known its meaning. They have no confidence in the social-revolutionary power of the proletariat — that is the alpha and omega of them all. One will embroider it with one colour thread and another with another, but at bottom that is it.

I claim to know whereof I speak because I know so many of them and know them so intimately — excuse me, knew them so intimately and know also what caused their renunciation of the struggle. They have been corrupted by that most ancient of corrupt ideas: that as for the lower class, there must always be one; that the lower class must always be exploited and oppressed; that there is not other way. That's their real feeling and that's what caused their renunciation of the struggle.

They are the Stalinists in reverse. They have lost their faith in the socialist faith for that reason and for that reason primarily and fundamentally.

They have lost their respect for the working class because for so long a period of time it can, and it has, and it does, lie dormant and stagnant and seems to be absolutely passive, immobilised in permanence. In other words, they have doomed it — this working class which has shown itself so capable of so many miracles in the past hundred and two hundred years of its struggle against the bourgeoisie and against oppression in general — doomed it to eternal servitude. That's why they are not Trotskyists; that's why they're not socialists; that's why they're not democrats; that's why they're not people with human integrity any longer.

Ask any of them point-blank (if you're on sufficiently good terms with them): do you believe that the working class can ever rule society and usher in a classless socialist regime? Do you believe that the working class has that capacity innate within it? Not one of them, if he is honest, will admit agreeing with it. You will notice everyone of them beginning to hedge and to hem and to haw and to talk about 25 other subjects — because in all of them the corrupt idea has taken sound and firm roots that the working class will always be oppressed and exploited by someone or another.

Look at Burnham and his "Machiavellians" — the whole theory is there, the whole snobbish bourgeois theory that goes back to feudalism and goes back to slavery before that: there have to be exploited workers and the best they can hope for is that the rulers fight among themselves and that in the interstices of this fight they may be able to promote their own interests just a little bit without ever changing their exploited status.

What is this at bottom but a variety of that notorious philosophy which the Stalinoid intellectuals and apologists used to whisper to us in justification of their support of the Kremlin: "You don't mean to say that you really believe that the working class can emancipate themselves, can themselves take power? ... They need a strong hand over them ..."

These people can't absorb the idea that the workers can free themselves. Take that diluted variety of these sceptics, the pro-war socialists (if you can call them socialists):

We would be for a Third Camp you see, if it existed. Show us a Third Camp and we would be the first ones to be for it — if it were big and powerful and had lots of dues-paying members. But there is no Third Camp now, so why be for it? But the minute it comes into being — we don't believe that it will ever happen, of course, but if despite our scepticism it should come into being against capitalism (which were are not really for) and against Stalinism (which we detest) — we will support it with all the power of speech and pen at our command. But until then allow us to be the snobs and careerists that we are.

Those who swoon with delight at being accepted nowadays in respectable society (of which, alas, we are not a part) have lost all respect for themselves — that's what it is with the cynics, with the somewhat milder version, the sceptics, the climbers, the turncoats and the veterans who never saw combat in the class struggle and who nevertheless have the effrontery to live off pensions from the bourgeoisie today in various institutions reserved for them exclusively.

For us who have nothing in common with such people and want nothing in common with such people, in all their 57 schools, the 25th anniversary comes after a quarter century of defeats and setbacks, yes, but defeats and setbacks accumulated only because men and movements left the working class in the lurch.

But although it is silent so often, and silent for so longer, and although it is disoriented, this proletariat — today's proletariat, or tomorrow's like yesterday's — will outlast this trial as it will outlast its old leaders and resume its iron march to socialist freedom. Our confidence in it, maintained these 25 years, is undiminished 25 years after we took up the banner of renewed faith in it and renewed willingness to learn from it, as well as to teach it what we know.

For the man who lives for himself, alone like a clod of mud in a ditch, like a solitary animal in a savage forest, 25 years of dedication to socialism is an incomprehensible as it is unendurable. But we are, thank god, not like the clods of mud, the careerists and the opportunists, the philistines of all sorts and varieties who have specially strong fountains of strength in this last trench of world capitalism, the United States. We are people who have been intellectually and spiritually emancipated by the great philosophical and cultural revolution in thought that Marx began and Trotsky so richly expanded. We are the fortunate ones who are not resigned and know that they need no resign ourselves, to the inevitability of advancing barbarism, to the decay and disintegration of society.

We know with scientific sureness that no reaction — not matter how strong at the moment, no matter how prolonged — can destroy that social force whose very conditions of existence force it into a revolutionary struggle against the conditions of its existence, the proletariat.

We know with scientific sureness that no matter how dark and powerful reaction may be at any given time, it not only generates but regenerates its gravedigger — that same proletariat, the only social force which class society has endowed with infinite capacity for recuperation from temporary defeat.

And we know with scientific sureness that the achievement of the fullest development of democracy which is socialism, is in safe hands when entrusted to the proletariat and in safe hands only when it is in its charge, for it alone must have democracy for its existence and it alone can realise it in full by its irrepressible aspiration for socialism and its unceasing fight for it.

For the man to whom the debasement and oppression of others is a mortal offence to himself, who cannot live as a free man while others are unfree, who understands that without resisting the decay of society there is no life worth living — for him the informed struggle against exploitation and social iniquity is the blood-stream of life. It is indispensable to the self-realisation of humanity and therefore to the attainment of his own dignity. It is the mark of his respect for his fellow man, of his yearning to gain the respect of others, and therewith to assure his respect for himself.

For such men, and we count ourselves as such, these turbulent 25 years are a long episode that has given richer and stouter meaning to the moral life of all who passed through it with their loyalties unimpaired, and it is in this life, the life of freedom, that the founder of our contemporary movement Leon Trotsky was a startling example.

It is to the grand vindication of this life that lies ahead that we renew our bond tonight — the oldest and noblest bond in history, the bond that will be redeemed only on the day when the last chain has been struck from the body and mind of man, so that he may walk for the first time among his equals erect.

The "mistakes" of the Bolsheviks

The causes for the decay of the Russian Revolution are often sought in the "mistakes of the Bolsheviks."

If only they had not suppressed freedom of speech and press! If only they had not suppressed the freedom of political organisation and all the non-Bolshevik parties! If only they had not established a one-party dictatorship! If only they had not set up a Communist International to split the Western European labor movement! Had they acted otherwise, we would have no fascism today, and no Stalinism, but instead a progressive development toward democratic socialism inside Russia and out.

That is the tenor of most of the criticism leveled at the Bolsheviks in the labor movement. Consistently thought out, they boil down to the idea that the real mistake was made in November 1917 when the Bolsheviks took power. This judgement is based essentially on the same factors that generated the fundamental theory of the Stalinist counter-revolution — "socialism in one country" — and differs from it only in that it is not on so high a level.

Class relations in Russia

The bonds by which Czarism held together the Russian Empire were brittle in the extreme. Under the stress of so minor a struggle as the Russo-Japanese war of 1904 and the revolutionary rehearsal a year later, the bonds almost shattered. Twelve years later, under the much heavier stress and pounding of the World War, they exploded beyond repair and tore Czarism to bits like the shot from shrapnel.

With Czarist despotism gone as an integrating force, who was left to keep the nation together and maintain it as a power, economic as well as political? One or two hundred years earlier in similar circumstances, it was the bourgeoisie. In one country after another, it united the nation on a new basis, eliminated or repressed the disintegrative forces, expanded the wealth and power of the country, and vouchsafed democracy to the masses in one measure or another. In Russia, however, the bourgeoisie had come too late. The solving of the problems of the democratic revolution had been too long postponed to permit a repetition of the French Revolution. This was the theory held in common by Lenin and Trotsky.

The period of the revolution in which Czarism was overturned tested the theory to the end. The bourgeoisie did come to power, but it was quite incapable of mastering the centrifugal tendencies which Czarism, in the comparatively peaceful days, had been able to hold in precarious check. The empire was falling apart. Be it in the person of Lvov, or Miliukov, or Kerensky, or Kornilov, the bourgeoisie made desperate, violent, but vain efforts to keep the subjected peripheral countries like Finland, the Ukraine, Poland, the Caucasus, inside the old empire with a new nameplate. It is unbelievable, but it is a fact, cried Lenin, that a peasant uprising is growing in a peasant country, "under a revolutionary republican government that is supported by the parties of the Social Revolutionists and the Mensheviks". The peasant rising did not come to strengthen the bourgeoisie and its pallid democracy, but was directed against it. The bourgeoisie was unable to deal with it in any better way than the Czar had discovered. At the same time, a proletarian power, the Soviets, not at all Bolshevik, grew up spontaneously by the side of the bourgeois power and threatened its existence.

The bourgeois democracy was incapable of seriously approaching a single one of the social and political problems at home. Given the collapse of Czarism, all the long-standing, outer-Russian imperialist tendencies to reduce Russia to a colony — tendencies most vigorously represented by the Germans, but not exclusive with them — received free rein. The country ruled by the bourgeois republicans was about to be overrun by foreign imperialism as a prelude to its partition among the great powers. This problem, too, the "revolutionary democracy" was unable to solve, or even undertake seriously to solve. The country faced complete economic ruin, political disintegration, chaos, dismemberment and subjugation from abroad, the imminent triumph of counter-revolution and reaction, with all the consequences flowing from them. The bourgeoisie, the bourgeois democracy, was impotent in dealing with the situation, notwithstanding the support it received from the Mensheviks and Social Revolutionists.

The "interference" of the Bolsheviks

To say that they might have solved these problems democratically if the Bolsheviks had not interfered is not only to ignore an overwhelming mass of facts, but to stand the question on its head. The "interference" of the Bolsheviks was made possible only because the bourgeois democrats, plus the social democrats, could not solve the problems.

Political action can be understood, not in the abstract, but in the concrete conditions in which it occurs. It cannot be rationally appraised by itself, but only in terms of the alternative. The alternative to the "risky" seizure of power by the working class under Bolshevik leadership was not the painless flowering of "democracy" but the triumph of savage counter-revolution and the partitioning and colonialisation of the country.

The "mistake" of the Bolsheviks in taking power when they did and where they did, not only saved the honor of international socialism and gave it a new and powerful lease on life, but it saved Russia. Without this "initial mistake", the greatest likelihood is that long ago German imperialism would have been ensconced in Petrograd and Moscow, French imperialism in Odessa, British imperialism in the Caucasus, Japanese imperialism throughout Siberia, Kerensky in a clerk's chair, with the Mensheviks running errands for them all.

The Bolsheviks cannot, and therefore must not, be judged as if they were uncontested masters of a situation in which they could calmly and undisturbedly plan a campaign of social reorganisation. The disdainful critics like to overlook the fact that they, or at least their friends and patrons, left no stone unturned

Cover of Communist International, no. 1, May 1919.
Design by B M Kustodiev.

or unhurled to prevent the new state power from working out its destiny. Class interest came before "scientific interest" in the "new social experiment."

Both Czar and bourgeoisie left the Bolsheviks, who took power almost without shedding a drop of blood, a heritage of chaos and violence and multitudinous unsolved problems.

The sabotage of the bourgeoisie, loyal patriots of the fatherland who were ready to sell it to foreign imperialism rather than have it ruled by the proletariat, forced the Bolsheviks to resort to the most radical socialist measures from the very beginning. The Bolsheviks were anything but Utopian. Their program was modest and realistic. If they took what would otherwise have been premature steps, it was done under the compulsions of the bitter class struggle immediately launched by the counter-revolution.

Decrees permitting capitalists to continue owning their factories under workers' control are impotent against shells loaded and fired at these factories by their departed owners. Terroristic attacks upon the government and its officials cannot be effectively met with sermons on the superiority of oral agitation and moral suasion. Freedom of the press cannot be extended by a government to "critics" who come to overthrow it with arms and battalions furnished by Czarists and foreign imperialists. Freedom must be defended from such critics, and with all available arms. Not only the bourgeois democrats like Kerensky, but the Mensheviks and SRs resorted to arms against the democratic Soviet power. Nor were they too finicky about the company they kept in their crusade against the Bolsheviks. Alliance with the Bolsheviks against their reaction was inadmissible in principle and beneath the integrity of these democrats. Alliance with reaction, with the Czarist generals, the Cossacks, the Clemenceaus and Churchills against the Bolsheviks, that was good, practical politics, realistic, tolerable by democracy.

In any country, such "practical politics" are commonly known as treason and treated accordingly. Against the Soviet power, this was not merely "treason to the nation," but treason to the working class and the working-class revolution. Those who tolerated the traitors, who even collaborated with them in a common party, who did not join the Bolsheviks in crushing them, were not much better. The Soviet power had no alternative but to outlaw these elements and their political institutions. This can be contested only by those who ignore facts — we say nothing of the class interests of the proletariat, of the interests of socialism! — including the fact that civil war is not conducted in accordance with the rules recommended in finishing schools for young ladies of good breeding.

What is downright outrageous is the impudence of the criticism of Bolshevism's dictatorial measures levelled by the very persons or groups which acted in such a manner as to leave the Soviet power no alternative but stern decisions of sheer self-defence.

The place of the Comintern

This holds true also for the organisation of the Communist International. The picture of Lenin as some sort of wild and irresponsible "professional splitter" is three-fifths myth and two-fifths abysmal misunderstanding. The social democracy during the war had led the working class into the cattle corrals of the bourgeoisie. The Communist International was organised to restore the class independence of the proletarian movement, of which it had been robbed by the leaders of the Second International. It was organised to unite the proletariat once more around a revolutionary socialist banner, to have it serve itself again, instead of serving the Kaiser, the French Steel Trust or the British Empire.

Above all, however, it was organised as an indispensable weapon of the Russian Revolution itself. The Comintern was the general staff of the world revolution. Its task was the organisation of the victory of the proletariat in the capitalist countries. This was assigned to it by the Bolsheviks, not out of considerations of abstract internationalism, but out of the thousand-times-repeated conviction that without the revolution in the West, the Russian workers' state could not hope to survive, much less solve its fundamental problems.

This fact is well known and widely acknowledged. Its full significance is not always grasped. The Russian Revolution was the first act of world revolution. That is how it was conceived by its authors. That was the starting point of all their policies. The heart of the question of the "mistakes of the Bolsheviks" is reached when this is thoroughly understood. Everything remains mystery and confusion if the question is studied from the standpoint of Stalin's nationalist theory.

The program of the Bolsheviks called for establishing the widest possible democracy. The Soviet regime was to be the most democratic known in history. If a state power, that is, coercion and dictatorship, was needed, it was to be directed only against a counter-revolutionary bourgeoisie.

Was so much concentration of dictatorial power and violence needed against the Russian bourgeoisie, that is, against a bourgeoisie described as helpless and hopeless? On a Russian national scale, the answer could easily have been in the negative. But as the world bourgeoisie understood, and immediately showed, the Russian Revolution was directed at international capitalism. Without world capitalism, the Russian bourgeoisie could have been disposed of by the Soviet power with a wave of the hand. With world capitalism behind it, the bourgeoisie of Russia, which is only another way of saying the danger of a victory for the counter-revolution, was a tremendous force against which the greatest vigilance was demanded.

Because the problem was only posed in Russia but could be solved only on a world scale, the Bolsheviks counted on the international revolution. Because they counted on the international revolution, the Bolsheviks allowed themselves all sorts of infringements upon the standards of political democracy, and even upon the standards of workers' democracy.

The suppression of democratic rights for other working-class organisations, even of those which were not directly engaged in armed insurrection against the Soviets, was first conceived as a temporary measure dictated by the isolation of the Russian Revolution and in virtue of that fact by the dangers to which it was immediately subject.

The victory of the revolution in the West would have meant a vast relaxation of suppressive measures. To this day the best of the Russian Mensheviks (if there are any left who have not gone over to Stalin) do not understand that the primary responsibility for their disfranchisement in Russia (and, more important, the degeneration of the revolution) falls upon the shoulders of their German co-thinkers, who so effectively prevented the German proletariat from coming to power.

In other words, if the October Revolution is looked at as a purely Russian revolution; if the world revolution on which the Bolsheviks reckoned is looked upon as a Utopia doomed in advance to failure; or if the world revolution is looked upon as a movement that should have been suppressed, as was done by the reaction and its social-democratic supporters; or if the world revolution is looked upon from the standpoint of the Stalinist theory of nationalist reaction — then the dictatorial and suppressive acts of the Bolsheviks (the Bolsheviks, not the Stalinists) become a series of mistakes and even crimes. If, however, these acts are regarded as measures imposed upon the Bolsheviks seeking to hold out at all costs while the world revolution was maturing — the world revolution on which they had every right to count — then their true nature is revealed. They are then understandable, not as something "inherent" in Bolshevism, as that which "unites" Bolshevism with Stalinism (or fascism!), or as that which produced the degeneration of the revolution; but as temporary measures aimed at overcoming the effects of an enforced isolation and superfluous to the extent that this isolation was relieved by socialist victory abroad.

Democracy in coming revolutions

However, if this is so, an important conclusion follows. The proletariat that triumphs in the next wave of socialist revolutions and triumphs in several of the advanced countries will have neither wish nor need to repeat all the measures of the Russian Revolution. It is absurd to think otherwise. It is much more absurd for the revolutionary movement to adopt a program advocating the universal repetition of all the suppressive measures of the Russian Bolsheviks. This injunction applies most particularly against the idea of a single, legal, monopolistic party, or as it is sometimes (and inaccurately) put, a "one-party dictatorship".

The workers' power in the advanced countries will be able to assure the widest genuine democracy to all working-class parties and organisations, and (given favourable circumstances, which mean, primarily, no attempt at counter-revolution) to bourgeois parties, and this assurance must be set down in advance. The assurance cannot be confined to a ceremonial pledge on holiday occasions, but must be reflected in the daily political practice of the revolutionary vanguard party. In the concrete case, the "daily practice" includes a critical re-examination of the Russian experience.

There were "mistakes" imposed upon the Bolsheviks by the actions of their opponents and by conditions in general. There were mistakes, without skeptical quotation marks, that cannot be sheltered under that heading. The most critical and objective reconsideration of the Bolshevik revolution does not, in our view, justify the attacks made upon Lenin and Trotsky for the violence they used against their violent, insurrectionary adversaries. Nor, even after all these years, can the excesses in repression and violence be regarded as having been weighty factors in the degeneration of the Soviet state. To condemn a revolution for excesses is to condemn revolution; to condemn revolution is to doom society to stagnation and retrogression.

But after having been compelled to overthrow all the non-Bolshevik parties, the leaders of the party in power made a virtue, and then a principle, out of a temporary necessity. "There is room for all kinds of parties in Russia", said one of them, Tomsky, if we rightly recall, "but only one of them in power and all the rest in prison". Tomsky merely expressed what had become the rule and principle for the other leaders.

The idea of one party in power is one thing, and not at all in violation of either bourgeois or workers' democracy. The idea that all other parties must be, not in opposition, with the rights of oppositions, but in prison, violates both bourgeois and workers' democracy, and it is with the latter that we are concerned here.

Even if every non-Bolshevik group, without exception, had resorted to armed struggle against the Soviet power, it was a disastrous mistake to outlaw them in perpetuity. From every point of view that may legitimately be held by a revolutionary party or a revolutionary government, it would have been wise and correct if the Soviet power had declared:

"Any political group or party that lays down its arms, breaks from the foreign imperialists and the counter-revolutionary bourgeoisie at home, adapts itself in word and deed to Soviet legality, repudiates armed struggle against the government and those who resort to armed struggle, will enjoy all democratic rights in the country, equal to those of the party in power."

The Bolsheviks made no such declaration. Instead, the kind made by Tomsky gained prevalence. There can be no question in our mind that the adoption and enforcement of the "Tomsky policy" contributed heavily to the degeneration of the revolution and the victory of Stalinism. From the prohibition of all parties but the Bolshevik, only a step was needed to the prohibition of all factions inside the Bolshevik Party at its tenth congress. Anyone acquainted with the history of the subsequent developments knows that this decision, also taken as an "emergency" measure, was a most powerful weapon in the hands of the bureaucracy against the Left Opposition. Disloyally construed, disloyally used, it smoothed the road to the totalitarian dictatorship of the bureaucracy.

The whole Bolshevik Party was politically miseducated and ideologically intimidated against the very idea of more than one party in the country, and for this miseducation none of its leaders can escape his share of responsibility. It is enough to recall that from the time of Zinoviev's first capitulation to Stalin in 1927 to the time of the last of the capitulators, every desertion from the Opposition was motivated to a considerable extent by the cry, "No two parties in the country!"

Proletarian revolution and democracy

The Bolshevik revolution was betrayed and crushed by the Stalinist counter-revolution. It is not right to say that nothing remains of the revolution. Much remains: its great tradition is still alive in millions of men; its ideas and teachings remain fundamentally sound for the much greater socialist revolution to come; its experiences are still before us and so are the lessons to be learned from them.

Not the least important lesson is the need to return to the principles set forth by Lenin in *The State and Revolution*. Especially in the light of what has happened the heaviest emphasis must be laid upon the dictatorship of the proletariat as the democratic rule of the workers; as the widest and most genuine democracy the workers have ever had; as the equitable enjoyment of democratic rights by small groups, political opponents of the government included, and military opponents alone excluded; as the safeguard of the principle of electivity of officials, above all of the trade unions and the soviets.

The revolutionary Marxists must learn, and then must teach, that the struggle for democratic rights is not just a clever device for embarrassing the undemocratic bourgeoisie, that the struggle is not confined to the days of capitalism. On the contrary: it is precisely when the new revolutionary power is set up that the struggle for democratic rights and democracy acquires its fullest meaning and its first opportunity for complete realisation. The revolutionists after the overturn of capitalism differ from the revolutionists before that overturn not in that they no longer need democratic rights and no longer demand them, but in the fact that they are for the first time really and fully able to promulgate them and to see to it that they are preserved from all infringement, including infringement by the new state or by bureaucrats in it. The right of free speech, press and assembly, the right to organise and the right to strike are not less necessary under the dictatorship of the proletariat, but more necessary and more possible.

Socialism can and will be attained only by the fullest realisation of democracy. The dictatorship of the proletariat must be counterposed to the dictator-ship of the bourgeoisie in this sphere because the latter denies the people access to and control over the very material bases whose monopoly by the bour-geoisie makes its "democracy" a formality not really enjoyed by the great masses.

That is what the revolutionary Marxists should teach. But first of all they must learn it, and thoroughly. It is one of the most important lessons of the Russian Revolution and its decay.

The Red Soldier at the Front is without Footwear and Clothing. By D S Moor (1920).

Bolshevism and gossip

My Life as a Rebel by Angelica Balabanoff

The memoirs of Angelica Balabanoff make up a sad book. Not because, like so many of her contemporaries in the radical movement, she has lost interest in the struggle or her socialist convictions, for she ends her recollections with a staunch re-affirmation of her ideals.

"My belief in the necessity for the social changes advocated by that [international labour] movement and for the realisation of its ideals has never been more complete than it is now when victory seems so remote... The experience of over forty years has only intensified my socialist convictions, and if I had my life to live over again, I would dedicate it to the same objective."

It is a sad book because it reveals that for all the passionately revolutionary spirit that animated her in four decades of activity in the working class movement, she did not succeed in mastering the simple lesson that Lenin tried to teach her friend Serrati at the Second Congress of the Comintern in 1920:

"Comrade Serrati said: we have not yet invented a sincerometer — this is a new French word which means an instrument for measuring sincerity — such an instrument has not yet been invented. We do not even need such an instrument, but we already do have an instrument for judging tendencies. It is a mistake on comrade Serrati's part — I should like to speak about this later — that he did not apply this long familiar instrument".

It is such a failure — or inability? — to replace subjective judgments, based on trifling personal incidents, by political judgments, that brought her own political life of the last two decades particularly to such a tragically futile conclusion. The same failure results in a deep discoloration of the pages of her memoirs.

Rebellious daughter of a wealthy and reactionary Russian family, she left her native Ukraine for Western Europe to take up studies which soon led her into active participation in the socialist movement. Moved by a genuine compassion for the exploited and oppressed, and a powerful spirit of indignation at all iniquity, she became, after joining the pre-war Italian Socialist Party, one of its most stirring and popular agitators. Her teachers, friends and associates were the Old Guard of Italian socialism — Antonio Labriola, Turati, Treves, Modigliani, Lazzari, and later, Serrati and Mussolini. Of the last-named, then a neurotic bitter young exile in Switzerland, she became the patron, nursing him politically into leading positions in the party, until he became a member of the Central Committee and editor of the official organ, *Avanti!* Her pictures of the later Duce, of his timorousness and braggadocio, his characterlessness and inspired mediocrity, are savage and telling.

Master of several languages — she was a talented translator at international assemblies — and associate of the internationalist left wing which, with Mussolini as its spokesman, effected the expulsion from the party of the patriots

Angelica Balabanoff

in the period of the Tripolitan war of 1912, she became, when the World War broke out, a central figure in the movement to reconstruct the collapsed Second International. The Zimmerwald and Kienthal anti-war conferences of the internationalist socialists — she was secretary of the Zimmerwald International Socialist Commission from its inception — were to a large extent due to her persevering work. She joined the Bolsheviks on the eve of the revolution and in 1919 was chosen by them as first secretary of the Communist International. She broke with the Comintern, and sided with the Serrati wing of the Italian Socialist Party, when the latter refused to adopt the famous 21 points and to break with the reformists of the Turati-Treves-Modigliani group. For the last 15 or more years, she has been the leader of that tiny fraction of Italian socialism which embraces all that is left of the once mighty "Maximalist" group, which its real leader, Serrati, abandoned before his death to rejoin the Communist International.

Her break with Lenin and the Bolsheviks — her estimate of them forms, next to her evaluation of Mussolini, the bitterest part of her volume of portraits and judgments — was inevitable; and the reason why remains, to her, uncomprehended down to the present day. There lies the nub of the tragedy of her political life and of her book. The torrential sweep of the Russian revolution sucked her into the Bolshevik party, but only for a brief period of time. Looking back upon it in 1938, and sitting in lofty judgment on the Bolsheviks, she explains the collapse of the Third International by the moral leprosy of Zinoviev — symbol of all that to her was inherently vicious in Bolshevism; the degeneration reflected in the recent trials "was developed under Zinoviev himself" and the frame-ups and confessions "were implicit in the development of the Bolshevik method, the Leninist strategy, since the Revolution ... the Bolshevik leaders were capable of anything to achieve their own political and factional ends ..."

Both the analysis and verdict have already served as the pathetic theme for reviews of the book in the petty bourgeois press, in which the dastardly immorality of the Bolsheviks is sanctimoniously and invidiously contrasted with what Professor Douglas calls the "fundamental sincerity" of Balabanoff, "one who believes that the means as well as the ends of economic action are important".

The explanation lies, however, elsewhere. Balabanoff's book is astoundingly devoid of political characterisations; it is filled with pictures of good men and bad men, honest men and crooks, blunderers and seers; and after the narration of all her experiences in various groups and movements, Balabanoff terminates her book without informing the reader of what are her specific political program and her political associations. Yet, while she does not apply political criteria to herself, it does not follow that such criteria are not applicable to her.

In international socialist politics, Balabanoff never was a communist but rather a representative of that wing of Menshevism led by Julius Martov. Its chief characteristic was a strong literary radicalism, which sometimes went so far as to bring it into peripheral touch with Lenin's thoroughgoing Marxism, but which rarely went so far as application in political life. The leaders of radical centrism could characterise the right wing with no lesser accuracy than did the Bolsheviks, but unlike the latter, who took seriously the proletarian revolution and the politics and methods leading to it, they could not bring themselves to a radical suspension of collaboration with the right wing. That is why even the most radical of Mensheviks, Martov, could "agree 95 percent" with the Bolsheviks, yet tax them with being "professional splitters", and devote 95 percent of his blows at Lenin and 5 percent at the right wing with which he scarcely agreed at all.

This is the reason — Balabanoff is not Martov, to be sure, but she suffers from the same political malady — why she could not remain in the Comintern, and not the intrigues, real or alleged, of Zinoviev. It is also the reason why her memoirs, even where they deal with personalities — and they deal with little else — are, with all respect to Professor Douglas' talk about "fundamental sincerity", hopelessly one-sided, splotched and distorted beyond balance and proportion. All the Bolsheviks are limned with splashes of black, shading off into blotches of mud; the social democrats, as a rule, are painted in nostalgic pastels.

Knowing his notorious weaknesses, one cannot be the advocate of Zinoviev; yet, throughout the early period of the Russian Revolution and the Comintern, he was the man, next to Lenin and Trotsky, who restored revolutionary Marxism to its rightful place in the world labour movement and who helped train up a whole generation — not excluding Balabanoff, for a time! — in its principles and traditions. Yet he emerges from her memoirs only as "the most despicable individual I have ever met".

On the other hand, however, Filippo Turati, leader of the Italian right wing, whose socialism Benedetto Croce aptly characterised as that of a "democrat à la Lombard", and who, by his politics, was more responsible than any other man in the movement for the paralysis of the Italian working class which made possible Mussolini's triumph, is very gently defended by Balabanoff.

"His approach was often misinterpreted in other countries because it was so typically Italian [!]. Many Italian intellectuals like to appear sceptical of theoretical axioms even if they are not ... Thus it was that Turati came to be considered [!] a theoretical sceptic and even [!!] an opportunist."

The author's approach, at any rate, cannot be misinterpreted...

Her description of events suffers also, and to such an extent, from her biased "approach", that stories calculated to be of telling significance about Bolshevik depravity end by having a significance only for evaluating her memoirs. In telling of the slowness of the Moscow courier in bringing her reports to Stockholm, where she was Bolshevik propagandist in 1917, she quotes a letter from Lenin:

"Dear Comrade: The work you are doing is of the utmost importance and I implore you to go on with it. We look to you for our most effective support. Do not consider the cost. Spend millions, tens of millions, if necessary. There is plenty of money at our disposal. I understand from your letters that some of the couriers do not deliver our papers on time. Please send me their names. These saboteurs shall be shot."

One gentleman-reviewer has already expressed his outraged horrification at this bloodthirsty despot who so lightly shot couriers merely for delaying with their dispatches. But surely Balabanoff is quoting from memory, and when it is borne in mind that she quoted quite a different letter eleven years ago, the conclusion is reached that — how shall we say it? — a political slant can play distressing tricks with one's memory. For in the German edition of her memoirs, *Erinnerungen und Erlebnisse* (Berlin 1927), the same incident is more innocently reported with the following letter from Lenin, also in quotation marks:

"Bravo, bravo! Your work, dear comrade, deserves the highest recognition. Please do not spare any means. That the material is furnished you in such an insufficient manner, is inexcusable. Please give me the name of the courier who is guilty of such gross, inexcusable negligence."

If letters are quoted from memory, it is still customary to omit quotation marks; and if they are quoted from a text at hand, not even the elapse of a decade permits a writer to quote it so differently on a second occasion as to include sentences about "tens of millions" (no trifle that!) and the summary execution of negligent but innocent couriers (also no trifle!) Otherwise, the author runs the risk not only of shocking the sensibilities of bourgeois reviewers, but also of arousing the feeling that her political objectives have, nolens volens, superseded her political objectivity, to say nothing of objectivity of the ordinary kind; and this is a failing which, we learn from the author, is the specific characteristic of the immoral Bolsheviks.

The feeling is deepened by other, and just as typical, discrepancies between the author's memoirs of 1938 and those of 1927.

According to the present edition, the Stockholm conference of the Zimmerwalders adopted a manifesto calling for a general strike in support of the Russian Revolution, an appeal which was not to be made public until endorsed by the constituent parties of the Allied countries. Radek, however, typical Bolshevik, began to insist that Balabanoff publish the appeal forthwith; "our mutual and unanimous understanding, our pledges and promises, and my own enormous responsibility meant nothing to Radek, and throughout the month of October he bombarded me with protests and demands". Meanwhile, Luise Zietz, representative of the German Independents, came to Stockholm to "prevent the premature publication of the manifesto in view of the precarious position of her party".

"Torn between the threatened extermination of left-wing socialism in Germany and the demands of those who spoke in the name of the Russian Revolution, I was utterly miserable, but I felt that there was only one course to pursue — to keep my pledge and obey the unanimous mandate of the Zimmerwald Commission. Shortly after I had given Radek my final decision the manifesto was published in the Finnish paper controlled by the Bolsheviks..."

Fortunately for Radek and the Bolsheviks, their moral turpitude does not stand out so heinously if we go by... the 1927 version of what happened. According to it, Radek did indeed insist upon the publication of the appeal in view of the terribly urgent situation in Russia and did threaten to publish it on his own responsibility. Zietz did indeed appeal, in a telegram to Balabanoff and then at a meeting of the Zimmerwald Commission, against its publication. But, we read in 1927, "however weighty were the reasons which Luise Zietz adduced, it was impossible for us to accept them, for political reasons on the one side and formal ones on the other. Disappointed and perhaps also enraged against me personally, comrade Zietz left Stockholm... At the same time I received a letter that comrade Ledebour [leader of the Independents] was not in agreement with the mission that Luise Zietz had undertaken in Stockholm in the name of the Independent Party."

Immediately thereafter, however, came the report that the Soviets had taken power. Balabanoff decided to make the appeal public telegraphically.

"All the obstacles that had stood in the way of the publication of the Zimmerwald manifesto only a few hours ago, had now fallen away with the great historical deed... That my collaborators in the International Socialist Commission would share my standpoint, of that I had no doubt; hardly had morning come than I telephoned them and obtained their complete consent."

And because the only member of the enlarged Commission who opposed immediate publication was Rakovsky — not Balabanoff! —

"I must say that my personal relations to Rakovsky from that time on were no longer as friendly and spontaneous as before."

Not less difficult to reconcile are the two distinctly different versions of the story that Trotsky complained about the difficulties put in the way of his return to Russia from the United States in 1917. In the English edition, we learn that on his arrival in Russia, Trotsky was particularly bitter because the Bolsheviks had tried to prevent or delay his return out of factional considerations, which would be just like the Bolsheviks, wouldn't it? "His interpretation seemed to me rather implausible then, but after my own later experiences with the Bolsheviks, I was not so sure of this."

In the 1927 version, however, Trotsky was, it is true, just as bitter; but his feelings were directed then at Robert Grimm, the Swiss social democrat, to whom Trotsky had turned with the request to have the Swiss government agree to let him pass through on his way home; and because it failed to grant Trotsky permission, "Trotsky hinted that good will was lacking on the part of Grimm or others". In 1928, Grimm receives the pardon of silence.

Numerous similar examples could be cited from the two conflicting sets of memoirs, and all of them pose the question of why an intervening decade of recollections sharpens so severely all judgments of Bolsheviks and moderates so charitably all judgments of social democrats, whose perfectly putrid role during and after the war — the period of the author's greatest activity — must surely have left a deep impression upon her. The answer is the one given by the late Henri Guilbeaux, who know her throughout the Zimmerwald period in Switzerland:

"Even though she flattered herself at being above the battle of the revolutionary Russian factions, she had a very clear point of view. Belonging to none of the factions, on the pretext of working for the restoration of unity, she was a Menshevik with all her soul."

It is, alas, this political reality that shapes and colours — more simply, misshapes and discolours — her memoirs, a tendency which is only strengthened by her literary collaboration with a duet of Mensheviks of California cultivation. But whatever its effects on the historical value of the book in relation to the points raised above — and the effects are disastrous — its value as an example of the "dialectical interdependence" of politics, morals, and the powers and tricks of memory, is not to be denied.

'In order to have more you must produce more', by A Zelinsky (1920).

October was a true working class revolution

The Independent Socialist League does not subscribe to any doctrine called Leninism. It does not have an official position on the subject and I am pretty certain that nobody could get the League to commit itself officially on a term which has been so varyingly and conflictingly defined as to make discussion of it more often semantic than ideological or political.

To me, and surely to most of our comrades, Leninism is a question primarily of historical importance in our time. Most often what is in people's minds is the Russian Revolution and democracy as the road and aim of socialism. In our view the Russian revolution has long ago been crushed. What is the fundamental and urgent political question is the relation between democracy and socialism. These questions concern socialists today and I want to outline my views on them.

We regard the Russian revolution of 1917, which Lenin led, as a socialist revolution that established a genuine workers' government. I have always defended this proposition and so have our comrades. You yourself have often in the past taken a similar view. I think it worth while here to note the fact that four years after the revolution, Morris Hillquit, a pretty severe critic of the Bolsheviks, wrote these interesting words:

"It is pretty idle cavilling to dispute the Socialist character of the Russian revolution... The Russian revolution has taken possession of the government in the name of the workers. It has effectively expropriated capitalist owners and nationalised the greater part of the industries. It has also written into its program the socialisation of the land. Measured by all practical tests it is therefore a Socialist revolution in character as well as intent. If it has not come as a result of the course of historic and economic development outlined by Marx, it has occurred through the working of another set of social conditions and forces, which have proved potent enough to create and maintain it. Its continued existence, year after year, in the face of almost incredible domestic difficulties and embittered foreign attacks, prove that we are not dealing with a mere freakish episode, but with a monumental historic event. This will remain true even if the Soviet government should not prove able to maintain itself indefinitely and should yield to another and substantially different form of government."

Now, I believe that the Soviet government finally yielded to "another and substantially different form of government" under the rise and consolidation of Stalin's power. I believe it to be as different as counter-revolution is from revolution, as different as the destruction of socialism is from the movement toward socialism. When you say that one grew out of the other "by natural processes," I would agree with that if it means "as a result of objective material forces." To that, I believe it important to add that Stalinism based itself to a considerable extent upon some of the ideas and institutions defended by Lenin. These it exploited or distorted to serve its own totalitarian and anti-socialist ends. Plainly, they were put forth originally in the desperate, groping attempt to get out of the blind alley formed around the revolution by the walls of the terrible backwardness of the country and the isolation of the republic.

I have in mind, most particularly, the decision of the Tenth Bolshevik Congress to prohibit factions inside the party, which played an enormous role in facilitating the rise of totalitarianism; and the point of view which became a principle defended by the Bolshevik leaders that all parties must be outlawed and kept outlawed. I must say that I unthinkingly accepted this proposition for years in the Communist and Trotskyist movements. But the grim realities of Stalinism forced a reconsideration of many questions. This one was not the least important. Fourteen years ago, I tried to re-examine this vital question, and I hope you will bear with a quotation from the article of 1943:

"The idea of one party in power is one thing, and not at all in violation of either bourgeois or workers' democracy. The idea that all other parties must be, not in opposition, with the rights of oppositions, but in prison, violates both bourgeois and workers' democracy, and it is with the latter that we are concerned here. Even if every non-Bolshevik group, without exception, had resorted to armed struggle against the Soviet power, it was a disastrous mistake to outlaw them in perpetuity ...

"The whole Bolshevik party was politically miseducated and ideologically intimidated against the very idea of more than one party in the country, and for this miseducation none of its leaders can escape his share of the responsibility ...

"The revolutionary Marxists must learn, and then must teach, that the struggle for democratic rights is not just a clever device for embarrassing the undemocratic bourgeoisie, that the struggle is not confined to the days of capitalism. On the contrary, it is precisely when the new revolutionary power is set up that the struggle for democratic rights and democracy acquires its fullest meaning and its first opportunity for complete realisation.

"The revolutionists after the overturn of capitalism differ from revolutionists before that overturn not in that they no longer demand them, but in the fact that they are for the first time really and fully able to promulgate them and to see to it that they are preserved from all infringement, including infringement by the new state or the bureaucrats in it. The right of free speech, press and assembly, the right to organise and the right to strike, are not less necessary under the dictatorship of the proletariat, but more necessary and more possible.

"Socialism can and will be attained by only the fullest realisation of democracy ... That is what the revolutionary Marxists should teach. But first of all they must learn it, and thoroughly. It is one of the most important lessons of the Russian revolution and its decay."

In the past fourteen years, I have expressed these views with increasing insistence and emphasis. I consider them today to be of fundamental importance to the coexistence and cooperation of all socialists whatever other matters they may differ on. It is from this socialist standpoint that I want to fight against the Stalinist regime, the Communist movement, their supporters, defenders and apologists. I am completely agreed that the regime is not just a "mistaken form of socialism" or any kind of socialism, but its betrayal and negation. And as you know, for years I defended the view that far from being some kind of socialism, the Russian regime represents a new form of totalitarian exploitation dominated by a new ruling class.

But I cannot see the political wisdom, or the factual foundation, for considering such an anti-socialist regime as the logical, inevitable and authentic continuation of a socialist revolution. This in precisely the main claim to socialist justification and legitimacy made by the Stalinists. I want to be able to say in any polemic it is necessary to conduct against them: You have not carried out the ideal and principles of the socialist revolution to a logical conclusion — you have betrayed and destroyed it. And I believe that the basic and relevant facts enable me to make that assertion honestly and sincerely. By defending everything that was said and done by Lenin or the other leaders of the revolution? Certainly not, but by emphasising the radical differences between the revolution and the present regime.

I have looked back on some of your own writings of fairly recent times and find them highly relevant to my point. You have written: "In Lenin's time the Communist Party was itself democratic." And: "Everybody knows that Lenin started with an extreme approach to equalitarianism." And: "It is true that in the very early days of the revolution the degree of workers' control in the factories was very great." And more along similar lines.

Now: I want to be able to say, in such debates as I have mentioned, that the Stalinist regimes have wiped out and betrayed all of that. To me, this is dictated by good political sense and is justified by ascertainable facts. I consider it of high political value and significance to say, as you do in your letter to me: "... if Lenin had lived, he might have repudiated Stalinism or been repudiated by it."

If anything, I would put it more emphatically, for it is my deep conviction. I say this without any thought of absolving Lenin or any other Bolshevik leader from their own responsibilities, excesses in the revolution, or of mistakes afterward. But also without any thought of making it mandatory upon all members of a democratic socialist party what you called "absolute identity of opinion" on a subject that is primarily of historical importance, and on which a pretty wide diversity of view exists — as it should — in every part of the socialist international with which I am familiar.

Only a sterile sect demands uniformity of opinion on all questions, historical, theoretical, philosophical, political and tactical. A political movement should and can be built only upon the degree of agreement that is necessary for its to carry out its political tasks effectively. Organisations like the Socialist Labor Party or the Socialist Workers Party are sorry examples of the former. I would like to see the SP-SDF as an encouraging model of the latter.[1]

---

1. This is from a letter to Norman Thomas in the negotiations about the entry of Max Shachtman and his co-thinkers to the Socialist Party. In fact, Shachtman went over very quickly to social democracy, though as far as we know he never repudiated the 1917 revolution.

Part 3: Lenin, Bolshevism, and Rosa Luxemburg

'A book is nothing but a person talking publicly' by S I Ivanov (1920)

Lenin and Rosa Luxemburg

Two legends have been created about the relationship between the views of Lenin and Rosa Luxemburg. Despite their antagonistic origins and aims, they supplement each other in effect. Neither one of the mythmakers approaches the extremely interesting and instructive subject from an objective historical standpoint. Consequently, the analysis made by each of them reduces itself to an instrument of factional politics which is, in both cases, the politics of reaction.

One school of thought, if such a term is permissible here, is headed by the faculty of Stalinist falsification. It covers up its reactionary objectives by posing as critics of Luxemburg and proponents of Lenin. A discussion of its arguments is rendered impossible by the very nature of its position, which formally prohibits both argument and discussion. Its scientific value is summarized in a few sentences from the papal bull issued by Stalin in 1932 in connection with the luckless Slutsky's study on Lenin's incorrect appraisal of Kautsky and Luxemburg: "You wish to enter into discussion against this Trotskyist thesis of Slutsky's? But what is there to discuss in this? Is it not plain that Slutsky is simply slandering Lenin, slandering the Bolsheviks? Slander must be branded, not transformed into a subject for discussion." The Stalinists have the Catholics' attitude toward their dogmas: they assume what is to be proved; their arbitrary conclusions are presented as their premises; their statement of the problem is at the same time their answer — and it brooks no discussion. "Bolshevism" is absolutely and at all points and stages irreconcilable with "Luxemburgism" because of the original sin of the latter in disputing the "organisational principles" of the former.

The other school of thought is less authoritarian in tone and form, but just as rigid in unhistorical dogma; and if, unlike the Stalinists, it is not wholly composed of turncoats from revolutionary Marxism, it has a substantial sprinkling of them. Their objectives are covered up by posing as critics of Lenin and defenders of Luxemburg. They include anachronistic philosophers of ultra-leftism and express-train travellers fleeing from the pestilence of Stalinism to the plague of social-democracy. Bolshevism, they argue, is definitely bankrupt. The horrors of Stalinism are the logical and inevitable outcome of Lenin's "supercentralism", or — as it is put by a recent critic, Liston Oak, who seeks the "inner flaws of Bolshevism" — of Lenin's "totalitarianism". Luxemburg, on the other hand, stressed the democratic side of the movement, the struggle, the goal. Hence, "Luxemburgism" is absolutely irreconcilable with "Bolshevism" because of the original sin, of the former in imposing its Jacobin, or bourgeois, or super-centralist, or totalitarian "organisational principles".

The use of quotation marks around the terms employed is justified and necessary, for at least in nine cases out of ten the airy analysts have only the vaguest and most twisted idea of what the disputes between Luxemburg and Lenin really were. In just as many cases they have revealed a cavalier indisposition to acquaint themselves with the historical documents and the actual writings of the two great thinkers. A brief survey will disclose, I believe, the superficiality of the arguments which, especially since the obvious putrescence of Stalinism, have gained a certain currency in the radical movement.

Nothing but misunderstanding can result from a failure to bear in mind the fact that Lenin and Luxemburg worked, fought and developed their ideas in two distinctly different movements, operating within no less different countries, at radically different stages of development; consequently, in countries and movements where the problems of the working class were posed in quite different forms. It is the absence of this concrete and historical approach to the disputes between Lenin, of the Social-Democratic Labor Party of Russia, and Luxemburg, of the Social-Democratic Party of Germany, that so surely brings most critics to grief.

The "organisational dispute" between Lenin and Luxemburg did not originate in the former's insistence on a break with Kautsky and the centrists before the war. When Stalin thunders against anyone "who can doubt" that the Bolsheviks brought about "a split with their own opportunists and centrist-conciliators long before the imperialist war (1904-1912) without at the same time pursuing a policy of rupture, a policy of split with the opportunists and centrists of the Second International" — he is simply substituting ukase for historical fact.

The truth is that Rosa Luxemburg reached a clear estimate of Kautsky and broke with his self-styled "Marxian centre", long before Lenin did. For many years after the turn of the century, Kautsky's prestige among all the factions of the Russian movement was unparalleled. The Menshevik Abramovich does not exaggerate when he writes that:

"A West European can hardly imagine the enormous authority which the leaders of the German social-democracy, the Liebknechts, the Bebels, the Singers, enjoyed in Russia. Among these leaders, Karl Kautsky occupied quite a special place... serving for all the Russian Marxists and social-democrats as the highest authority in all the theoretical and tactical questions of scientific socialism and the labor movement. In every disputed question, in every newly-arisen problem, the first thought always was: What would Kautsky say about this? How would Kautsky have decided this question?"

Lenin's much-disputed *What to Do?* held up, as is known, the German so-cial-democracy and its leader, Bebel, as models for the Russian movement. When Kautsky wrote his famous article, after the 1905 revolution in Russia, on the Slavs and the world revolution, in which, Zinoviev writes, under Lux-emburg's influence, he advanced substantially the Bolshevik conception, Lenin was highly elated. "Where and when," he wrote in July 1905, in a polemic against Parvus, "have I characterised the revolutionism of Bebel and Kautsky as 'opportunism'? Where and when have I presumed to call into existence in the international social-democracy a special tendency which was not identi-cal with the tendency of Bebel and Kautsky?" A year and a half later, Lenin wrote that "the vanguard of the Russian working class knows Karl Kautsky for some time now as its writer", and a month later, in January 1907, he de-scribed Kautsky as "the leader of the German revolutionary social-democrats". In August 1908, Lenin cited Kautsky as his authority on the question of war and militarism as against Gustave Hervé, and as late as February 1914, he in-voked him again as a Marxian authority in his dispute with Rosa Luxemburg on the national question. Finally, in one of his last pre-war articles, in April 1914, *Wherein the German Labor Movement Should Not Be Imitated*, speaking of the "undoubted sickness" of the German social-democracy, he referred exclu-sively to the trade union leaders (specifically to Karl Legien) and the parlia-mentary spokesmen, but did not even mention Kautsky and the centrists, much less raise the question of the left wing (also unmentioned) splitting with them.

It is this pre-war attitude of Lenin towards the German centre — against which Luxemburg had been conducting a sharp frontal attack as early as 1910 — that explains the vehemence and the significant terminology of Lenin's stric-tures against Kautsky immediately after the war broke out, for example, his letter to Shliapnikov on October 27, 1914, in which he says: "I now despise and hate Kautsky more than all the rest... R. Luxemburg was right, she long ago understood that Kautsky had the highly-developed 'servility of a theoretician' ..." In sum, the fact is that by the very nature of her milieu and her work before the war, Rosa Luxemburg had arrived at a clearer and more correct apprecia-tion of the German social-democracy and the various currents within it than had Lenin. To a great extent, this determined and explained her polemic against Lenin on what appeared to be the "organisational questions" of the Russian movement.

The beginning of the century marked the publication of two of Lenin's most audacious and stirring works, *One Step Forward, Two Steps Backward*, and its forerunner, *What to Do?* The Russian movement was then in no way compara-ble to the West-European, especially the German. It was composed of isolated groups and sections in Russia, more or less autonomous, pursuing policies at odds with each other and only remotely influenced by its great revolutionary

Marxists abroad — Plekhanov, Lenin, Martov, Potressov, Trotsky and others. Moreover, the so-called "Economist" tendency was predominant; it laid the greatest stress on the element of spontaneity in the labour struggle and under-rated the element of conscious leadership.

Lenin's *What to Do?* was a merciless criticism of "Economism", which he identified with "pure-and-simple trade unionism", with khvostism (i.e., the policy of dragging at the tail of events, or of the masses), with opportunism. Social-democracy, he argued, is not a mere outgrowth of the spontaneous economic struggles of the proletariat, nor is it the passive servant of the workers; it is the union of the labor movement with revolutionary socialist theory which must be brought into the working class by the party, for the proletariat, by itself, can only attain a trade-union and not a socialist consciousness. In view of the dispersion of the movement in Russia, its primitive and localistic complexion, an all-Russian national party and newspaper had to be created immediately to infuse the labour movement with a socialist, political consciousness and unite it in a revolutionary struggle against Czarism. The artificers of the party, in contrast with the desultory agitators of the time, would be the professional revolutionists, intellectuals and educated workers devoting all their time and energy to revolutionary activity and functioning within an extremely centralised party organisation. The effective political leadership was to be the editorial board of the central organ, edited by the exiles abroad, and it would have the power to organise or reorganise party branches inside Russia, admit or reject members, and even appoint their local committees and other directing organs. I differ with the Mensheviks in this respect, wrote Lenin in 1904:

"The basic idea of comrade Martov ... is precisely a false 'democratism', the idea of the construction of the party from the bottom to the top. My idea, on the contrary, is 'bureaucratic' in the sense that the party should be constructed from above down to the bottom, from the congress to the individual party organisations."

It should be borne in mind that, despite subsequent reconsideration, all the leaders of the *Iskra* tendency in the Russian movement warmly supported Lenin against the Economists. "Twice in succession," wrote A.N. Potressov, later Lenin's furious enemy, "have I read through the booklet from beginning to end and can only congratulate its author. The general impression is an excellent one — in spite of the obvious haste, noted by the author himself, in which the work was written." At the famous London Congress in 1903, Plekhanov spoke up in Lenin's defence: "Lenin did not write a treatise on the philosophy of history, but a polemical article against the Economists, who said: We must wait until we see where the working class itself will come, without the help of the revolutionary bacillus." And again: "If you eliminate the bacillus, then there remains only an unconscious mass, into which consciousness must be brought from without. If you had wanted to be right against Lenin and if you had read through his whole book attentively, then you would have seen that this is just what he said."

It was only after the deepening of the split between the Bolsheviks and the Mensheviks (Plekhanov included) that the latter launched their sharp attacks on Lenin's polemical exaggeration — that is what it was — of the dominant role of the intellectuals as professional revolutionists, organisers and leaders of the party, and of the relationship between spontaneity and the element of socialist consciousness which can only be introduced into the labour movement from without. Lenin's defence of the ideas he expressed in 1902 and 1904 on these questions and on centralism, is highly significant for an understanding of the concrete conditions under which they were advanced and the concrete aims they pursued.

In "The Fruits of Demagogy", an article written in March 1905 by the Bolshevik V. Vorovsky (read and revised by Lenin), the author quotes Plekhanov's above-cited praise of Lenin's *What to Do?* and adds:

"These words define perfectly correctly the sense and significance of the Lenin brochure and if Plekhanov now says that he was not in agreement, from the very beginning, with its theoretical principles, it only proves how correctly he was able to judge the real significance of the brochure at a time when there was no necessity of inventing 'differences of opinion in principle' with Lenin. In actuality, *What to Do?* was a polemical brochure (which was entirely dedicated to the criticism of the khvostist wing in the then social-democracy, to a characterisation and a refutation of the specific errors of this wing). It would be ridiculous if Lenin, in a brochure which dealt with the 'burning questions of our movement', were to demonstrate that the evolution of ideas, especially of scientific socialism, has proceeded and proceeds in close historical connection with the evolution of the productive forces (in close connection with the growth of the labour movement in general). For him it was important to establish the fact that nowhere has the working class yet worked itself up independently to a socialist ideology, that this ideology (the doctrine of scientific socialism) was always brought in by the social-democracy..."

In 1903, at the Second Congress itself, Lenin had pointed out that "the Economists bent the staff towards the one side. In order to straighten it out again, it had to be bent towards the other side and that is what I did", and almost two years later, in the draft of a resolution written for the Third Congress, he emphasised the non-universality of his organisational views by writing that "under free political conditions our party can and will be built up entirely upon the principle of electability. Under absolutism, this is unrealisable for all the thousands of workers who belong to the party." Again, in the period of the 1905 revolution, he showed how changes in conditions determined a change in his views:

"At the Third Congress I expressed the wish that in the party committees there should be two intellectuals for every eight workers. How obsolete is this wish! Now it would be desirable that in the new party organisations, for every intellectual belonging to the social-democracy there should be a few hundred social-democratic workers."

Perhaps the best summary of the significance of the views he set forth at the beginning of the century is given by Lenin himself in the foreword to the collection, *Twelve Years*, which he wrote in September 1907:

"The basic mistake of those who polemise against *What to Do?* today, is that they tear this work completely out of the context of a definite historical milieu, a definite, now already long past period of development of our party... To speak at present about the fact that Iskra (in the years 1901 and 1902!) exaggerated the idea of the organisation of professional revolutionists, is the same as if somebody had reproached the Japanese, after the Russo-Japanese war, for exaggerating the Russian military power before the war, for exaggerating concern over the struggle against this power. The Japanese had to exert all forces against a possible maximum of Russian forces in order to attain the victory. Unfortunately, many judge from the outside, without seeing that today the idea of the organisation of professional revolutionists has already attained a complete victory. This victory, however, would have been impossible if, in its time, this idea had not been pushed into the foreground, if it had not been preached in an 'exaggerated' manner to people who stood like obstacles in the way of its realisation... *What to Do?* polemically corrected Economism, and it is false to consider the contents of the brochure outside of its connection with this task".

The ideas contained in *What to Do?*, which should still be read by revolutionists everywhere — and it can be read with the greatest profit — cannot, therefore, be understood without bearing in mind the specific conditions and problems of the Russian movement of the time. That is why Lenin, in answer to a proposal to translate his brochure for the non-Russian parties, told Max Levien in 1921:

"That is not desirable; the translation must at least be issued with good commentaries, which would have to be written by a Russian comrade very well acquainted with the history of the Communist Party of Russia, in order to avoid false application."

Just as Lenin's views must be considered against the background of the situation in Russia, so must Luxemburg's polemic against them be viewed against the background of the situation in Germany. In her famous review in 1904 of Lenin's *One Step Forward, Two Steps Backward* (an extension of the views of *What to Do?*), Luxemburg's position was decisively coloured by the realities of the German movement. Where Lenin stressed ultra-centralism, Luxemburg stressed democracy and organisational flexibility. Where Lenin emphasised the dominant role of the professional revolutionist, Luxemburg countered with emphasis on the mass movement and its elemental upsurge.

Why? Because these various forces played clearly different roles in Russia and in Germany. The "professional revolutionists" whom Luxemburg encountered in Germany were not, as in Russia, the radical instruments for gathering together loose and scattered local organisations, uniting them into one national party imbued with a firm Marxian ideology and freed from the opportunistic conceptions of pure-and-simple trade unionism. Quite the contrary. In Germany, the "professionals" were the careerists, the conservative trade union bureaucrats, the lords of the ossifying party machine, the reformist parliamentarians, the whole crew who finally succeeded in disemboweling the movement. An enormous conservative power, they weighed down like a mountain upon the militant-minded rank and file. They were the canal through which the poison of reformism seeped into the masses. They acted as a brake upon the class actions of the workers and not as a spur. In Russia the movement was loose and ineffectual, based on circles, as Lenin said, "almost always resting upon the personal friendship of a small number of persons". In Germany, the movement was tightly organised, conservatively disciplined, routinised, and dominated by a semi-reformist, centralist leadership. These concrete circumstances led Luxemburg to the view that only an appeal to the masses, only their elemental militant movement could break through the conservative wall of the party and trade union apparatus. The "centralism" of Lenin forged a party that proved able to lead the Russian masses to a victorious revolution; the "centralism" that Luxemburg saw growing in the German social-democracy became a conservative force and ended in a series of catastrophes for the proletariat. This is what she feared when she wrote against Lenin in 1904:

"... the role of the social-democratic leadership becomes one of an essentially conservative character, in that it leads to working out empirically to its ultimate conclusions the new experience acquired in the struggle and soon to converting it into a bulwark against a further innovation in the grand style. The present tactic of the German social-democracy, for example, is generally admired for its remarkable manifoldness, flexibility and at the same time certainty. Such qualities simply mean, however, that our party has adapted itself wonderfully in its daily struggle to the present parliamentary basis, down to the smallest detail, that it knows how to exploit the whole field of battle offered by parliamentarism and to master it in accordance with given principles. At the same time, this specific formulation of tactics already serves so much to conceal the further horizon that one notes a strong inclination to perpetuate that tactic and to regard the parliamentary tactic as the social-democratic tactic for all time."

But it is a far cry from the wisdom of these words, uttered in the specific conditions of Luxemburg's struggle in Germany, to the attempts made by syndicalists and ultra-leftists of all kinds to read into her views a universal formula of rejection of the idea of leadership and centralisation. The fact of the matter is that the opportunistic enemies of Luxemburg and her closest collaborator, Leo Jogisches (Tyzsko), especially in the Polish movement in which she actively participated, made virtually the same attacks upon her "organisational principles" and "regime of leadership" as were levelled against Lenin. During the war, for example, the Spartakusbund was highly centralised and held tightly in the hands of that peerless organiser, Jogisches. The Social-Democracy of Poland and Lithuania, which she led, was, if anything, far more highly centralised and far more merciless towards those in its ranks who deviated from the party's line than was the Bolshevik party under Lenin. In his history of the Russian movement, the Menshevik Theodore Dan, who did not spare Lenin for his "organisational regime", and sought to exploit Luxemburg's criticism of Lenin for his own ends, nevertheless wrote that the Polish social-democracy of the time shared in its essentials the organisational principles of Lenin, against which Rosa Luxemburg had polemised at the birth of Bolshevism; it also applied these principles in the practice of its own party, in which a rigid, bureaucratic centralism prevailed and people like Radek, Zalevsky, Unschlicht and others, who later played a leading role in the Communist party, were expelled from the party because of their oppositional stand against the party executive.

"Bureaucratic centralism", was (and is) the term generally applied by Dan and Mensheviks of all stripes to Lenin and Luxemburg and all others who seriously sought to build up a purposeful party of proletarian revolution, in contrast to that "democratic" looseness prevalent in the Second International which only served as a cover behind which elements alien to the revolution could make their way to the leadership of the party and, at crucial moments, betray it to the class enemy. The irreconcilable antagonism which the reformists felt towards Lenin and Luxemburg is in sharp and significant contrast to the affinity they now feel towards the Stalinist international, in which full-blooded and genuine bureaucratic centralism has attained its most evil form. It is not difficult to imagine what Rosa Luxemburg would have written about the Stalin regime had she lived in our time; and by the same token it is not difficult to understand the poisonous campaign that the Stalinists have conducted against her for years.

The years of struggle that elapsed since the early polemics in the Russian movement, the experiences that enriched the arsenal of the great revolutionists of the time, and above all the Russian Revolution itself, undoubtedly served to draw the political tendency of Rosa Luxemburg closer to that represented with such genius by Lenin. Had she not been cut down so cruelly in the prime of her intellectual power, there is little doubt in my mind that she would have become one of the greatest figures and champions of the Communist International — not of the horribly twisted caricature that it is today, but as it was in the early years. It does not even occur to me, wrote Karl Kautsky, her bitter foe, in 1921, "to deny that in the course of the war Rosa drew steadily closer to the communist world of thought, so that it is quite correct when Radek says that 'with Rosa Luxemburg there died the greatest and most profound theoretical head of communism'."

The judgment is a correct one and doubly valid because it comes from a political opponent who knew her views so well. It is worth a thousand times more than all the superficial harpings on the theme of the irreconcilability of Marxism's greatest teachers in our time.

New International, **May 1938**

Part 4: Appendices

Shachtman confronts Kerensky

Kerensky, head of the government that Lenin ousted, debates Max Shachtman

Seldom does history record the former head of a government, deposed by social revolution, facing up in an open debate 34 years later to a modern representative of the same ideological current which swept him from power. This was the situation in the February 8 [1951] debate at the University of Chicago where Max Shachtman confronted Alexander Kerensky, the head of the régime which was overthrown by the great Russian Revolution.

To recall to consciousness all the relevant facts of that vast revolution and vindicate its democratic and socialist aims and achievements, Shachtman, national chairman of the Independent Socialist League, brought a clearly defined and thoroughly Marxist appreciation of the meaning of democracy.

Alexander Kerensky, erstwhile president of the short-lived Russian Provisional Government and self-styled "arch-democrat", brought no understanding whatsoever of democracy, substituting for that lack his own garbled version of historical facts and a relentless penchant for reiterating fraudulent quotations from Lenin. Indeed, how could a "democrat" proceed otherwise, who could not even explain publicly that he was not put into office by popular election!

The intervening years since the revolution have witnessed the rise in Russia of the totalitarian bureaucratic oligarchy of Stalinism. Grabbing onto this bare historical fact, Kerensky sought to bury the anti-democratic crimes of his own régime by pointing an accusing finger at Lenin and the Bolsheviks as those responsible for Stalin's monstrous despotism. Shachtman thus faced a double task in this debate, one familiar enough to genuine socialists: that of establishing historical truth against the combined opposition of both capitalist and Stalinist falsifiers of the past 34 years.

This is the reason that Shachtman, in opening the discussion, found it necessary to remark: "The Stalinist régime never slackens in its efforts to portray itself as the legitimate successor of the Bolshevik Revolution. It needs this great authority to help befuddle the thinking of people and to maintain itself in power… It came into power as the result of a counter-revolution which systematically destroyed not only every single one of the great achievements of the Bolshevik Revolution but likewise exterminated all its founders, builders and defenders".

Scouting the idea that the evening's discussion on "Was the Bolshevik Revolution of 1917 democratic?" was of merely historical interest, Shachtman indicated its vital relationship to the most important social and political question of our time, the answer to which will determine conclusively the fate of society. Formulated by Lenin, the leader of Bolshevism, that question is: the working class cannot attain socialism except through the fight for democracy, and democracy cannot be fully realised without the fight for socialism.

Following is a running summary and digest of the presentations and rebuttals of the two speakers. The digest of Shachtman's presentation is based on his written notes.

One must judge a revolution out of the circumstances from which it sprang. The social structure of Czarism, the most reactionary and outlived in Europe, was in a state

March on International Working Women's Day February 1917 marks the start of the Russian Revolution.

of complete collapse. The imperialist war was bleeding the country white; a consciousness of the futility of continuing it deepened not only among the people at home but also among the soldiers at the front. At the top in official and court circles, bigotry, corruption and every conceivable form of social and intellectual leprosy was eating into the régime. At the front, a blood-letting that was as useless as it was incredible; at home a veritable orgy of war-profiteering among the capitalist classes and an unendurable growth of hunger among the working classes.

In February 1917 the Czarist régime appeared to be the most powerful in the world, with the world's biggest army at its disposal, with a subject people at once docile and impotent. Shortly after, the régime was overthrown by the same people and the same army.

It was an imposing example to all statesmen and politicians that the patience of the people is not inexhaustible and that, once they are determined to rise in the struggle for liberty, for their aspirations, they stand on no ceremony, on no formalities. They take action directly and stop waiting for the promises of their well-wishers to be fulfilled in some distant and indefinite future. The example was also instructive to statesmen and politicians capable of learning from the people. As it soon turned out, not many of them are capable of learning very much.

Tonight we are discussing democracy, the rule of the sovereign people.

Democracy does not consist in imposing upon the people what their rulers, by themselves, decide is a good thing for the people. It consists in the free expression of the desires of the people and their ability to realise these desires through institutions manned by their freely-chosen representatives. What then did the people who had just put an end to czarist rule want? It would be a bold man who contended that two opinions are possible on this score.

They wanted:

(1) an end to the imperialist war;

(2) the convocation of a national, democratically-representative Constituent Assembly;

(3) an end to the rule of the predatory landlords and a distribution of land among the peasants;

(4) a radical change in industry, beginning with the 8-hour work day and the assuring of the beginning of the end of completely arbitrary rule of industry by the capitalist class by the establishment of workers' control in industry;

(5) the right of national self-determination for the nationalities oppressed by czarism.

Not a single one of these desires is, by itself, the equivalent of socialism. Every single one of the demands of the Russian people was democratic through and through. And yet, as we shall see, they required a socialist revolution for their realisation.

Virtually from the first day the revolution established what were tantamount to two governments, two powers, contesting with one another for political supremacy.

One was the soviets; in 1917, as in 1905 they were spontaneously established. More democratic institutions it would be hard to imagine. They were directly and freely elected and sat in permanent session as direct representatives of the workers, peasants and soldiers. They were not the creation or invention of the Bolsheviks. While they were spontaneously formed without waiting for instructions from anybody, they were dominated by the right-wing socialists and the Socialist-Revolutionists. The Bolsheviks started as a tiny minority in the soviets.

While the soviets were the only elected body on a nation-wide basis in the land, and only they could thus speak authoritatively for the people, being referred to even by Kerensky as the "revolutionary democracy," they did not seek to become the government of Russia under their compromising leadership.

But they were the real power, recognised by all: by the czarist generals who wanted to crush them and restore reaction; by all the provisional governments; by the Bolsheviks who wanted them to take all governmental power; and above all by the people. Not a single significant political or military step could be taken by the official government without their support.

Appearing to stand above the soviets were the various provisional governments. These were not democratic, if by that term is understood a government elected by popular suffrage in regularly fixed elections and submitting its conduct to the control of any popularly elected democratic body.

The provisional government was constructed exclusively from the top, bureaucratically, by agreements among party leaders, self-constituting and self-perpetuating. Unstable by its very nature, it had no independent power of its own. It depended for its existence on the unpreparedness, and therefore the tolerance, of the reactionary forces on the one side and the revolutionary forces on the other.

While the soviets mistakenly thought the government could be the vehicle for the advancement of the revolution, they watched its every step, particularly its reactionary wing and allies, and tried to control each step, reflecting the attitude of the whole people. The provisional government tried to maintain itself by satisfying both the real social and political forces, the reaction and the revolution.

This aim was utopian; the two forces could not be reconciled. Both forces realised their life and future depended on the other's destruction. The governments became more and more governments of chaos, sure to produce nothing but that.

The 8 months' record of provisional governments in this stormy period when the desires of the people were urgent and manifest consisted of the following:

(1) The main body of the czarist officialdom remained intact, only few changes being made at the top. Czarist officers primarily remained at the head of the army, doing everything to undermine the soldiers' soviets, soldiers' rights, and even keeping enough power to threaten this same government. Cossack regiments, symbol of the czarist knout, were kept intact.

(2) The Constituent Assembly was not convened, on the basis of all kinds of pretexts. The real reason for this, as the bourgeoisie openly declared, was that the election results would not be acceptable to them and would mean that the régime might refuse to con-

tinue the imperialist war.

(3) While the people wanted peace, the provisional government, in obedience to Czarist commitments made to the Anglo-French allies, drove the army into the June offensive at a horrible cost in lives and against conservative military opinion that it would be doomed. The people did not want to fight for the Czar's secret treaties, authentic agreements made among imperialist pirates. While Kerensky had been told by Miliukov about them, he never repudiated them and refused to publish them, since such would be a "discourtesy to the Allies".

(4) While the rule of the landlords continued, the peasants who wanted the land received promises. But they were taking the land, carrying out the revolution themselves in the traditional style of every great agrarian revolution. The provisional government forbade them to act, instead of carrying out its own reforms. It sent Cossacks against the peasants, who had never seen a Bolshevik in their lives but who were taking things into their own hands.

(5) No changes in industry. While the capitalists sabotaged production by locking out workers, the government failed to intervene. The 8-hour work day decreed by the government was not enforced. Everything was promised for after the "Constituent Assembly" met, but its convocation was constantly delayed. Workers saw that their soviets' influence in the government declined as that of the capitalists and Czarists grew.

(6) The treatment of oppressed nationalities represents the "acid test" for a democrat. The Finnish social-democrats obtained a majority in early June and declared for their autonomy, enjoyed previously under the Czars. The provisional government dissolved the Finnish parliament, barring its doors with Russian soldiers… In June Kerensky prohibited the holding of the Ukrainian Soldiers Congress called by the nationalist Rada. Vinnichenko, head of that body and an anti-Bolshevik, attacked the provisional government for being "imbued with the imperialist tendencies of the Russian bourgeoisie". In October Kerensky demanded an explanation of alleged criminal agitation started there for a Ukrainian Constituent Assembly and an investigation of the Rada was ordered. On the basis of this record of failing to meet the continuing demands of the revolution, the provisional government of Kerensky fell. It also explains why the power of the compromiser Menshevik-SR leadership in the Soviets likewise fell. They had urged confidence in the provisional government, which showed it did not deserve the masses' confidence.

After the Kornilov affair, the Bolsheviks won uninterrupted victories in the soviets, while the Mensheviks and SRs split up and declined. Bolshevik influence was won fairly, openly, democratically, in spite of huge handicaps. Their leaders were arrested or driven underground, presses and headquarters smashed, press outlawed, forbidden entry to the garrisons and a lynch spirit aroused against them as German agents.

On November 7 the soviet congress, whose convocation had been delayed by its compromising leadership, was called together by that same leadership. The Bolsheviks had a clear majority. The congress endorsed the uprising led by the Military Revolutionary Committee of the Petrograd Soviet under Trotsky by electing a new government of Bolsheviks holding soviet power. Two weeks later the Peasant Soviet Congress, called by the compromisers, gave a majority to the Left SRs and the Bolsheviks, and the Left SRs entered the new soviet government.

In a few days the soviet government did all the things the provisional government had failed to do:

(1) gave the land to the peasants;

(2) offered peace by broadcast to all governments and peoples, starting with proposals for a three months' armistice;

(3) inaugurated workers' control of production to stop bourgeois sabotage of industry;

(4) decreed freedom for all nationalities, beginning with Finland and the Ukraine;

(5) denounced and published all secret treaties and Czarist rights in China and Persia;

(6) wiped out all Czarist power in the army and began creation of new workers' and peasants' army;

(7) abolished special Cossack privileges and caste position;

(8) inaugurated the new soviet régime of direct representation, with full right of recall.

The Constituent Assembly finally met in January; and because of its then unrepresentative character, big changes having occurred in mass thinking since its lists were drawn and the election held, and its refusal to recognise that the revolution had conferred full power on the soviets, it was dissolved.

No champions could be found among the people for it — only reaction supported it. The country rallied to the soviet power as the only guarantee of the great democratic achievements consolidated by the Bolshevik Revolution.

The future proved to be a difficult one. The country was plunged into civil war by the dispossessed classes, landlords, bankers, bondholders, monarchist and reactionary scum in general who sought to arouse the wealthier peasants against the régime, and by all the imperialist powers who forgot their differences in the face of the socialist enemy.

This civil war brought devastation to the country from which it took years to emerge. It forced upon the soviets a harsh régime, and laid the basis for the eventual rise and triumph of a counter-revolutionary bureaucracy which is in power today.

But in spite of that these achievements are immortal; nothing that happened afterwards can eradicate that from history or from the thoughts of mankind. They are a monument and a guidepost.

The road out of the blind alley into which society is being driven more and more, lies in the struggle for democracy. The struggle for democracy receives its clarity, purpose and guarantee in the struggle for socialism; the struggle for socialism lies in the hands of the working class — the beast of burden, the despised of the earth — whose will to victory was forever underlined by their first great revolution, the Bolshevik Revolution in Russia.

Kerensky's presentation followed Shachtman, who had devoted his time to developing the whole picture of the unfolding revolution in Russia, in its historical context and in a rounded interpretation. Kerensky devoted his time to picking holes in this interpretation, from the viewpoint of a government official of narrow social vision.

He based himself on the necessity for the provisional government to "defend Russia" during the war, opposing the elements of extreme monarchist reaction who favoured a separate peace with Germany and likewise opposing the desire of the people to get out of the disastrous war.

He took the stand that the social reforms demanded by the people must be postponed until the war was over. The government could legitimately adopt measures such as its land reforms, the 8-hour day, the need for a Constituent Assembly, the right of self-determination for oppressed nationalities — but (and it was a very big but) nothing could really be done until the Constituent Assembly met, and it would be better for that body to meet only after the conclusion of the war.

After all, the organisation of a Constituent Assembly is a "big job". The Germans were advancing, and the "Lenin crisis in the rear" forced the Constituent Assembly commission to cease its never-ending labour after only three weeks. The provisional government was "in direct contact with all forces — exception: the Bolsheviks".

This section of Kerensky's presentation had already been anticipated in Shachtman's speech, which had made clear in advance the garbled version of history which Kerensky was presenting. Nor did Kerensky even try to meet Shachtman on the ground of the meaning of democracy and the role of the masses.

Instead he spent the major part of his time plucking out and attacking quotations from Lenin's writing, with a view to proving their conspiratorial, treasonous and totalitarian nature.

According to Kerensky's story, Lenin foresaw that Kerensky's proposals would win the support of the peasantry — after the victory of Russia's noble but crumbling armies. Therefore Lenin had to act fast, before this happened.

He had to marshal his Bolsheviks to organise army deserters in the countryside and to steer a course toward armed insurrection, before the provisional government had a sporting chance to show its sterling mettle to the peasants on some indeterminate future date after the equally indeterminate conclusion of hostilities.

The aim of Bolshevism, according to Kerensky, was to exploit the country in totalitarian fashion. The real question here, he announced, is what happened after the revolution — but he abruptly stopped at this point, apparently remembering that the subject of the discussion was the revolution itself; however, he picked up this theme from time to time later.

Lenin, Trotsky and Stalin, he said, were playing a double game of trickery on the country and the government. Lenin sent various "secret instructions" to his central committee. (Kerensky, without pointing it out, was referring to the period when his own government had jailed Trotsky and other Bolsheviks and had forced Lenin to go into hiding!)

In one of those "instructions" Lenin committed the heinous crime of saying that the soviets would be of value to the people only if they carried through the needs of the revolution.

Another aim of Bolshevism, Kerensky charged, was "to distract the freest country in the world from preparing a base for the future world socialist movement." So, Lenin concluded, the provisional government had to be stopped.

"For this they ruined Russian democracy," he cried, after having made clear that he understood nothing about the urgent desire of the Russian masses for the democratic and socialist reforms which only the Bolsheviks were fighting for.

Striking a personal note, Kerensky drew some applause when he cried: "Maybe my government was unpopular but I needed no bodyguards. In Kiev when I took a walk the people liked to gather around me and speak to me". Kerensky was presumably referring to Stalin's secluded and guarded living habits (and it is a safe bet that he was not referring to [US President] Truman's bodyguard); but while he was supposed to be discussing Lenin and the days of the Russian Revolution, he made no mention of the fact that Lenin and the other Bolshevik leaders continually mingled with the workers at all kinds of meetings and elsewhere, guarded at other times as the crisis neared only against the police vengeance of Kerensky himself.

He concluded his presentation by quoting an attack by Proudhon on... Marx. The French petty bourgeois radical had denounced Marx's Communist Manifesto with the cry that "Communism is nothing more than inequality, subjugation, and slavery".

The fight in 1917, said Kerensky, was "not a fight between capitalism and socialism, but between freedom and slavery." And "Stalin is the most faithful, most able, most talented disciple of Lenin."

Shachtman opened his rebuttal with a reminder to the audience that he had initially stated that the Stalinists have the biggest lie factory against the Bolshevik Revolution, but they by no means have a monopoly on the business.

He proceeded to discuss Kerensky's garbled quotations — that is, forgeries — purporting to prove that Lenin favoured "treason", discussing in particular Lenin's opposition to the czar's war and the world-wide imperialist war and his views on the so-called "revolutionary defeatism".

The ISL chairman demanded to know "who elected" the supposedly "democratic" provisional government — which, of course, had been put into power by no popular vote of any kind. In contrast, he pointed out, the Bolshevik government took power with the support of a free vote of the broadest and most representative body ever assembled in Russia or for that matter in the world — the soviets (councils) of the workers, peasants, and soldiers of the country — in a congress organised and prepared by enemies of the Bolsheviks.

It will be a curious spectacle for future historians to picture the president of a government whom no people had elected contesting the democratic character of the only revolutionary regime in the history of the world's revolutions which did come to power with the recorded, freely voted support of the broad masses!

Shachtman presented the documentation of the recent book on *The Election to the Russian Constituent Assembly of 1917* by O H Radkey as even more conclusive proof that the compromising leadership of the Mensheviks and SRs "no longer commanded the allegiance" of the masses.

He stressed the absurdity, not to speak of the slanderousness, of Kerensky's claim that the Bolsheviks were able to lead a vast, tumultuous, surging mass revolution of the people through "trickeries."

How many insurrections, he asked, had Kerensky ever organised in which he gave public instructions (not "secret instructions") so that the reaction would know the time, place, and forces at his disposal?

"Whom did the Bolsheviks suppress during the civil war? White guards, czarists and Mensheviks who had taken up arms against the government and the revolution... Did that 'maniac' Lincoln ever permit the Confederate States in the US Civil War to open up a recruiting station in Chicago?"

Kerensky had referred in rapturous terms to the president of the first provisional government in 1917, Prince Lvov, one of the biggest landowners in Russia, as "one of the most extraordinary democrats in the world". Shachtman stated his regret that he had no time to take up this democratic idol of Kerensky's properly, but it is worthwhile to mention Kerensky's estimate for the light it casts on his own conceptions of democracy.

Kerensky had argued that while his provisional government had denied self-determination to Finland and the Ukraine, it had granted immediate freedom to Poland. Shachtman had only to point out that this was done when (and because) Poland was under the German sword at the time!

Kerensky was magnanimously giving freedom to a people whom he no longer controlled, while ruthlessly maintaining Russian control over the Finns and Ukrainians whom the Germans did not have in their power.

As reported above, Kerensky had also waved the flag of the Kronstadt revolt against the Bolsheviks, which took place in 1921 during the civil war of the White Guards and foreign armies against the revolution.

It was "ill-advised" for Kerensky to mention the word Kronstadt on his lips, Shachtman said. The provisional government — in 1917 — had "merely" ordered submarines to blow up the ships of the pro-Bolshevik Kronstadt sailors to compel their submission to the government!

In his rebuttal, Kerensky differentiated his own attack on Lenin as a "German agent" (one of the crudest of all the slanders against Lenin) from that of others in that he did not accuse Lenin of being a vulgar agent for German gold. It was "Lenin's point of view", he said, that coincided with German interests.

Taking up the question of why he had denied self-determination to the Ukrainians, he gave as his excuse the Ukrainians' "excessive" territorial demands, which for him could be solved only by the same Constituent Assembly which he was continually post-

'The Menagerie of the Future by M M Cheremnykh (1918). In the first part of the 20th century human zoos sometimes existed, of different types of human beings, for example, pygmies or other racial types.

poning.

His main appeal was "Why was it necessary to organise the uprising?", implying that it is "always possible" for things to be worked out.

As is also reported elsewhere, Shachtman, by the terms of the debate, was then supposed to have a surrebuttal, but he did not get the opportunity since the chairman adjourned the meeting due to the lateness of the hour. But even without this last word, there is little doubt that the solid, fact-buttressed, cogent picture of the Russian Revolution that he had presented clearly lighted up the socialist inspiration and democratic heritage of the great revolutionary struggle.

An anti-Bolshevik eye-witness
Hal Draper

[In his book *In the Workshop of the Revolution*, Steinberg, a Left SR leader of 1917-8] presents the 1917 upheaval not as a conspiracy but as a real people's revolution. And he is very inconsiderate of the myths about the "democratic" Kerensky regime which the bad Bolsheviks overthrew, as well as the Menshevik and Right Socialist Revolutionary allies of Kerensky.

Actually Steinberg's language about the "moderate socialist parties" (Menshevik and Right SRs) is very mild, but the outline of the picture he pains is damning enough. That picture is of an elemental revolutionary upsurge of the masses from blow, determined to throw off all oppression and equally determined to end the war, which the rights and moderates tried to oppose, and which the Bolsheviks (and left SRs) supported. This was the simple difference between the historic reality and the anti-Bolshevik myth of a "conspiracy".

Of the right wing socialists, Steinberg writes that they believed "that the necessary conditions were not yet in evidence to realise the programme of the people. They conceived it impossible to end the war without the co-operation of the Allied powers. They thought it utopian to transfer political power to the working classes since, in their view the capitalist order in Russia was inevitable. Their interpretation of the revolution as only a democratic bourgeois succession to Tsarism, demanded, of course, a corresponding strategy — the strategy of class compromise and political compliance. This strategy put the moderate two parties (Mensheviks and Right Socialist-Revolutionaries) halfway between the bourgeois and the working-class programmes, gave their activities an air of vacillation and, in fact, fortified the position of the bourgeois camp."

Now to be sure, the anti-Bolsheviks argue strenuously that anything beyond a bourgeois revolution was indeed impossible, but what Steinberg point up sharply is that this line meant that the right-wingers had to set themselves against and get ready to suppress the revolutionary dynamism of the people. It is because the anti-Bolsheviks have to get around this inconvenient fact that the myth of a "conspiracy" was born.

By the time of the new Kerensky government of 10 July, Steinberg relates, "Kerensky had lost hold of the ties of confidence which once had bound him to the people." Discreditment rebounded not only against Kerensky but also against the Menshevik and Right SR ministers who joined his cabinet.

"The main speaker for and exponent of this rootless coalition", writes Steinberg "was the Social-Democrat (Menshevik) Tseretelli. As minister of the interior, he dispatched a circular to the whole country designed to redouble the power of the government commissars against the active local soviets. He ordered these commissars to block the 'illegal distribution of landed properties,' the 'appropriation, ploughing and sowing of other people's lands.' He thus sustained the policy of his predecessor, Prince Lvov. Every circular of this kind was like a match thrown into the powder keg of the revolution."

Being highly concerned with the democratic forms of the revolution, Steinberg especially emphasises the transformation of the Kerensky regime into a "quasi-dictatorship" — with the consent and support of the very democratic Mensheviks and S-Rs who were later to issue howling blasts of anguish at every step the Soviet government took even to defend itself against armed insurrection.

Steinberg's general sketch of the whole development, of course, contributes nothing new to historical knowledge, its main interest lying the character of the narrator. There are vignette touches here and there.

• In August, at the State Conference organised by Kerensky we see the scene where Bublikov, a leading industrial capitalist, steps up to shake Tseretelli's hand before the assemblage, an impressive piece of symbolism while at the same moment a general strike of workers in Moscow was going on.

• While we all know that Kerensky and a few die-hard slanderers still preserve the chestnut about the Bolsheviks being "German agents" we can read in Steinberg that the Kerensky government itself was thus accused. In the manifesto of the Kornilov revolt, the reaction declared "The Provisional Government standing as it does under the pressure of the Bolsheviks in the soviets, works in full agreement with the German General Staff…"

With regard to the seizure of power itself, Steinberg is typically ambivalent. "The Left Social-Revolutionaries" he relates "did not think it advisable to precipitate such a rebellion. In their opinion it would be sufficient for the [Soviet] Congress to maintain the positions of the people and lead the revolution to the Constituent Assembly. But they felt that, if the masses were to rebel, they would not stand against them."

No initiative toward revolution — and no opposition to it: you just go along with the surge of the people. The Left S-Rs could never have been leaders of the revolution, the role that had to be played by the Bolsheviks, and on the other hand they could never have been enemies of the revolution. They combined the fuzziest of ideologies with real revolutionary sentiments and combination which doomed them to be simple fellow travellers of the revolution.

They had no political compass of their own, but as sincere revolutionists they could feel which way the revolutionary aspirations of the people where blowing. When the wind stopped blowing in one clear direction, they were lost.

Steinberg does not link up his above-quoted reference to the Left S-R's coolish opinion on the seizure of power with what he describes later as the great result of the "inadvisable" rebellion. Left S-Rs like Steinberg never could orient themselves in the criss cross of events and policies but they could respond like sensitive barometers to revolutionary élan.

"The October Revolution brought tremendous expectations, there was now a profound sense of relief. It is true that there was also great bitterness about the past, great anxiety for the future; but the deepest sensation which October aroused in the people was joy. In city, village and army, people rejoiced in the fullness of the their liberation, in the limitless freedom that now summoned their creative efforts. It was as if the walls of Jericho had crumbled before their eyes. A new life called to them with a thousand voices: from now on 'everything is possible to man'. 'Everything is possible' did not mean licence and wilful destruction, but full freedom to satisfy the constructive urges and the noblest ideals of man.

"All aspects of existence, social economic, political, spiritual, moral, familial, were opened to purposeful fashioning by human hands. Ideas for social betterment and progress that had been gathering for generations in Russia and elsewhere seemed to wait on the threshold of the revolution ready to put forth and permeate the life of the Russian people. The issues were not only social and economic reforms and thoroughgoing political changes; with equal zeal the awakened people turned to the fields of justice and education, to art and literature. Everywhere the driving passion was to create something new, to effect a total difference with 'the old world' and its civilisation. It was one of those uncommon moments of self perception and self assertion. The storm passed nobody by: neither those who hailed it as a blessing nor those who spurned it as a curse."

It was this climate of a world reborn which in the first place doomed the Constituent

Assembly as a vestigial remnant of the "old world". When the Constituent Assembly was swept away in the tide, it scarcely created even a ripple. It had ceased to have any significance.

What played a greater role at the time was a different question: coalition cabinet or one-party cabinet. And here, in Steinberg's account, we come to another reason why determined anti-Bolsheviks will not like this book. The reason is this: even when Steinberg is doing his best to be as "anti-Bolshevik" as they come, he just can't seem to squeeze out any facts to give colour to his strictures. The trouble it would seem, is that he had old-fashioned prejudices against simply inventing suitable "facts" to fit anti-Bolshevik specifications.

Steinberg and the Left S-Rs were enthusiastically in favour of constituting the first Soviet government as a coalition of all the socialist parties, including the Mensheviks and Right S-Rs. But the latter made it impossible, for a simple and straightforward reason: they were against the revolution and would enter its government only to behead it. Steinberg uses some language blaming "extremists" on both sides (Bolsheviks as well as the rightists, presumably) but every fact in his account speaks one way only:

"Protesting violently, the Mensheviks and Right Socialist-Revolutionaries quit the Second Soviet Congress when it proclaimed the Soviet Republic. Thus, the moderates caused the final split in the camp of the working classes and facilitated the establishment eventually of a purely Bolshevik government."

So the Left S-Rs set out to be the honest brokers who would bring the right-wing socialists back into the coalition. After all, these right-wing socialists had lived more or less happily in a coalition government dominated by imperialists and capitalists; why should they be so intransigent about entering a coalition with revolutionary socialists? It disconcerted the honest brokers no end.

On the day the first cabinet was established, the Bolsheviks formally invited the Left S-Rs to name three representatives. At this point the Left S-Rs refused, on the ground that they wanted an all-around coalition. So the Bolsheviks had to set up the cabinet themselves.

Negotiations for the inclusion of the rightists continued, but uselessly; for the condition which the Mensheviks and Right S-Rs set for their participation was breathtaking: nothing more than that Lenin and Trotsky (by name) should be kicked out of the government! Fantastic as it seemed, they were not even clever enough to try to undermine the revolution by stealth: they openly demanded just as if they had not been defeated and discredited, that the revolution behead itself in order to obtain as a reward their own worthy personages, now a little shopworn from being kicked around by Kerensky but still willing to "save" the revolution for capitalism and war.

"It was amazing" writes Steinberg. "During the February period, the Mensheviks and the Right Socialist-Revolutionaries had countenanced all possible coalitions with bourgeois parties, even when they were openly reactionary. But the same leaders now rejected indignantly the idea of a socialist coalition, that is, co-operation with the Bolshevik Party, which at that time was still weak and still sought support in other related elements. Lenin's face for them seemed to eclipse all of the revolution. And again they unwittingly helped prepare the ground for his future dictatorship."

So Steinberg complains that "Lenin's secret political purpose" was a "dictatorship" all the while, but whereas Steinberg was clever enough to mind-read Lenin's secret thoughts, no one else in the country had to be half so clever in order to see that it was in fact the right wing socialists who were torpedoing any unity.

The Left S-Rs finally joined the coalition themselves, and their course afterwards is another story.

Steinberg's account of the Constituent Assembly adds nothing new to the question. What he choose to emphasise, however, is that it was the right wing socialists (again) who excluded any possible compromise.

When the Constituent Assembly met, Chernov (Right S-R leader) was elected president and

"Of all possible attitudes toward the Soviets, Chernov (and the Right Socialist Revolutionary Party that stood behind him) chose the most dangerous, if not the most foolish tactic: he simply ignored the Soviets as if they did not exist at all. His major speech, which naturally encompassed all cardinal issues of the revolution, was delivered with the incredible pretence that the Constituent Assembly had convened in a social vacuum. He announced that negotiations for peace would be started with the Allied powers, that the socialisation of land would be carried through; that the federative rights of all nationalities would be proclaimed. Not with a single word did he mention that all these vital tasks were already being realised in the country and followed with intense interest in the whole world.

"What did all this mean? By implication it was a challenge to the Soviets and the masses that stood by them. For the Constituent Assembly, the only chance of survival lay in some compromise with the revolutionary forces that had already struck roots. It would have been easy to find some legal, constitutional and political form for such understand. But this one way of averting civil war within the camp of the working people was ignored by the majority [of the assembly]. Did it then hope that the Soviets would simply capitulate?"

Like all the others in the mainstream of the revolution the Left SRs now looked on the Constituent Assembly as an obsolete reminder of pre-October Russia.

In chapter 13 of Steinberg's book we find him in jail! What has happened? For a whole chapter our honest author goes through description of some local prison colour, ponderings about the French Revolution, tales about prisoners etc… and not a word about why he and a whole group of Left S-R's have been imprisoned. He barely manages to mention casually even that the Left S-Rs had left the government: why? Not a word.

At one point, he pictures himself as wondering "What this the final break-up of the once common front?" The reader naturally must suppose that this break up has taken place because of the Bolsheviks' action in jailing their ex-partners.

It is well-nigh incredible but Steinberg drags the reader through three more whole chapters before he even discusses his own version of what had happened to the coalition.

The reason for this peculiar structure is no mystery or personal idiosyncrasy.

Steinberg is deliberately engaged in recreating the impression — without deliberately lying at all — that he and his Left S-R party broke with the Bolsheviks over questions of democracy and terrorism, that is, over questions which today are "respectable" ones for anti-Bolsheviks.

And of course, the indisputable historic fact is that his party broke with the Bolsheviks over an entirely different issue… because of their intransigent and violent opposition to the Brest-Litovsk peace with Germany and for no other reason.

The Tsarist army had disintegrated, the whole land in revolution was in turmoil, the German army was threatening on the borders, whole regiments were deserting the lines, the front could not be held; better yield to the German' robber demands for a peace than have the revolution crushed: a revolutionary war against the German invasion could not be sustained; there was no choice… So Lenin argued, not only against the Steinbergian phrasemongers but also against a strong minority of the Bolsheviks themselves, a minority which publicly campaigned for its position outside of the party and against its majority.

The Left S-Rs advocated war, not peace; but this position was defeated at the Congress of Soviets which met to ratify the Brest Treaty. Thereupon, right there, Steinberg announced for the Left S-Rs that they were withdrawing from the government — "to the consternation of all present" he adds.

One searches the six meagre pages which he devotes to the whole issue… for Steinberg's statement of reasons in favour of his position of continuing the war rather than accepting the forced peace. This is what one finds.

Continuing the war by partisan warfare "might encourage the German people to resistance against their own masters. But 'peace' would automatically strengthen the German imperialist forces both at home and abroad."

Now, as a matter of record it was the Brest peace which did play an important part in stimulating revolutionary discontent in Germany; but the 1917 general strike wave in Germany and Austria proved that the revolution there was not yet ripe; and it was not at all necessary for the Left S-Rs to agitate Lenin about the quintessential need for the German revolution to come to the aid of the Soviets; and… But all this is really beside the point.

Steinberg and the Left S-Rs did not adopt their position out of overweening anxiety for the German revolution. It is transparent rationalisation. If not, the Left S-R position would have been merely tactical opposition to the Brest treaty, as indeed was the case with the dissident Bolsheviks. On the contrary, as Steinberg makes clear, for the Left S-Rs the surrender of Russian territory to the German robbers was a principled "capitulation" of the revolution.

Steinberg quotes himself from an 1918 article: If we sign, "no trace will remain of the meaning and content of the [Soviet] republic." At the end of this chapter we also find that the Brest treaty "broke the moral backbone of the coalition."

Why? Why were the Left S-Rs so frenetically and principledly outraged by this peace signed at the point of Germany's guns? Was it perhaps, the infusion of sheer national-chauvinism in their fuzzy ideology which prevented them from accepting the loss of Russian territory, even in order to save the revolution? Yes.

Well, then… had the Bolsheviks thrown the [Steinberg] in jail just because he and his Left S-Rs had left the government? Of course, no…

The Left S-R party decided to make up for their defeat in the Soviet Congress by embroiling Russia in war with Germany by their own organised provocation.

On 6 July, two Left S-R agents assassinated the German ambassador, count von Mirbach. Steinberg does not boggle over the question of the party Central Committee's responsibility for this move. The party's leaders, Spiridonova, proudly claimed full responsibility for the act, in the traditions of S-R individual terrorism. The last thing the party wanted was to have the assassination treated as merely involving two individual murderers. It was the party that had ordered Mirbach's assassination in order to provoke Germany into renewing its assault on Russia.

Steinberg writes that "In actuality the Left S-Rs *at that time* had had no intentions of staging a revolt." [Italics added]. That is at that time they were "merely" trying to get a war started against the country so as to bring about the "revolutionary war" which the majority of the country had rejected. (Democrats they are, you see.)

Perhaps some readers will not believe that a man can be so naïve as to tell this story on himself and still continue to write as if his party was engaged in nothing out of the ordinary from the point of view of its democratic rights. But the fact is that Steinberg actually writes the following fantastic and almost unbelievable words:

"but Lenin and Trotsky could not forgive another party for acting independently and thus challenging their dictatorship."

He actually writes this after himself recounting his party's war plot! The Left S-Rs were just "acting independently" of the government, is that a crime? Doesn't a party have the democratic right to assassinate an ambassador in order to get a war started against its own country! You call this a revolt? And so what if our "revolutionary war" program has been voted down by the Soviet Congress? So what if I, Steinberg, nowhere in my own book even claim that a majority of the country was really for it? Is it not another proof of Lenin's "dictatorship" that he could not "forgive" this little innocent plan to start a war? Are we not great democrats and he a dictator?

Like a character straight out of Wonderland, Steinberg continues to write about the Left S-Rs' "deep shock" when the Bolshevik government reacted sharply.

But this was July 1918. It was not until seven months later that Steinberg found himself wondering about things in prison. What had happened in this interval?

Specifically we have already seen that Steinberg had written that "the Left S-Rs at that time [July 1918] had had no intentions of staging a revolt". How did their intentions develop?

This brings us to the question of the Left S-Rs going over to the programme of armed struggle against the Soviet government.

This party, which had broken with the Bolsheviks over the issue of war rather than peace, which proclaimed that it regarded this issue as involving the whole content of the revolution, which was so frantically anxious to blow up the Brest peace that it reverted to its terrorist-assassination methods in order to embroil the country in a war on the vote of its own narrow Central Committee as against the vote of the Soviet Congress — was it true or wasn't it true that this party then moved to a programme of armed insurrection against the government.

Steinberg not only states but documents the party position.

Steinberg first summarises the thinking of the party on attitude to the government.

"Almost unwittingly, a policy of 'war on two fronts' evolved," he writes — one war against the White interventionists and simultaneous war against the Soviet government. "They might have said: We shall fight the bourgeois counter-revolution as if the Bolshevik state did not exist, and we shall fight Bolshevism as if social reaction did not stand poised to stab us in the back."

But — Steinberg continues in summary of his comrades' thinking — could such a two-front war succeed? Perhaps we should table our quarrel with Bolshevik policy in order to defend the revolution's future?

"Should the Left S-Rs then inform the regime of their decision, so that they might be released and take part in the battle?" The conclusion seemed logical, but — "it did not satisfy the moral conscience of the prisoners."

(Note that Steinberg implicitly demonstrates that he has no doubt that the Left S-Rs would have been freed from jail if they had been willing to adopt this position against armed insurrection.)

How did the party divide on the question? It was debated among the Left S-R prisoners and "argued in the secret correspondence with the illegal Central Committee of the Left S-Rs outside."

One faction (the "moderates") declared "We… reject *for the time being* any armed struggle against the Bolshevik government because it might play into the hands of forces hostile to the revolution." (Italics added.) The other faction ("intransigents") argued for the two-front war: "you cannot destroy one without the other." They were for armed struggle now and against calling for a "fight against the Denikins under Bolshevik leadership."

Who won? All Steinberg reports is that the "moderates" were "restrained" and their will "paralysed" by their fear that they might be regarded by the others as selling out to the Bolsheviks. Is it fair to conclude then that the "intransigents" dominated the party councils? At this point Steinberg simply ignores the obvious question. And of course it should be remembered the "moderates" were those who did not favour *immediate* organisation of armed struggle.

In any case, Steinberg next presents the text of a document hitherto unknown to me, which is decisive by itself.

It proves to the hilt without any possibility of doubt whatsoever that the Bolshevik government asked only, as the condition for releasing the Left S-Rs that they state publicly that they were against "armed action" to overthrow the government. This the S-Rs refused to do. Hence there is no possibility of dispute over why they were in prison.

All this emerges from the text of the document itself. Steinberg does not point it up one way or the other. One can even wonder whether he realised the meaning of the document which he quotes!

In August 1919 the Left S-R Central Committee with the agreement of "all party circles" decided on negotiations with the Bolsheviks for an agreement which would legalise their party and free their prisoners. Kamenev, Beloborodoff and Stassova represented the Bolsheviks, the Left S-R delegation of three included Steinberg. There are six pages of direct quotations from the discussion as selected by Steinberg himself.

The Left S-Rs proposed to "transfer the centre of our political operations to the provinces occupied by the Whites." In exchange for this they demanded: legality in these provinces after liberation and immediate release of all jailed party members.

Immediately what Kamenev wanted to know was: "Will you give up your tactics of armed struggle against us!"

Of course, the Bolsheviks also kept pointing out that it was impossible and absurd to have a situation where a party was illegal in (say) Moscow because of its programme of armed insurrection while it was legal in a recently liberated province, insecurely held, with the same programme.

The Bolsheviks kept hammering away at the main point.

In his very first speech Kamenev said "Can we ever come to an agreement with you, as clear and decisive as our split has been? Back in October (1917) we had differences of opinion too, yet we were able to work together." And in this framework he posed the decisive question of "armed action".

Turning and twisting the Left S-Rs refused to say yes or no. One of them evaded by merely saying that "You have no proofs whatsoever of our participation in any plots." But the Bolsheviks were not asking for proof one way or the other. They were merely asking for a public statement of party policy against armed action.

The second S-R evaded with the following phrase: "And anyway we have been refraining from armed action against you for some time" — apparently not even realising what he was admitting with this formulation.

Steinberg didn't even refer to Kamenev's insistent question (according to the text of the first conference as given in his book).

The send and last negotiation conference took place in September. Here the Bolsheviks were even more insistent in narrowing the issue down to insurrection.

It is not enough for you S-Rs, the Bolsheviks explained quite patiently, to say that you reject armed insurrection because at the present time you don't actually have the means for it. That only convinces us that as soon as you can gather your forces, locally or nationally, you will act as before. What we are asking for is a statement of party policy against it…

Steinberg replied: "Our party has not, so far, officially proclaimed any armed struggle against the Bolsheviks. You will not be able to find a single such decision in our party conferences. That is why we do not need a paragraph about it."

Of course the party had not yet proclaimed any armed struggle. That was in question at no point. As to a statement of party policy not a word could be elicited from the Left S-R delegation other than what we have quoted above.

And that settles that.

'Knowledge will Tear apart the Chains of Slavery' by A A Radakov

What is Trotskyism?

Our criticism of Trotsky's later theory of the "workers' state" introduces into it an indispensable correction. Far from "demolishing" Trotskyism, it eliminates from it a distorting element of contradiction and restores its essential inner harmony and continuity. The writer considers himself a follower of Trotsky, as of Lenin before him, and of Marx and Engels in the earlier generation.

Such has been the intellectual havoc wrought in the revolutionary movement by the manners and standards of Stalinism, that "follower" has come to mean serf, worshipper, or parrot. We have no desire to be this kind of "follower."

There are "followers" who seem to think that the whole of Trotskyism (that is, the revolutionary Marxism of our time) is contained in the theory that Russia is still a workers' state and in the slogan of "unconditional defence of the Soviet Union." They merely prove that they have retired from a life of active and critical thought, and from the realities of life in general, and confine themselves to memorising by heart two pages of an otherwise uncut and unread book. They would be the first to deny, by the way, that the whole of Leninism is contained in Lenin's theory of the "democratic dictatorship of the proletariat and peasantry" or in his strictures against Trotsky and the theory of the permanent revolution.

The whole of Trotsky, for the new generation of Marxists that must be trained up and organised, does not lie in his contradictory theory of the class character of Russia; it is not even a decisively important part of the whole. Trotskyism is all of Marx, Engels and Lenin that has withstood the test of time and struggle — and that is a good deal! Trotskyism is its leader's magnificent development and amplification of the theory of the permanent revolution. Trotskyism is the defence of the great and fundamental principles of the Russian Bolshevik revolution and the Communist International, which it brought into existence. Trotskyism is the principle of workers' democracy, of the struggle for democracy and socialism.

In this sense — and it is the only one worth talking about — *The New Course* is a Trotskyist classic. It was not only a weapon hitting at the very heart of decaying bureaucratism in revolutionary Russia. It was and is a guide for the struggle against the vices of bureaucratism throughout the labour and revolutionary movements.

Bureaucratism is not simply a direct product of certain economic privileges acquired by the officialdom of the labour movement. It is also an ideology, a concept of leadership and of its relationship to the masses, which is absorbed even by labour and revolutionary officialdoms who enjoy no economic privileges at all. It is an ideology that reeks of its bourgeois origin. Boiled down to its most vicious essence, it is the kind of thinking and living and leading which says to the rank and file, in the words Trotsky once used to describe the language of Stalinism: "No thinking! Those at the top have more brains than you."

We see this ideology reflected in the every-day conduct of our own American trade union bureaucracy: "We will handle everything. Leave things to us. You stay where you are, and keep still." We see it reflected throughout the big social-democratic (to say nothing of the Stalinist) parties: "We will negotiate things. We will arrange everything. We will manoeuvre cleverly with the enemy, and get what you want without struggle. You sit still until further orders. That is all you are fit for." We even see it in those smaller revolutionary groups which are outside the reformist and Stalinist movements and which consider that this fact alone immunises them from bureaucratism. We repeat, it is a bourgeois ideology through and through. It is part of the ideas that the bourgeoisie, through all its agencies for moulding the mind of the masses, seeks to have prevail: "Whatever criticism you may have to make of us, remember this: The masses are stupid. It is no accident that they are at the bottom of the social ladder. They are incapable of rising to the top. They need a ruler over them; they cannot rule themselves. For their own good, they must be kept where they are."

The New Course does more than dismiss this odious ideology that fertilises the mind of the labour bureaucracy. It analyses its source and its nature. It diagnoses the evil to perfection. It indicates the operation needed to remove it, and the tools with which to perform the operation. It is the same tool needed by the proletariat for its emancipation everywhere. Its name is the democratically organised and controlled, self-acting, dynamic, critical, revolutionary political party of the working class.

The counter-revolution in Russia was made possible only because Stalinism blunted, then wore down, then smashed to bits this indispensable tool of the proletariat. The bureaucracy won. "If Trotsky had been right," says the official iconographer of Stalin, Henri Barbusse, "he would have won." How simple! What a flattering compliment to ... Hitler. The bureaucracy not only won, but consolidated its power on a scale unknown in any country of the world throughout all history. Stalin himself is now the Pope-Czar of the Russian Empire.

But that is only how it seems on the surface; that is how it is only for a very short while, as history counts. "Any imbecile can rule with a state of siege," said Rochefort. Only the really powerful and confident can rule by establishing peaceful relations in the country. That, the new bureaucracy, without a past and without a future, cannot do. The combined efforts of world capitalism cannot do that nowadays, still less the efforts of the Stalinist nobility. The latter has succeeded in establishing "socialism," for itself and "in a single country." It will not live long to enjoy it. Together with all modern rulers, it is doomed to perish in the unrelenting world crisis that it cannot solve, or to perish at the hands of an avenging socialist proletariat.

Cromwell's Roundheads marched with Bibles in their hands. The militant proletariat needs no divine revelations or scriptural injunctions, no Bibles or saviours. But it will march to victory only if its conscious vanguard has assimilated the rich and now-more-timely-than-ever lessons to be learned from the classic work of the organiser of the first great proletarian revolution.

Blanquism and social democracy

Rosa Luxemburg

Comrade Plekhanov has published in the [Polish newspaper] *Kuryer* a detailed article in which he accuses the so-called Bolsheviks of Blanquism.

It is not our job to defend the Russian comrades against whom comrade Plekhanov brings up the artillery of his erudition and his dialectics. For sure they can do that themselves. But the problem in question calls forth a few remarks which may be of interest to our readers, and so we give them some space.

Comrade Plekhanov, to characterise Blanquism, quotes Engels on Blanqui, a French revolutionary of the 1840s from whose name the designation of a whole tendency derives. Engels says:

"In his political activity he was mainly a 'man of action', believing that a small and well organised minority, who would attempt a political stroke of force at the opportune moment, could carry the mass of the people with them by a few successes at the start and thus make a victorious revolution... From Blanqui's assumption, that any revolution may be made by the outbreak of a small revolutionary minority, follows of itself the necessity of a dictatorship after the success of the venture. This is, of course, a dictatorship, not of the entire revolutionary class, the proletariat, but of the small minority that has made the revolution, and who are themselves previously organised under the dictatorship of one or several individuals".

Frederick Engels, Karl Marx's comrade-in-arms, is indubitably a great authority, but the question of whether this characterisation of Blanqui is completely just can nevertheless be discussed. For, in 1848, Blanqui was not at all obliged to presume that his group would be a "small minority"; on the contrary, then, in that time of powerful revolutionary stirrings, he reckoned with certainty that the whole body of the working people, in Paris at least if not in France, would respond to his appeal for struggle against the deceitful and iniquitous policy of a bourgeois ministry seeking to "cheat the people of their victory". However, the basic question is elsewhere: that comrade Plekhanov tries to demonstrate that the characterisation of Blanqui made by Engels applies exactly to the so-called "Bolsheviks" (whom Plekhanov, with no more ado, calls "the minority" because they found themselves in a minority at the reunification congress [in April 1906, of the RSDLP, of which the Bolsheviks and Mensheviks were factions]).

He says, in so many words: "This whole characterisation applies completely to our present minority".

He justifies this statement in the following way: "The relation of the Blanquists to the popular masses was utopian in the sense that they had not understood the significance of the revolutionary autonomy of those masses. According to their schemes, only the conspirators were properly speaking active, and the masses merely supported them, carried along by the well organised minority.

Comrade Plekhanov affirms that the Russian "Bolshevik" comrades (we prefer to stick to the usual terminology) have succumbed to this "original sin of Blanquism". In our opinion, comrade Plekhanov does not prove his case. For comparison with the members of [the 19th century populist movement] Narodnaya Volya [People's Will], who really were Blanquists, proves nothing, and the malevolent remark that Zhelyabov, the hero and leader of Narodnaya Volya, was endowed with a more astute political instinct than the "Bolshevik" leader Lenin is in too bad taste to detain us. As we have already said, it is not our business to break lances in defence of the "Bolsheviks" and comrade Lenin, for they let nothing past them from anyone. We are concerned with the root of the matter. The question is: in the Russian revolution of today, is Blanquism possible? If such a tendency could exist, could it exert any influence?

We think that it is enough to pose the question thus for anyone with any awareness of the current revolution, anyone who has had direct contact with it, to give a negative reply. The whole difference between the French situation of 1848 and the current situation in the Russian empire consists exactly in the fact that the relationship between the "organised minority", that is the proletarian party, and the masses has fundamentally changed. In 1848 the revolutionaries, to the degree that they were socialists, made desperate efforts to bring socialist ideas to the masses, to stop them supporting empty bourgeois liberalism. That socialism was nebulous, utopian, and petty-bourgeois.

Today in Russia, things present themselves differently: neither our [Poland's] rancid old liberal democrats, nor the Cadet organisation, the Tsarist constitutionalists of Russia, nor any other national "progressive" bourgeois party in another part of the [Tsarist] empire, has been able to win the broad working masses. Today, those masses are gathering under the socialist banner: when the revolution broke out, they placed themselves, of their own initiative, almost spontaneously, under the red flag. And that is the best evidence in favour of our party.

We do not want to hide the fact that in 1903 we were still only a handful, that as a party, in the strict sense of the term, as effectively organised comrades, we were at the most a few hundred, and when we appeared publicly or demonstrated only a small band of workers joined us. Today, we as a party are counted in tens of thousands.

Where does the difference come from? Is it because we have so many genius leaders in our party? Because we are, perhaps, such celebrated conspirators? Not at all. For sure, none of our leaders, that is, none of those to whom the party has confided responsibilities, would want to expose themselves to ridicule by inviting a comparison between themselves and old Blanqui, yesteryear's lion of the revolution. Few of our agitators can match the old conspirators of the Blanquist group as regards their personal catchment and their organisational abilities. How is our success, and the lack of success of the Blanquists, explained? Quite simply by the fact that the "masses" are different "masses": the groups of workers who today are fighting Tsarism; people whom life itself has made into socialists; people who absorbed hate of the established

Karl Kautsky

order with their mothers' milk; people whom necessity has taught to think socialistically.

That is the difference. Neither the leaders, nor even the ideas, but social and economic conditions, have given rise to it — conditions which are such that they exclude any possibility of a common class struggle uniting the proletariat and the bourgeoisie.

Since the masses are different, since the proletariat is different, we cannot speak today of Blanquist or conspiratorial tactics. Blanqui and his heroic comrades made superhuman efforts to draw the masses into class struggle; they did not succeed, because they had to do with workers who had not yet broken with the guild system, who were still immersed in petty bourgeois ideology.

We social democrats [i.e. Marxists] have a much simpler and easier task: today we just have to work to lead the class struggle which has ignited with an inexorable necessity. The Blanquists tried to pull the masses behind them, while we, the social democrats [Marxists] today, are almost pushed along by the masses. The difference is great, as great as that between a pilot who with great efforts gets his boat to move against the stream and one who steers a boat carried along by the torrential current. The first has not enough strength and he will not achieve his goal, while the second has only the task of making sure that the boat does not deviate from its route, does not shatter on a reef or get stuck on a sandbank.

Comrade Plekhanov should calm down as regards the "revolutionary autonomy of the masses". That autonomy exists, nothing will hold it back, and all the bookish sermons (we ask pardon for this expression, but we can find no other) about its necessity can only bring smiles from those who are working among and with the masses.

We contest the idea that the comrades of the so-called "majority" in today's Russian were victims of Blanquist deviations in the course of the revolution, as comrade Plekhanov reproaches them for having been. It is possible that there were traces in the organisational plan which comrade Lenin drafted in 1902, but that lies in the past — the distant past, since today life is moving quickly, dizzyingly quickly. Those errors have been corrected by life itself, and there is no danger of them being repeated. There is nothing scary about the ghost of Blanquism, because it cannot come back to life these days. The danger we face, on the contrary, is that comrade Plekhanov and his "minority" supporters who fear Blanquism so much will fall into the opposite extreme and ground the boat on a sandbank. We see that opposite extreme in the fact that these comrades so fear being in the minority that they reckon on masses outside the proletariat.

From that flow the calculations about the Duma [fake Parliament convened by the Tsar]; the false slogans in the Central Committee's "directives" about supporting the Cadet gentlemen; the attempt to raise the demand "Down with the bureaucrat ministry" [i.e. the Menshevik slogan "For a Duma ministry"], and similar errors. The boat will not remain grounded on the sandbank; there is no danger of that; the tumultuous stream of the swelling revolution will quickly push along the boat of the workers; but it would be a pity if these mistakes lose us even one moment of time.

In the same way, the notion of the "dictatorship of the proletariat" has taken a different significance from before. Frederick Engels justly emphasises that the Blanquists envisaged not a dictatorship of "the entire revolutionary class, the proletariat, but of the small minority that has made the revolution". Today, the question is presented quite differently. It is not an organisation of conspirators that "makes the revolution" and can envisage establishing its dictatorship. Even the Narodnaya Volya types and their claimed heirs, the Russian Social Revolutionaries [populists], have long ceased dreaming of any such thing.

If the "Bolshevik" comrades today speak of the dictatorship of the proletariat today, they have never given it the old Blanquist meaning, and nor have they ever fallen into the mistake of Narodnaya Volya, which dreamed of "taking power for itself". On the contrary, they have affirmed that the current revolution can find its culmination when the proletariat, the whole revolutionary class, has seized the state machine. The proletariat, as the most revolutionary element, will perhaps take on its role of liquidator of the old regime by "taking power for itself" in order to oppose the counter-revolution and stop the revolution being weakened by the bourgeoisie which is reactionary in its very nature. No revolution has ever been concluded other than by the dictatorship of a class, and all the indications are that at the present time the proletariat can become that liquidator of the old regime.

Obviously, no social-democrat [Marxist] indulges in the illusion that the proletariat can maintain itself in power: if it could maintain itself in power, it would then bring its class ideas to dominion, it would realise socialism. Its forces are not sufficient for that today, since the proletariat, in the strict sense of the word, constitutes a minority of society in the Russian Empire. The realisation of socialism by a minority is unconditionally excluded, since the very idea of socialism excludes the domination of a minority. Thus, on the day after the political victory of the proletariat over Tsarism, the majority will withdraw from it the power that it will have conquered.

To speak concretely: after the fall of Tsarism, power will pass into the hands of the most revolutionary part of society, the proletariat, because the proletariat will seize all the posts and will remain on its guard as long as power is not in the hands of those legally called upon to hold it, in the hands of the new government, which only the Constituent Assembly, as the legislative organ elected by the whole population, can determine. The plain fact is that in society not the proletariat, but the petty bourgeoisie and the peasantry, constitutes the majority, and that consequently, in the Constituent Assembly, not the Social Democrats but the peasant and petty bourgeois democrats will form the majority. We can deplore this fact, but not change it.

Such is, in broad outline, the situation, according to the assessment of the "Bolsheviks", and all the Social Democratic organisations and parties outside Russia itself [i.e. in the non-Russian bits of the Tsarist empire] share this vision. Where the Blanquism is in here, is difficult to see.

To justify his assertion even superficially comrade Plekhanov is obliged to rip the words of comrade Lenin and his supporters from their context. If we wanted to do that sort of thing, we could also show that the "Mensheviks" have recently been "Blanquists", starting with comrade Parvus and finishing with comrade... Plekhanov! But that would be a sterile game in scholasticism. Comrade Plekhanov's article is embittered and embittering, which is a bad thing: "When Jupiter rages, it is that Jupiter errs."

It is high time to finish with this scholasticism and with all this fuss about who is "Blanquist" and who is "orthodox Marxist". Today, it is a matter of whether, in current conditions, the tactic recommended by comrade Plekhanov and the Menshevik comrades is correct — a tactic which aims to work with the Duma as much as possible, and outside the Duma with the elements which are represented there; or whether on the contrary the tactic is correct which we like the Bolshevik comrades are pursuing, a tactic based on the principle that the centre of gravity is situated outside the Duma, in the active presence on the scene of the revolutionary popular masses. So far the Menshevik comrades have been able to persuade no-one of the correctness of their views, and no-one will be any more persuaded by them linking the "Blanquist" label to their adversaries.[1]

---

[1] Originally published as "Blanquism and social-democracy", *Czerwony Sztander* [Red Flag], Cracow, no. 82, 27 June 1906. Translated at bit.ly/rl-bsd from a German version in the volume *Die Polnische Schriften*, edited by Jurgen Hentze, pp.298ff. The German text gives the title of Plekhanov's article as "Wo bleibt die Rechte?", roughly, where is the right wing? But it seems to be the same article as referred to in Lenin's article, "How comrade Plekhanov argues about social-democratic tactics", *Collected Works* volume 10, p.460: the Plekhanov article is there referred to as "On Tactics and Tactlessness".

Hands off Rosa Luxemburg!

Leon Trotsky

Stalin's article, "Some Questions Concerning the History of Bolshevism," reached me after much delay. After receiving it, for a long time I could not force myself to read it, for such literature sticks in one's throat like sawdust or mashed bristles. But still, having finally read it, I came to the conclusion that one cannot ignore this performance, if only because there is included in it a vile and barefaced calumny about Rosa Luxemburg. This great revolutionist is enrolled by Stalin into the camp of centrism! He proves — not proves, of course, but asserts — that Bolshevism from the day of its inception held to the line of a split with the Kautsky centre, while Rosa Luxemburg during that time sustained Kautsky from the left. I quote his own words: "... long before the war, approximately since 1903-04, when the Bolshevik group in Russia took shape and when the Lefts in the German Social Democracy first raised their voice, Lenin pursued a line toward a rupture, toward a split with the opportunists both here, in the Russian Social Democratic Labour Party, and over there, in the Second International, particularly in the German Social-Democratic Party." That this, however, could not be achieved was due entirely to the fact that "the Left Social Democrats in the Second International, and above all in the German Social-Democratic Party, were a weak and powerless group ... and afraid even to pronounce the word 'rupture', 'split'."

To put forward such an assertion, one must be absolutely ignorant of the history of one's own party, and first of all, of Lenin's ideological course. There is not a single word of truth in Stalin's point of departure. In 1903-04, Lenin was, indeed, an irreconcilable foe of opportunism in the German Social Democracy. But he considered as opportunism only the revisionist tendency which was led theoretically by Bernstein.

Kautsky at the time was to be found fighting against Bernstein. Lenin considered Kautsky as his teacher and stressed this everywhere he could. In Lenin's work of that period and for a number of years following, one does not find even a trace of criticism in principle directed against the Bebel-Kautsky tendency. Instead one finds a series of declarations to the effect that Bolshevism is not some sort of an independent tendency but is only a translation into the language of Russian conditions of the tendency of Bebel-Kautsky. Here is what Lenin wrote in his famous pamphlet, *Two Tactics*, in the middle of 1905: "When and where did I ever call the revolutionism of Bebel and Kautsky 'opportunism'?... When and where have there been brought to light differences between me, on the one hand, and Bebel and Kautsky on the other?... The complete unanimity of international revolutionary Social Democracy on all major questions of program and tactics is a most incontrovertible fact". Lenin's words are so clear, precise, and categorical as to entirely exhaust the question.

A year and a half later, on December 7, 1906, Lenin wrote in the article "The Crisis of Menshevism": "... from the beginning we declared (see *One Step Forward, Two Steps Back*): We are not creating a special 'Bolshevik' tendency; always and everywhere we merely uphold the point of view of revolutionary Social Democracy. And right up to the social revolution there will inevitably always be an opportunist wing and a revolutionary wing of Social Democracy".

Speaking of Menshevism as the opportunistic wing of the Social Democracy, Lenin compared the Mensheviks not with Kautskyism but with revisionism. Moreover he looked upon Bolshevism as the Russian form of Kautskyism, which in his eyes was in that period identical with Marxism. The passage we have just quoted shows, incidentally, that Lenin did not at all stand absolutely for a split with the opportunists; he not only admitted but also considered "inevitable" the existence of the revisionists in the Social Democracy right up to the social revolution.

Two weeks later, on December 20, 1906, Lenin greeted enthusiastically Kautsky's answer to Plekhanov's questionnaire on the character of the Russian revolution: "He has fully confirmed our contention that we are defending the position of revolutionary Social Democracy against opportunism, and not creating any 'peculiar' Bolshevik tendency. . ." ("The Proletariat and Its Ally in the Russian Revolution," December 10, 1906).

Within these limits, I trust the question is absolutely clear. According to Stalin, Lenin, even from 1903, had demanded a break in Germany with the opportunists, not only of the right wing (Bernstein) but also of the left (Kautsky). Whereas in December 1906, Lenin as we see was proudly pointing out to Plekhanov and the Mensheviks that the tendency of Kautsky in Germany and the tendency of Bolshevism in Russia were — identical. Such is part one of Stalin's excursion into the ideological history of Bolshevism. Our investigator's scrupulousness and his knowledge rest on the same plane!

Directly after his assertion regarding 1903-04, Stalin makes a leap to 1916 and refers to Lenin's sharp criticism of the war pamphlet by Junius, i.e., Rosa Luxemburg. To be sure, in that period Lenin had already declared war to the finish against Kautskyism, having drawn from his criticism all the necessary organisational conclusions. It is not to be denied that Rosa Luxemburg did not pose the question of the struggle against centrism with the requisite completeness — in this Lenin's position was entirely superior. But between October 1916, when Lenin wrote about Junius's pamphlet and 1903, when Bolshevism had its inception, there is a lapse of thirteen years; in the course of the major part of this period Rosa Luxemburg was to be found in opposition to the Kautsky and Bebel Central Committee, and her fight against the formal, pedantic, and rotten-at-the-core "radicalism" of Kautsky took on an ever increasingly sharp character.

Lenin did not participate in this fight and did not support Rosa Luxemburg up to 1914. Passionately absorbed in Russian affairs, he preserved extreme caution in international matters. In Lenin's eyes Bebel and Kautsky stood immeasurably higher as revolutionists than in the eyes of Rosa Luxemburg, who observed them at closer range, in action, and who was much more directly subjected to the atmosphere of German politics. The capitulation of German Social Democracy on August 4 was entirely unexpected by Lenin. It is well known that the issue of the Vorwärts with the patriotic declaration of the Social-Democratic faction was taken by Lenin to be a forgery by the German general staff. Only after he was absolutely convinced of the awful truth did he subject to revision his evaluation of the basic tendencies of the German Social Democracy, and while so doing he performed that task in the Leninist manner, i.e., he finished it off once for all.

On October 27, 1914, Lenin wrote to A. Shlyapnikov: "I hate and despise Kautsky now more than anyone, with his vile, dirty, self-satisfied hypocrisy... Rosa Luxemburg was right when she wrote, long ago, that Kautsky has the 'subservience of a theoretician' — servility, in plainer language, servility to the majority of the party, to opportunism".

Were there no other documents — and there are hundreds — these few lines alone could unmistakably clarify the history of the question. Lenin deemed it necessary at the end of 1914 to inform one of his colleagues closest to him at the time that "now," at the present moment, today, in contradistinction to the past, he "hates and despises" Kautsky. The sharpness of the phrase is an unmistakable indication of the extent to which Kautsky betrayed Lenin's hopes and expectations. No less vivid is the second phrase, "Rosa Luxemburg was right when she wrote, long ago, that Kautsky has the 'subservience of a theoretician'..." Lenin hastens here to recognise that "verity" which he did not see formerly, or which, at least, he did not recognise fully on Rosa Luxemburg's side.

Such are the chief chronological guideposts of the questions, which are at the same time important guideposts of Lenin's political biography. The fact is indubitable that his ideological orbit is represented by a continually rising curve. But this only means that Lenin was not born Lenin full-fledged, as he is pictured by the slobbering daubers of the "divine," but that he made himself Lenin. Lenin ever extended his horizons, he learned from others and daily drew himself to a higher plane than was his own yesterday. In this perseverance, in this stubborn resolution of a continual spiritual growth over his own self did his heroic spirit find its expression. If Lenin in 1903 had understood and formulated everything that was required for the coming times, then the remainder of his life would have consisted only of reiterations. In reality this was not at all the case. Stalin simply stamps the Stalinist imprint on Lenin and coins him into the petty small change of numbered adages.

In Rosa Luxemburg's struggle against Kautsky, especially in 1910-14, an important place was occupied by the questions of war, militarism, and pacifism. Kautsky defended the reformist program: limitations of armaments, international court etc. Rosa Luxemburg fought decisively against this program as illusory. On this question Lenin was in some doubt, but at a certain period he stood closer to Kautsky than to Rosa Luxemburg. From conversations at the time with Lenin I recall that the following argument of Kautsky made a great impression upon him: just as in domestic questions, reforms are products of the revolutionary class struggle, so in international relationships it is possible to fight for and to gain certain guarantees ("reforms") by means of the international class struggle. Lenin considered it entirely possible to support this position of Kautsky, provided that he, after the polemic with Rosa Luxemburg, turned upon the right-wingers (Noske and Co.). I do not undertake now to say from memory to what extent this circle of ideas found its expression in Lenin's articles; the question would require a particularly careful analysis. Neither can I take upon myself to assert from memory how soon Lenin's doubts on this question were settled. In any case they found their expression not only in conversations but also in correspondence. One of these letters is in the possession of Karl Radek.[2]

I deem it necessary to supply on this question evidence as a witness in order to attempt in this manner to save an exceptionally valuable document for the theoretical biography of Lenin. In the autumn of 1926, at the time of our collective work over the platform of the Left Opposition, Radek showed Kamenev, Zinoviev, and me — probably also other comrades as well — a letter of Lenin to him (1911?) which consisted of a defence of Kautsky's position against the criticism of the German Lefts. In accordance with the regulation passed by the Central Committee, Radek, like all others, should have delivered this letter to the Lenin Institute. But fearful lest it be hidden, if not destroyed, in the Stalinist factory of fabrications, Radek decided to preserve the letter till some more opportune time. One cannot deny that there was some foundation to Radek's attitude. At present, however, Radek himself has — though not very responsible — still quite an active part in the work of producing political forgeries. Suffice it to recall that Radek, who in distinction to Stalin is acquainted with the history of Marxism, and who, at any rate, knows this letter of Lenin, found it possible to make a public statement of his solidarity with the insolent evaluation placed by Stalin on Rosa Luxemburg. The circumstance that Radek acted thereupon under Yaroslavsky's rod does not mitigate his guilt, for only despicable slaves can renounce the principles of Marxism in the name of the principles of the rod.

However the matter we are concerned with relates not to the personal characterisation of Radek but to the fate of Lenin's letter. What happened to it? Is Radek hiding it even now from the Lenin Institute? Hardly. Most probably, he entrusted it, where it should be entrusted, as a tangible proof of an intangible devotion. And what lay in store for the letter thereafter? Is it preserved in Stalin's personal archives alongside with the documents that compromise his closest colleagues? Or is it destroyed as many other most precious documents of the party's past have been destroyed?

In any case there cannot be even the shadow of a political reason for the concealment of a letter written two decades ago on a question that holds now only a historical interest. But it is precisely the historical value of the letter that is exceptionally great. It shows Lenin as he really was, and not as he is being recreated in their own semblance and image by the bureaucratic dunderheads, who pretend to infallibility. We ask, where is Lenin's letter to Radek? Lenin's letter must be where it belongs! Put it on the table of the party and of the Comintern!

If one were to take the disagreements between Lenin and Rosa Luxemburg

---

1. Lenin's letter to Radek (30 September 1910) is at bit.ly / 100930-vil

in their entirety, then historical correctness is unconditionally on Lenin's side. But this does not exclude the fact that on certain questions and during definite periods Rosa Luxemburg was correct as against Lenin. In any case, the disagreements, despite their importance and at times their extreme sharpness, developed on the bases of revolutionary proletarian policies common to them both.

When Lenin, going back into the past wrote in October 1919 ("Greetings to Italian, French, and German Communists") that "... at the moment of taking power and establishing the Soviet republic, Bolshevism was united; it drew to itself all that was best in the tendencies of socialist thought akin to it...", I repeat, when Lenin wrote this he unquestionably had in mind also the tendency of Rosa Luxemburg, whose closest adherents, e.g., Markhlewsky, Dzerzhinsky, and others were working in the ranks of the Bolsheviks.

Lenin understood Rosa Luxemburg's mistakes more profoundly than Stalin; but it was not accidental that Lenin once quoted the old couplet in relation to Luxemburg: Although the eagles do swoop down and beneath the chickens fly, chickens with outspread wings never will soar amid clouds in the sky. Precisely the case! Precisely the point! For this very reason Stalin should proceed with caution before employing his vicious mediocrity when the matter touches figures of such stature as Rosa Luxemburg.

In his article "A Contribution to the History of the Question of the Dictatorship" (October 1920), Lenin, touching upon questions of the Soviet state and the dictatorship of the proletariat already posed by the 1905 revolution, wrote: "While such outstanding representatives of the revolutionary proletariat and of unfalsified Marxism as Rosa Luxemburg immediately realised the significance of this practical experience and made a critical analysis of it at meetings and in the press," on the contrary, "... people of the type of the future 'Kautsky-ites' proved absolutely incapable of grasping the significance of this experience..." In a few lines, Lenin fully pays the tribute of recognition to the historical significance of Rosa Luxemburg's struggle against Kautsky — a struggle which Lenin himself had been far from immediately evaluating at its true worth. If to Stalin, the ally of Chiang Kai-shek, and the comrade-in-arms of Purcell, the theoretician of "the worker-peasant party" of "the democratic dictatorship," of "non-antagonising the bourgeoisie," etc. — if to him Rosa Luxemburg is the representative of centrism, to Lenin she is the representative of "unfalsified Marxism." What this designation meant coming as it does from Lenin's pen is clear to anyone who is even slightly acquainted with Lenin.

I take the occasion to point out here that in the notes to Lenin's works there is among others the following said about Rosa Luxemburg: "During the florescence of Bernsteinian revisionism and later of ministerialism (Millerand), Luxemburg carried on against this tendency a decisive fight, taking her position in the left wing of the German party... In 1907 she participated as a delegate of the SD of Poland and Lithuania in the London congress of the RSDLP, supporting the Bolshevik faction on all basic questions of the Russian revolution. From 1907, Luxemburg gave herself over entirely to work in Germany, taking a left-radical position and carrying on a fight against the centre and the right wing... Her participation in the January 1919 insurrection has made her name the banner of the proletarian revolution."

Of course the author of these notes will in all probability tomorrow confess his sins and announce that in Lenin's epoch he wrote in a benighted condition, and that he reached complete enlightenment only in the epoch of Stalin. At the present moment announcements of this sort — combinations of sycophancy, idiocy, and buffoonery — are made daily in the Moscow press. But they do not change the nature of things: What's once set down in black and white, no axe will hack nor all your might. Yes, Rosa Luxemburg has become the banner of the proletarian revolution!

How and wherefore, however, did Stalin suddenly busy himself — at so be-lated a time — with the revision of the old Bolshevik evaluation of Rosa Lux-emburg? As was the case with all his preceding theoretical abortions so with this latest one, and the most scandalous, the origin lies in the logic of his strug-gle against the theory of permanent revolution. In this "historical" article, Stalin once again allots the chief place to this theory. There is not a single new word in what he says. I have long ago answered all his arguments in my book *The Permanent Revolution*. From the historical viewpoint the question will be sufficiently clarified, I trust, in the second volume of *The History of the Russian Revolution* (the October Revolution), now on the press. In the present case the question of the permanent revolution concerns us only insofar as Stalin links it up with Rosa Luxemburg's name. We shall presently see how the hapless theoretician has contrived to set up a murderous trap for himself.

After recapitulating the controversy between the Mensheviks and the Bol-sheviks on the question of the motive forces of the Russian revolution and after masterfully compressing a series of mistakes into a few lines, which I am com-pelled to leave without an examination, Stalin writes: "What was the attitude of the German Left Social Democrats, of Parvus and Rosa Luxemburg, to this controversy? They invented a utopian and semi-Menshevik scheme of perma-nent revolution... Subsequently, this semi-Menshevik scheme of permanent revolution was seized upon by Trotsky (in part by Martov) and turned into a weapon of struggle against Leninism." Such is the unexpected history of the origin of the theory of the permanent revolution, in accordance with the latest historical researches of Stalin. But, alas, the investigator forgot to consult his own previous learned works. In 1925 this same Stalin had already expressed himself on this question in his polemic against Radek. Here is what he wrote then: "It is not true that the theory of the permanent revolution... was put for-ward in 1905 by Rosa Luxemburg and Trotsky. As a matter of fact this theory was put forward by Parvus and Trotsky." This assertion may be consulted on page 185, Problems of Leninism, Russian edition, 1926. Let us hope that it ob-tains in all foreign editions.

So, in 1925, Stalin pronounced Rosa Luxemburg not guilty in the commission of such a cardinal sin as participating in the creation of the theory of the per-manent revolution. "As a matter of fact this theory was put forward by Parvus and Trotsky." In 1931, we are informed by the identical Stalin that it was pre-cisely "Parvus and Rosa Luxemburg... who invented a utopian and semi-Men-shevik scheme of permanent revolution." As for Trotsky he was innocent of creating the theory, it was only "seized upon" by him, and at the same time by... Martov! Once again Stalin is caught with the goods. Perhaps he writes on questions of which he can make neither head nor tail. Or is he consciously shuffling marked cards in playing with the basic questions of Marxism? It is incorrect to pose this question as an alternative. As a matter of fact both the one and the other are true. The Stalinist falsifications are conscious insofar as they are dictated at each given moment by entirely concrete personal interests. At the same time they are semi-conscious, insofar as his congenital ignorance places no impediments whatsoever to his theoretical propensities.

But facts remain facts. In his war against "the Trotskyist contraband," Stalin has fallen foul of a new personal enemy, Rosa Luxemburg! He did not pause for a moment before lying about her and vilifying her; and moreover, before proceeding to put into circulation his giant doses of vulgarity and disloyalty, he did not even take the trouble of verifying what he himself had said on the same subject six years before.

The new variant of the history of the ideas of the permanent revolution was indicated first of all by an urge to provide a dish more spicy than all those preceding. It is needless to explain that Martov was dragged in by the hair for the sake of the greater piquancy of theoretical and historical cookery. Martov's attitude to the theory and practice of the permanent revolution was one of unalterable antagonism, and in the old days he stressed more than once that Trotsky's views on revolution were rejected equally by the Bolsheviks and the Mensheviks. But it is not worthwhile to pause over this.

What is truly fatal is that there is not a single major question of the international proletarian revolution on which Stalin has failed to express two directly contradictory opinions. We all know that in April 1924, he conclusively demonstrated in *Problems of Leninism* the impossibility of building socialism in one country. In autumn, in a new edition of the book, he substituted in its place a proof — i.e., a bald proclamation — that the proletariat "can and must" build socialism in one country. The entire remainder of the text was left unchanged. On the question of the worker-peasant party, of the Brest-Litovsk negotiations, the leadership of the October Revolution, on the national question, etc., etc., Stalin contrived to put forward, for a period of a few years, sometimes of a few months, opinions that were mutually exclusive. It would be incorrect to place the blame in everything on a poor memory. The matter reaches deeper here. Stalin completely lacks any method of scientific thinking, he has no criteria of principles. He approaches every question as if that question were born only today and stood apart from all other questions. Stalin contributes his judgments entirely depending upon whatever personal interest of his is uppermost and most urgent today. The contradictions that convict him are the direct vengeance for his vulgar empiricism. Rosa Luxemburg does not appear to him in the perspective of the German, Polish, and international workers' movement of the last half-century. No, she is to him each time a new, and, besides, an isolated figure, regarding whom he is compelled in every new situation to ask himself anew, "Who goes there, friend or foe?" Unerring instinct has this time whispered to the theoretician of socialism in one country that the shade of Rosa Luxemburg is irreconcilably inimical to him. But this does not hinder the great shade from remaining the banner of the international proletarian revolution.

Rosa Luxemburg criticised very severely and fundamentally incorrectly the policies of the Bolsheviks in 1918 from her prison cell. But even in this, her most erroneous work, her eagle's wings are to be seen. Here is her general evaluation of the October insurrection: "Everything that a party could offer of courage, revolutionary farsightedness, and consistency in a historic hour, Lenin, Trotsky, and the other comrades have given in good measure. All the revolutionary honour and capacity which the Social Democracy of the West lacked were represented by the Bolsheviks. Their October uprising was not only the actual salvation of the Russian Revolution; it was also the salvation of the honour of international socialism." Can this be the voice of centrism?

In the succeeding pages, Luxemburg subjects to severe criticism the policies of the Bolsheviks in the agrarian sphere, their slogan of national self-determination, and their rejection of formal democracy. In this criticism we might add, directed equally against Lenin and Trotsky, she makes no distinction whatever between their views; and Rosa Luxemburg knew how to read, understand, and seize upon shadings. It did not even fall into her head, for instance, to accuse me of the fact that by being in solidarity with Lenin on the agrarian question, I had changed my views on the peasantry. And moreover she knew these views very well since I had developed them in detail in 1909 in her Polish journal. Rosa Luxemburg ends her criticism with the demand, "in the policy of the Bolsheviks the essential must be distinguished from the unessential, the fundamental from the accidental." The fundamental she considers to be the force of the action of the masses, the will to socialism. "In this," she writes, "Lenin and Trotsky and their friends were the first, those who went ahead as an example to the proletariat of the world; they are still the only ones up to now who can cry with Hutten, 'I have dared!'"

Yes, Stalin has sufficient cause to hate Rosa Luxemburg. But all the more imperious therefore becomes our duty to shield Rosa's memory from Stalin's calumny that has been caught by the hired functionaries of both hemispheres, and to pass on this truly beautiful, heroic, and tragic image to the young generations of the proletariat in all its grandeur and inspirational force.

A group of POUM fighters in Barcelona, 1937.

Trotskyism and the PSOP

Leon Trotsky

It was with real interest that I turned to Marceau Pivert's article, *The PSOP and Trotsky-ism*, appearing in the June 9, 1939, issue of the PSOP organ.[1] I had supposed that Pivert would finally submit the differences separating him from the Fourth International to a concrete analysis. Regrettably, from the very first lines I was disappointed. Pivert does not make even an attempt to venture into the field of Marxian theory and class politics. His entire criticism of "Trotskyism" remains on the level of psychology, moralising, and the rules of politeness. Pivert manifestly avoids any serious discussion of the funda-mental problems of the labor movement. This I shall try to demonstrate through patient analysis of all the ideas and even nuances of ideas contained in Pivert's article, which in its theme is programmatic.

"Claims to Hegemony"
    Pivert is ready to collaborate with "Trotskyism," provided only that the latter aban-dons all claims to "hegemony" and takes the pathway of "trustful collaboration with all elements that have courageously broken with social patriotism and national com-munism." The very counterposing of collaboration to "claims to hegemony" is enough to arouse suspicion. The participation of different tendencies within a party unques-tionably presupposes trust in the possibility of convincing one another, learning from one another. If differences arise, every tendency confident of its views will seek to win a majority. Precisely this constitutes the mechanics of party democracy. What other "hegemony" is possible within a democratic party save that of winning a majority to one's views? After all did not Marceau Pivert and his friends strive to gain a majority at the last congress of the PSOP? And didn't they obtain it? Didn't they thereby install their "hegemony" in the party? Was that to their discredit? Pivert's line of argumenta-tion shows that he considers the "hegemony" of his own tendency as the norm and the law, and any attempt of another tendency to win a majority a violation of the norm, a crime, worse yet — Trotskyism. Where then is democracy?

"Factional Methods"
    Having thus proclaimed "hegemony" to be his private monopoly in the party, Pivert thereupon demands that the Trotskyites "abandon factional methods." This demand, repeated several times, comes somewhat incongruously from the pen of a politician who constantly underscores the democratic nature of his organisation. What is a fac-tion?
    It is a temporary non-statutory and voluntary grouping of closest co-thinkers within a party, whose aim is to convince the party of the correctness of their viewpoint in the shortest possible period of time. The appearance of factions is unavoidable even in the most mature and harmonious party, owing to the extension of its influence upon new layers, the cropping up of new problems, sharp turns in the situation, errors of the lead-ership, and so on. From the standpoint of monolithism a factional struggle is an "evil";

---

1. Trotsky wrote this article (*New International*, October 1939) in response to moves by the leader-ship around Marceau Pivert against French Trotskyists who had joined the PSOP (Workers' and Peasants' Socialist Party), a group formed by left-wingers expelled from the French Socialist Party (led by Léon Blum) at the end of 1938. The PSOP collapsed soon after the start of World War 2. Pivert was active after World War 2 as a leftish Third Camp socialist within the Socialist Party.

but it is an unavoidable evil and, in any event, a far lesser evil than the prohibition of factions. True enough, attempts at the formation of factions lacking an adequate principled basis in consequence of political immaturity, personal ambition, careerism, etc. are frequently observable, especially in young parties. In all such cases it is the task of the leadership to expose, without recourse to police measures, the hollowness of these enterprises and in that way to discredit them before the party membership. Only in this way is it possible to create profound attachment for the party so that episodic conflicts, no matter how sharp, do not threaten its unity. The existence of factions, in the nature of things, provokes friction and involves an expenditure of energy, but this is the inevitable overhead expense of a democratic regime. A capable and authoritative leadership strives to reduce factional friction to a minimum. This is achieved by a correct policy tested by collective experience; by a loyal attitude toward the opposition; by the gradually increasing authority of the leadership; but never by prohibition of factions, something which cannot fail to invest the struggle with a hypocritical and poisonous character. Whoever prohibits factions thereby liquidates party democracy and takes the first step toward a totalitarian regime.

Building "Cells"

Pivert next demands of the "Trotskyites" that they renounce "building cells commanded from outside." The possibility itself of such a "demand" arises from a glaring confusion of concepts. Pivert himself doubtlessly considers it the duty of every PSOP member to organise cells in the trade unions to win over the majority of the workers. To the extent that these cells are attacked by the Jouhaux clique, Stalin's spies and the Sureté Nationale, they are compelled to lead an undercover existence. The PSOP, as a party, retains, I believe, the leadership of these cells in its hands "from outside". Were the PSOP to renounce such methods of work within the trade unions, within Blum's party and Stalin's party, it would thereby abandon the struggle for "hegemony" of the working class, that is to say, its revolutionary mission. I hope that is not the case! Where then are the differences? Pivert is simply scaring himself and scaring the party with the bogie of the Bolshevik method of "cells" without having reflected upon the gist of the problem.

But perhaps it is not a question of that at all, but rather of "Trotskyite" cells within the PSOP itself? We are then merely confronted with a restatement of the charge of factionalism. In this case, however, it is altogether wrong to speak of building cells, since it is open political collaboration which is involved, and an equally open ideological struggle between two tendencies. Assuredly, if the ideological struggle were to be replaced by bureaucratic repressions, then the "Trotskyites" would not only be justified but duty bound in my opinion to resort to the method of undercover cells. À la guerre comme à la guerre! But the responsibility for the existence of undercover cells would in that case fall squarely upon the shoulders of the totalitarian bureaucracy.

"Commanded from Outside"

Just what is implied by "commanded from outside?" Here, too, Pivert mentions no persons, no institutions and no facts (apparently in the interests of politeness). We may assume, however, that he wished to say: "Commanded by Trotsky." Many, for lack of serious arguments, have resorted to this insinuation. But just what does the term "command" signify in this case? The Stalin bureaucracy commands by dint of power and money. Blum's machine commands by dint of its ties with the bourgeoisie. The Trotskyites have neither money, nor a GPU, nor ties with the bourgeoisie. How then can they "command?" It is simply a question of solidarity on fundamental questions. Why then the insinuation?

Nor is the expression "from outside" in any way happier. Is that an allusion to people outside the party? Or foreigners? Of what crime are these foreigners guilty? Of expressing their opinions and offering advice? When a serious struggle occurs within a revolutionary party, it inevitably engenders international repercussions. The representatives of one and the same tendency in various countries naturally seek to support each other. What is malicious or criminal about it? On the contrary, it is a manifestation of internationalism. Instead of chiding the "Trotskyites," one should learn from them!

An Example of "Comradely" Tone

Pivert then goes on to demand of the Trotskyites that they abandon their "means of pressure (?) or corruption (??) or systematic denigration ..." What is implied by the expression "means of pressure?" The apparatus of the party is in Pivert's hands, and the methods of pressure permitted by that apparatus are by no means alien to Pivert. The opposition has nothing at its disposal, save its ideas. Does Pivert wish to prohibit the exercise of ideological pressure? The term "corruption" has a very precise meaning in the language of politics: bribery, careerism, etc. In my opinion the Fourth International is the last organisation one could possibly accuse of such sins. There remains "systematic denigration." Experience has demonstrated that the vaguer the views of a politician, and the less he endures criticism, the more readily does a trenchant argument seem to him "denigration." An excess of sensitivity is a symptom of inner lack of confidence. As a party leader, Pivert should set an example of "trustful collaboration" and yet he permits himself to speak of "corruption." Let us hold that Pivert's pen slipped and that he himself will find occasion to make the correction.

Bolshevism and Factions

After refusing the opposition the right to struggle for a majority ("hegemony") in the party, and in accordance with this prohibiting factions, that is, trampling underfoot the elementary principles of a democratic régime, Pivert is imprudent enough to counterpose the democracy of the PSOP to Bolshevik centralism. A risky contraposition! The entire history of Bolshevism was one of the free struggle of tendencies and factions. In different periods, Bolshevism passed through the struggle of pro and anti-boycottists, "otzovists," ultimatists, conciliationists, partisans of "proletarian culture," partisans and opponents of the armed insurrection in October, partisans and opponents of the Brest-Litovsk treaty, left communists, partisans and opponents of the official military policy, etc., etc. The Bolshevik Central Committee never dreamed of demanding that an opponent "abandon factional methods," if the opponent held that the policy of the Central Committee was false. Patience and loyalty toward the opposition were among the most important traits of Lenin's leadership.

It is true that the Bolshevik party forbade factions at the Tenth Party Congress in March 1921, a time of mortal danger. One can argue whether or not this was correct. The subsequent course of development has in any case proved that this prohibition served as one of the starting points of the party's degeneration. The bureaucracy presently made a bogie of the concept of "faction," so as not to permit the party either to think or breathe. Thus was formed the totalitarian régime which killed Bolshevism. Is it not astonishing that Pivert who so loves to talk about democracy, freedom of criticism, etc., should borrow not from the vital, vigorous and creative democracy of young Bolshevism, but rather from the home of decadent Bolshevism take his bureaucratic fear of factions?

Discipline in Action

The corrective for factional struggle is discipline in action. The party is not a social club but a combat group. If Pivert had stated that the "Trotskyites" were violating discipline in action, that would have been a serious argument. But Pivert makes no such claim, which means that this is not the case.

Pivert's Faction

The demand to "abandon factional methods" is all the more inadmissible since Pivert himself has wholly at his disposal "hegemony," without doubt his own faction also, his own undercover meetings (for example, in the struggle against Trotskyism), etc. The only difference lies in this, that "Trotskyism" deals its blows against the right and Pivert against the left.

The Fourth International and Factions

In complete contradiction with reality, Pivert depicts the régime in the Fourth International as a régime of monolithism and blind submission. It would be hard to invent a caricature more fantastic and less scrupulous. The Fourth International has never prohibited factions and has no intention of doing so. Factions have existed and do exist among us. Controversy occurs always over the content of the ideas of each faction, but never over its right to existence. From the standpoint of Bolshevik ideas on party democracy I would consider it an outright scandal to accuse an opponent, who happened to be in the minority, of employing "factional" methods, instead of engaging in discussion with him over the gist of the question. If the differences are serious ones then factional methods are justified. If the differences are not serious then the adversary will find himself discredited. The factional struggle can result only in a more profound principled fusion or a split. No one yet has invented another alternative, if we leave aside the totalitarian regimes.

Verification of a Concrete Question

On the question of entry into the PSOP, for example, one could least of all discover among the "Trotskyites" "monolithism" or "blind submission." Our French comrades for a long time passionately discussed the question and in the end they split. What was my personal attitude in the matter? Let me state frankly — I hesitated. A few months ago I expressed myself in a private letter rather negatively. This did not prevent an influential group of French comrades under the leadership of Rous from entering the PSOP. I believe they have been proved correct.

A part of our French section has obviously revealed organisational conservatism and sectarianism. It would be an astonishing thing if under present political conditions such tendencies did not manifest themselves among those of the hounded and persecuted extreme left. Irrefutable facts testify that the Fourth International struggles against sectarianism and moreover with increasing success. A split is of course a regrettable episode, but nothing more than an episode. If the PSOP continues to evolve in a revolutionary direction (and we heartily hope that it will), it will draw into its ranks the dissident section of the "Trotskyites." If under the pressure of the bourgeoisie, social patriots and Stalinists the PSOP expels the "Trotskyites", unity will be reestablished outside of the PSOP.

"Leader Party"

Generalising his views on the party Pivert writes: "To the conception of a leader-party, a kind of centralised staff which prepares under cover of conspiracies a so-called (?) revolutionary action we prefer the conception of a party wide open to the real mass movement and offering the revolutionary vanguard all the possibilities of direct contact with the widest possible layers of the worker and peasant proletariat." As always, Pivert remains in the realm of abstractions and nebulous formulas. What "leader party" is referred to here? Is it the old Bolshevik Party? If so, why isn't this stated openly? Is it possible to educate workers by anonymous allusions? Furthermore, these allusions are false to the core. There has never been a party in history with a profound internal democracy which was distinguished by such awareness, boldness and flexibility in approaching the masses as the Bolshevik Party. Pivert still can only promise to establish contact with "the broadest possible layers;" while the Bolshevik Party united millions in action for victory. Incidentally, of what "undercover conspiracies" does Pivert speak so contemptuously? Is it perhaps the preparation of the October insurrection? But in that case he is merely repeating what has always been maintained by liberals, Mensheviks, and Social Revolutionaries.

Bolshevism Alone Built the Revolutionary Party

Organisational conceptions do not of course possess an independent character. But through them, and through them alone, is the programmatic and tactical position completely expressed. To dilettantes of the former Parisian magazine, *Masses*, and their ilk, organisational questions are reduced to assuring their "hegemony" over a little magazine and of protecting themselves from disagreeable criticism — further than that they do not go. The organisation of the Social Democracy was and remains entirely adapted to electoral tasks. To this day Bolshevism alone has been able to find organisational forms suitable for the revolutionary struggle for power. To wave Bolshevism aside by means of clichés without having behind one any other revolutionary experience is inadmissible, frivolous, and ignoble. That is not the way to educate workers!

Rosa Luxemburg

To prop up his organisational views (more exactly, their absence) Pivert of course cites Luxemburg. But that does not advance us greatly. Much can be learned from Rosa; but her organisational views were the weakest points in her position, for in them was summed up her errors in the sphere of theory and politics. In Germany, Rosa was unable to build a revolutionary party or a faction and this was one of the causes for the foundering of the 1918-1919 Revolution (on this point see the article of Walter Held in *Unser Wort*). As for the Polish party of Rosa Luxemburg under the influence of the events of the revolution it was compelled to reconstruct itself on the Bolshevik model. These historical facts are far more important than quotations!

Trotskyism in 1904

In 1904 I wrote a brochure, *Our Political Tasks*, which in the organisation sphere developed views very close to those of Rosa Luxemburg (Souvarine quotes this brochure with sympathy in his biography of Stalin). However, all subsequent experience demonstrated to me that Lenin was correct in this question as against Rosa Luxemburg and me. Marceau Pivert counterposes to the "Trotskyism" of 1939, the "Trotskyism" of 1904. But after all since that time three revolutions have taken place in Russia alone. Have we really learned nothing during these thirty five years?

"Libertarian" Promises

The better to recommend his spirit of democracy, Pivert promises that his "method of building socialism will not be authoritarian but libertarian." It is impossible not to smile sadly at this pompous and vague phrase. Does this formula of "libertarian" socialism signify anarchy, that is, rejection of the dictatorship of the proletariat? But Pivert considers himself a Marxist and not a Proudhonist or Bakuninist. The dictatorship of the proletariat by its very essence is "authoritarian," otherwise it would not be a dictatorship. It goes without saying that there are limits to "authoritarianism", that is, within the régime of the dictatorship there are differences. If Pivert wishes to say that he would strive to have the Soviets as the organs of dictatorship preserve the broadest possible internal democracy, then he will only be repeating what the "Trotskyites" have struggled for since 1923. However, for Pivert's promise to ring more convincingly he should not now be trampling internal party democracy underfoot in the manner of Leon Blum and Paul Faure, refusing the minority its most legitimate rights, prohibiting oppositional factions and preserving "hegemony" as a monopoly for his own faction; he should, in other words, establish at least one tenth of that democracy which distinguished the Bolshevik Party under Czarist illegality and during the first years of the Soviet régime. As long as this is not so, the promise of "libertarian" benefits in the indefinite future carries little value. It recalls somewhat the promise of recompense beyond the grave for sufferings in this world.

Such are the organisational views of Pivert. They signify in effect a break with party democracy and the substitution of bureaucratic centralism for democratic centralism, that is, the hegemony of the apparatus over ideas. We shall see presently that in the sphere of doctrine, program and politics, things do not go much better.

A Unilateral Demand

Pivert demands, as we know already, "trustful" collaboration with all those elements which have "courageously" broken with social patriotism and national communism. In principle we are prepared to accept such a demand. But unfortunately Pivert himself violates it and in a fashion that cries out. Bolshevism broke with all species of patriotism a quarter of a century before the PSOP. Pivert, however, doesn't at all reveal a "trustful" attitude towards Bolshevism. The Trotskyites, who have demonstrated the revolutionary character of their internationalism through a long struggle and with innumerable victims, are duty bound to trust Pivert; but Pivert is not at all obliged to trust the Trotskyites. Pivert's rule is — trust for the right — threats and repressions for the left. But this is the rule of Léon Blum, shifted only a few degrees.

The Break with Social Patriotism

Internationalism is indubitably the fundamental premise for collaboration. Our French comrades have taken into account very seriously the PSOP's break with the social patriotic party of Blum, otherwise they would not have entered the PSOP But to depict the matter as if a split with a putrified party automatically solves all questions is incorrect. After the break it is necessary to elaborate a revolutionary program and to determine exactly who are one's friends and one's enemies. The leadership of the PSOP has not yet done this. And this is not accidental. It is still a long way from having cut completely the old umbilical cord.

Freemasonry

The misfortune is that the leaders of the PSOP have not broken "courageously" with social patriotism, for they have not broken with Freemasonry, that important reservoir of imperialist patriotism. The other day I received the excellent pamphlet of Pierre Bailly *Yes, Freemasonry Is a Danger*. Rejecting all psychological and philosophical hogwash, which hasn't the slightest value since in the course of its entire development Freemasonry has contributed nothing either to science or philosophy, the author approaches the question in a Marxist manner, that is, from the class standpoint. On the basis of the documents of Freemasonry itself he has irrefutably demonstrated its imperialist, reactionary and demoralising rôle.

Bailly's pamphlet is, incidentally, the best proof of the fact that in contrast to all other factions and groups our comrades know how to approach a complex problem as proletarian revolutionists. Even the minor fact that Nikolitch's pamphlet, hollow and loaded with bourgeois sentimentality, is very well printed while Bailly's serious work is mimeographed illustrates well enough the social position of centrist and revolutionary ideas.

Social Pacifism

No, Pivert hasn't at all broken "courageously" with social patriotism and its variation, social pacifism — otherwise he would not have concluded an alliance against us with Maxton, the leader of the British Independent Labour Party. Between revolutionary Marxism and the imperialist pacifism of Maxton there is an abyss. Fenner Brockway is slightly to the left of Maxton. But, as the entire experience of the Independent Labour Party has demonstrated, Maxton at every critical occasion threatens to resign and Fenner Brockway immediately flops on hands and knees before Maxton. One may shut one's eyes to this. But the facts remain. Let Pivert explain to the workers just what links him with Maxton against the Fourth International. "Tell me who your friends are and I will tell you who you are."

Sneevliet

Pivert marches hand in hand with Sneevliet, whose entire politics in recent years has been — with God's help (!) not to provoke the anger of the Dutch government and not to deprive his sectarian trade union organisation of government subsidies. Dozens of times we demanded that Sneevliet's party elaborate a political platform, that Sneevliet as a member of parliament advance fighting slogans, that agitation among the masses be conducted in a revolutionary spirit. Sneevliet systematically equivocated so as not to break with his conservative government. It is best not to recall the "tone" which this democrat employed in discussions with young comrades. When the Conference of the Fourth International finally convened and at last took up the question of the Dutch section, Sneevliet quit our organisation and naturally began complaining about our bad "methods." Beyond doubt, Pivert's methods are much better: he keeps silent about the capitulatory politics of Sneevliet and directs his blows against the Trotskyites.

The POUM

Pivert strives to defend the personal memory of Andres Nin against base calumnies and this is of course excellent. But when he depicts Nin's politics as a revolutionary model then it is impermissible to call this anything but a crime against the proletariat. In the heat of revolutionary war between the classes Nin entered a bourgeois government whose goal it was to destroy the workers' committees, the foundation of proletarian government. When this goal was reached, Nin was driven out of the bourgeois government. Instead of recognising after this the colossal error committed, Nin's party demanded the reestablishment of the coalition with the bourgeoisie. Does Pivert dare deny this? It is not words which decide but facts. The politics of the POUM were determined by capitulation before the bourgeoisie at all critical times, and not by this or that quotation from a speech or article by Nin. There can be no greater crime than coalition with the bourgeoisie in a period of socialist revolution.

Instead of mercilessly exposing this fatal policy Pivert reprints in its justification all the old articles of Kurt Landau. Like Nin, Landau fell victim to the GPU. But the most ardent sympathy for the victims of Stalin's executioners does not free one from the obligation of telling the workers the truth. Landau, like Nin, represented one of the varieties of left Menshevism, was a disciple of Martov and not of Lenin. By supporting Nin's mistakes, and not our criticism of these mistakes, Landau, like Victor Serge, like Sneevliet, like Pivert himself, played a regrettable role in the Spanish revolution. Within the POUM a left opposition is now beginning to raise its head (José Rebull and his friends). The duty of Marxists is to help them draw the final conclusions from their criticisms. Yet Pivert supports the worst conservatives in the POUM of the Gorkin type. No, Pivert has not drawn the conclusions of his break with Blum!

"Practical Results"

It is with a disdain sufficiently out of place that Pivert speaks of the "practical results" achieved by Trotskyism as far too insignificant to force him to change his point of view. But just how in our epoch of universal reaction can a revolutionary party become a mass movement? At the present time owing to the avowed bankruptcy of the two former Internationals, the situation is becoming more favourable for the revolutionists. One of the signs is the split of the PSOP from Blum's party. But we began the struggle a long time before that. If Pivert thought in a critical manner he would understand that without the long preparatory work of the "Trotskyists" in all probability he would not yet have broken with Blum. From the broad historical outlook, the PSOP as a whole is only a by-product of the struggle of Trotskyism. Can it be that Pivert considers this "practical result" insignificant?

Reaction and "Trotskyism"

The fact that the Stalinists, as well as the bourgeois police, label every leftward tendency as Trotskyism shows that in the last analysis the entire force of world reaction beats down upon the Fourth International. The GPU maintains a large staff of agents on the one hand for espionage, frame ups and murders and on the other for provoking conflicts and splits in our ranks. Never before in history has there been a revolutionary tendency subjected to such persecution as ours. Reaction understands only too well that the danger is the Fourth International. Only thanks to the relentless criticism and propaganda of the Fourth International have the centrists begun to stir, the left centrists to separate themselves from the right centrists, the latter to demarcate themselves from the avowed social patriots. Several years ago Pivert stated correctly that struggle against Trotskyism was a certain sign of reaction. Sad to say, however, this reaction is drawing him into its ranks.

The Inner Power of the Fourth International

The international organisation of Brandler, Lovestone, etc., which appeared to be many times more powerful than our organisations has crumbled to dust. The alliance between Walcher of the Norwegian Labor Party and Pivert himself burst into fragments. The London Bureau has given up the ghost. But the Fourth International, despite all the difficulties and crises, has grown uninterruptedly, has its own organisations in more than a score of countries, and was able to convene its World Congress under the most difficult circumstances, the terror of the GPU (murder of Klement!), and to elaborate its program, to which no one has yet counterposed anything equivalent. Let Pivert attempt to enumerate Marxist publications which in their theoretical level can be placed alongside *The New International, Clave, Unser Wort,* and other organs of the Fourth International.

All the left groupings which gravitate in the orbit of the London Bureau or thereabouts represent heterogeneous splinters of the past without a common program, with senile routine and incurable maladies. The Fourth International is developing as a grouping of new and fresh elements on the basis of a common program growing out of the entire past experience, incessantly checked and rendered more precise. In the selection of its cadres the Fourth International has great advantages over the Third. These advantages flow precisely from the difficult conditions of struggle in the epoch of reaction. The Third International took shape swiftly because many "lefts" easily and readily adhered to the victorious revolution. The Fourth International takes form under the blows of defeats and persecutions. The ideological bond created under such conditions is extraordinarily firm. But the tempo of growth, at all events in the initial period, remains a slow one.

A Dilettante's Criterion

Victor Serge says: "You cannot create a workers' International worthy of the name just by wanting it." What a smug and at the same time hollow statement! One might imagine that Serge carries in his back pocket all the measurements for an International, exactly as for a pair of trousers. And can a national party "worthy of the name" be built "just by wanting it"? Does the PSOP, for example, correspond to Serge's measurements? People who approach the matter with such superficial criteria thereby demonstrate that for them the International is a solemn and pompous institution, something in the nature of a temple. When the magnificent edifice shall have been built (By whom? How?), then they will enter its arch. We approach it in a different manner. The International is for us, like a national party, an indispensable instrument of the proletariat. This instrument must be constructed, improved, sharpened. This is just what we are doing. We do not wait for someone else to do this work for us. We call upon all revolutionists to take part — right now, immediately, without losing an hour. When the Fourth International becomes "worthy of the name" in the eyes of Messrs. Litterateurs, Dilettantes, and Sceptics, then it will not be difficult to adhere to it. A Victor Serge (this one, or another) will then write a book in which he will prove (with lyricism and with tears!) that the best, the most heroic period of the Fourth International was the time, when bereft of forces, it waged a struggle against innumerable enemies, including petty-bourgeois sceptics.

Our Section in the USA

Pivert should beware of hasty conclusions! The PSOP is still far from being a mass party and has not yet had the opportunity of testing the power of its resistance to the pressure of imperialism. On the other hand our various sections have not only proved their viability but have also entered the arena of mass struggles. In the United States, the most powerful capitalist country in the world, the Socialist Workers Party, from a propaganda circle, which it had been for a number of years, is turning before our very eyes into a militant factor in working class politics. The struggle against fascism and the struggle against war are headed by the American section of the Fourth International. One of the chief fascist agitators, Father Coughlin, was recently compelled to devote one of his radio speeches to our American section and its struggle to build workers defence guards. The Socialist Workers Party is engaged in serious work in the trade unions, publishes an excellent twice weekly newspaper, a serious monthly journal, a newspaper for the youth (issued twice monthly) and renders important ideological and material assistance to other sections.

In Belgium

Our section in Belgium, almost wholly proletarian in composition, received some 7,000 votes in the last elections. Each vote in the present background of reaction and chauvinism is worth a hundred votes cast for reformist parties. Let Pivert not be too hasty in drawing a balance sheet! Let him rather attentively read the declaration issued by our Belgian comrades elected at Flénu. But alas! Instead of seeking ties with the Belgian Revolutionary Socialist Party, Pivert lends his ear to bankrupts and sectarians. Is it Vereecken together with Sneevliet and Victor Serge who will hew a highway to the masses?

A Voice From Saigon

In connection with the elections to the Colonial Council held April 30 of this year, the Bolshevik Leninists have written me from Saigon (Indo-China): "Despite the infamous coalition between the Stalinists and bourgeois of all colours, we have gained a brilliant victory. This victory was all the more hard won because the minds of the voters were befuddled for months by the foggy propaganda of a centrist group called 'October'... We marched into the struggle with the banner of the Fourth International fully unfurled... Today, more than ever before," the letter continues, "we understand the significance not only of the program of the Fourth International, but also of the struggle of 1925, 1926, 1927, and 1928 against the theory and practice of socialism in one country, the struggle against the Anti-Imperialist League and other pompous parade committees, Amsterdam Pleyel and tutti quanti."

This voice of the revolutionary workers from Saigon is infinitely more important than the voices of all the London Bureaus and pseudo-"Marxist centres." The advanced workers of an oppressed country rally to a persecuted International. From the experiences of their own struggles they have come to understand our program and they will know how to champion it. Especially precious and important is the declaration that the advanced Saigon workers understand the meaning of the struggle of the Left Opposition during the years from 1925 to 1928. Only continuity of ideas creates a revolutionary tradition, without which a political party sways like a reed in the wind.

In England and France

In the old colonising countries, England and France, the labour bureaucracy, directly interested in colonial superprofits, is more powerful and conservative than anywhere else in the world, and the revolutionary masses find it very difficult there to raise their heads. This is the explanation for the extremely slow development of the sections of the Fourth International in these countries. Upon the evolution of the PSOP depends to a large extent whether the revolutionists will succeed there in forcing a serious breach in the wall of betrayal and treachery in the coming months. But no matter how things turn out in this respect, the general course of development leaves no room for doubt. When the most oppressed strata in England and France erupt to the surface, they will not tarry at halfway positions but will adopt that program which gives an answer to the profundity and sharpness of the social contradictions.

"Dogma"

Pivert either refuses or is unable to understand that our invincible strength lies in our theoretical thoroughness and irreconcilability. "Trotsky allows in his organisation," writes Pivert, "only those members who accept as dogma (?), and consequently without discussion (?) a systematic reference to the principles elaborated in the first four congresses of the Communist International. Our conception of the party is altogether different." Subject to all sorts of dubious influences, Pivert attempts to reduce the movement of the Fourth International to a single individual; "Trotsky allows in his organisation..." Pivert couldn't possibly be ignorant of the fact that the Left Opposition from the very first embraced the flower of the Bolshevik Party; revolutionists tempered in illegality, heroes of the civil war, the best representatives of the younger generation — hundreds upon hundreds of exemplary Marxists who would have done honour to any party. Tens of thousands of "Trotskyites" died a lingering death. Was it really only because "Trotsky allows" or doesn't allow? Such gibberish should be left to Brandler, Walcher, Lovestone, Sneevliet and other cynics... but let us return to "dogma." In the Bolshevik Party differences arose after the first four congresses of the Comintern whose decisions were elaborated with the most direct participation of the future leaders of "the Left Opposition." A sharp turn towards opportunism was sanctioned by the Fifth Congress. Without renouncing the revolutionary tradition, the greatest in the annals of history, we have nevertheless not made of the first four congresses more than our starting point, nor have we restricted ourselves to them. We have observed, studied, discussed, criticised, worked out slogans and marched ahead. I might cite as proof our theoretical journals, internal bulletins, scores of programmatic books and pamphlets issued in the last fifteen years. Perhaps Pivert can mention a single serious critical work of our opponents which remains unanswered by us? Perhaps Pivert himself and his friends have a criticism of the decisions of the first four congresses not considered by us? Where is it?

In the very same article, Pivert demands of Trotskyists "that they accept the charter (of the PSOP), its structure, its statutes, the decisions of the majority, and oblige themselves to fulfill them without remiss." This demand is legitimate in itself but does this mean that the charter of the PSOP, its structures, statutes, etc. are "dogma"? Or is it solely the programmatic decisions of the first four congresses that are "dogma"?

Make Believe

Pivert reasons as follows; we must find, uncover, and reject those traits, those peculiarities and shortcomings of classic Bolshevism which Stalinism subsequently seized upon. This reasoning is formalistic and lifeless. Stalinism didn't at all formally seize always upon the worst traits of Bolshevism. Self-sacrifice is a magnificent quality of a revolutionist. Some of the defendants in the Moscow trials were undoubtedly guided by the spirit of self-sacrifice: to give their lives and even their reputations for the sake of "defence of the USSR." Does this imply that in place of self-sacrifice it is necessary to inculcate egoism? To this one might reply "it is necessary to develop critical insight". But that is a commonplace. The Bolsheviks were by no means less capable of critical insight than their latter day critics. But objective historical conditions are more powerful than the subjective ones. When a new bureaucracy in an isolated and backward country rises above the revolutionary class and strangles its vanguard, of necessity it utilises the formulas and traditions of Bolshevism, qualities and methods inculcated by it, but it charges them with a diametrically opposite social content. Lenin following Marx taught us that during the first stages of socialism elements of inequality will inevitably still remain. The bureaucracy transformed this idea into justification of its gangster privileges. Must we, because of this abuse, unconditionally reject the correct idea of Marx?

The dialectic of the class struggle throughout the length of history has accomplished similar transformations, substitutions, and transfigurations. This was the fate of Christianity, Protestantism, democracy, etc. This in particular was the fate of Freemasonry. It originated in the 17th Century as a reaction of the petty bourgeoisie against the decomposing spirit of capitalist individualism and attempted to resurrect the idealised morality of guild "brotherhood." In the course of the class struggle it later became an instrument in the hands of the big bourgeoisie for disciplining and subordinating the petty bourgeoisie to its own aims. It is impermissible to approach principles outside of social reality, outside of those classes which support them.

The criticism of Bolshevism which Pivert develops in the wake of Victor Serge and others does not contain an iota of Marxism. It substitutes for a materialistic analysis a game of make believe.

For the hegemony of scientific thought

A serious revolutionist who foresees the grave decisions which the party must make in critical times, feels acutely his responsibility in the preparatory period, painstakingly, meticulously analyses each fact, each concept, each tendency. In this respect a revolutionist resembles a surgeon who cannot rest content with commonplaces concerning anatomy but must know exactly the articulation of the bones, the muscles, the nerves and the tendons and their interconnection, so as not to make a single false movement with his scalpel. The architect, the physician, and the chemist would regard indignantly any proposal against rendering scientific concepts and formulas more precise, against claiming "hegemony" for the laws of mechanics, physiology, or chemistry, in favour of a conciliatory attitude toward other views no matter how erroneous. Yet this is precisely Pivert's position. Without plumbing the gist of programmatic differences, he repeats commonplaces on the "impossibility" of any one tendency "claiming to incorporate in itself all truth." Ergo? Live and let live. Aphorisms of this type cannot teach an advanced worker anything worthwhile; instead of courage and a sense of responsibility they can only instill indifference and weakness. The Fourth International wages a struggle against quackery for a scientific attitude toward the problems of proletarian politics. Revolutionary ardour in the struggle for socialism is inseparable from intellectual ardour in the struggle for truth.

Bolshevism or Menshevism?

To Pivert it seems that we are the representatives of dogmatism and routine whereas he is a proponent of critical thought. As a matter of fact in his criticism of "Trotskyism" Pivert merely repeats the hoary formulas of the Mensheviks without adding to them one original syllable. But Menshevism was also put to a test, and not a minor one. The Bolshevik Party victoriously led the greatest revolution in history. Finding itself isolated, it was unable to withstand the pressure of hostile historical forces. In other words Russian Bolshevism found it beyond its powers to substitute itself for the international working class. Menshevism on the contrary contributed nothing to the revolution except prostration and perfidy. Left Menshevism in the person of Martov signified sincere perplexity and impotence. The historic task posed by October has not been resolved. The fundamental forces participating in the struggle are the same. The choice is not between "Trotskyism" and the PSOP but between Bolshevism and Menshevism. From the starting point of Bolshevism we are ready to march forward. We refuse to crawl back.

The Program of the Fourth International

Pivert finds it necessary in June 1939 to return to the "four congresses," but we have succeeded already in marching far ahead. Last autumn, a year ago, our international conference adopted the program of Transitional Demands corresponding to the tasks of our epoch. Is Pivert familiar with this program? What is his attitude towards it? For our part we are desirous of nothing so much as criticism. In any "tone" you please, but getting to the heart of the matter!

Here is a concrete proposal which I take the liberty to make "from outside"; to proceed immediately to a discussion and an elaboration of an international program of the proletariat and to create a special publication for an international discussion on this question. As a basis for this discussion I propose the program of the Fourth International, "The Death Agony of Capitalism and the Tasks of the Fourth International." It goes without saying, however, that our International is prepared to accept as a basis for discussion another draft if it is forthcoming. Perhaps Pivert and his friends will accept this proposal? It would undoubtedly be a great step forward!

I have analysed Pivert's article with a meticulousness which might appear to some as superfluous and tiresome. To others the "tone" might again appear too sharp. But I believe, nevertheless, that a detailed explanation, precise and clear, is far greater evidence of a desire for collaboration than diplomatic equivocations supplemented by threats and insinuations. I should like not only Marceau Pivert but also Daniel Guérin to reflect on this. It is necessary to cease feeding on the empty formulas of yesterday. It is necessary to take the road of serious and honest discussion of the program and strategy of the new International.

The class, the party, and the leadership

Leon Trotsky

The extent to which the working class movement has been thrown backward may be gauged not only by the condition of the mass organisations but by ideological groupings and those theoretical inquiries in which so many groups are engaged. In Paris there is published a periodical *Que Faire* (What To Do) which for some reason considers itself Marxist but in reality remains completely within the framework of the empiricism of the left bourgeois intellectuals and those isolated workers who have assimilated all the vices of the intellectuals.

Like all groups lacking a scientific foundation, without a program and without any tradition this little periodical tried to hang on to the coat-tails of the POUM — which seemed to open the shortest avenue to the masses and to victory. But the result of these ties with the Spanish revolution seems at first entirely unexpected: The periodical did not advance but on the contrary retrogressed. As a matter of fact, this is wholly in the nature of things. The contradictions between the petty bourgeoisie, conservatism and the needs of the proletarian revolution have developed in the extreme. It is only natural that the defenders and interpreters of the policies of the POUM found themselves thrown far back both in political and theoretical fields.

The periodical *Que Faire* is in and of itself of no importance whatever. But it is of symptomatic interest. That is why we think it profitable to dwell upon this periodical's appraisal of the causes for the collapse of the Spanish revolution, inasmuch as this appraisal discloses very graphically the fundamental features now prevailing in the left flank of pseudo-Marxism.

Que Faire Explains

We begin with a verbatim quotation from a review of the pamphlet *Spain Betrayed* by comrade Casanova:

"Why was the revolution crushed? Because, replies the author (Casanova), the Communist Party conducted a false policy which was unfortunately followed by the revolutionary masses. But why, in the devil's name, did the revolutionary masses who left their former leaders rally to the banner of the Communist Party? 'Because there was no genuinely revolutionary party.' We are presented with a pure tautology. A false policy of the masses; an immature party either manifests a certain condition of social forces (immaturity of the working class, lack of independence of the peasantry) which must be explained by proceeding from facts, presented among others by Casanova himself; or it is the product of the actions of certain malicious individuals or groups of individuals, actions which do not correspond to the efforts of 'sincere individuals' alone capable of saving the revolution. After groping for the first and Marxist road, Casanova takes the second. We are ushered into the domain of pure demonology; the criminal responsible for the defeat is the chief Devil, Stalin, abetted by the anarchists and all the other little devils; the God of revolutionists unfortunately did not send a Lenin or a Trotsky to Spain as He did in Russia in 1917."

The conclusion then follows: "This is what comes of seeking at any cost to force the ossified orthodoxy of a chapel upon facts." This theoretical haughtiness is made all the more magnificent by the fact that it is hard to imagine how so great a number of banalities, vulgarisms and mistakes quite specifically of conservative philistine type could be compressed into so few lines.

The author of the above quotation avoids giving any explanation for the defeat of the Spanish revolution; he only indicates that profound explanations, like the "condition of social forces" are necessary. The evasion of any explanation is not accidental. These critics of Bolshevism are all theoretical cowards, for the simple reason that they have nothing solid under their feet. In order not to reveal their own bankruptcy they juggle facts and prowl around the opinions of others. They confine themselves to hints and half-thoughts as if they just haven't the time to delineate their full wisdom. As a matter of fact they possess no wisdom at all. Their haughtiness is lined with intellectual charlatanism.

Let us analyse step by step the hints and half-thoughts of our author. According to him a false policy of the masses can be explained only as it "manifests a certain condition of social forces," namely, the immaturity of the working class and the lack of independence of the peasantry. Anyone searching for tautologies couldn't find in general a flatter one. A "false policy of the masses" is explained by the "immaturity" of the masses. But what is "immaturity" of the masses? Obviously, their predisposition to false policies. Just what the false policy consisted of, and who were its initiators: the masses or the leaders — that is passed over in silence by our author. By means of a tautology he unloads the responsibility on the masses. This classical trick of all traitors, deserters and their attorneys is especially revolting in connection with the Spanish proletariat.

Sophistry of the Betrayers

In July 1936 — not to refer to an earlier period — the Spanish workers repelled the assault of the officers who had prepared their conspiracy under the protection of the People's Front. The masses improvised militias and created workers' committees, the strongholds of their future dictatorship. The leading organisations of the proletariat on the other hand helped the bourgeoisie to destroy these committees, to liquidate the assaults of the workers on private property and to subordinate the workers' militias to the command of the bourgeoisie, with the POUM moreover participating in the government and assuming direct responsibility for this work. What does "immaturity" of the proletariat signify in this case? Self-evidently only this, that despite the correct political line chosen by the masses, the latter were unable to smash the coalition of socialists, Stalinists, anarchists and the POUM with the bourgeoisie. This piece of sophistry takes as its starting point a concept of some absolute maturity, i.e. a perfect condition of the masses in which they do not require a correct leadership, and, more than that, are capable of conquering against their own leadership. There is not and there cannot be such maturity.

But why should workers who show such correct revolutionary instinct and such superior fighting qualities submit to treacherous leadership? object our sages. Our answer is: There wasn't even a hint of mere subordination. The workers' line of march at all times cut a certain angle to the line of the leadership. And at the most critical moments this angle became 180 degrees. The leadership then helped directly or indirectly to subdue the workers by armed force.

In May 1937 the workers of Catalonia rose not only without their own leadership but against it. The anarchist leaders — pathetic and contemptible bourgeois masquerading cheaply as revolutionists — have repeated hundreds of times in their press that had the CNT wanted to take power and set up their dictatorship in May, they could have done so without any difficulty. This time the anarchist leaders speak the unadulterated truth. The POUM leadership actually dragged at the tail of the CNT, only they covered up their policy with a different phraseology. It was thanks to this and this alone that the bourgeoisie succeeded in crushing the May uprising of the "immature" proletariat. One must understand exactly nothing in the sphere of the inter-relationships between the class and the party, between the masses and the leaders in order to repeat the hollow statement that the Spanish masses merely followed their leaders. The only thing that can be said is that the masses who sought at all times to blast their way to the correct road found it beyond their strength to produce in the very fire of battle a new leadership corresponding to the demands of the revolution. Before us is a profoundly dynamic process, with the various stages of the revolution shifting swiftly, with the leadership or various sections of the leadership quickly deserting to the side of the class enemy, and our sages engage in a purely static discussion: Why did the working class as a whole follow a bad leadership?

The Dialectic Approach

There is an ancient, evolutionary-liberal epigram: Every people gets the government it deserves. History, however, shows that one and the same people may in the course of a comparatively brief epoch get very different governments (Russia, Italy, Germany, Spain, etc.) and furthermore that the order of these governments doesn't at all proceed in one and the same direction: from despotism to freedom as was imagined by the evolutionist liberals. The secret is this, that a people is comprised of hostile classes, and the classes themselves are comprised of different and in part antagonistic layers which fall under different leadership; furthermore every people falls under the influence of other peoples who are likewise comprised of classes. Governments do not express the systematically growing "maturity" of a "people" but are the product of the struggle between different classes and the different layers within one and the same class, and, finally, the action of external forces — alliances, conflicts, wars and so on. To this should be added that a government, once it has established itself, may endure much longer than the relationship of forces which produced it. It is precisely out of this historical contradiction that revolutions, coup d'états, counterrevolutions, etc. arise.

The very same dialectic approach is necessary in dealing with the question of the leadership of a class. Imitating the liberals our sages tacitly accept the axiom that every class gets the leadership it deserves. In reality leadership is not at all a mere "reflection" of a class or the product of its own free creativeness. A leadership is shaped in the process of clashes between the different classes or the friction between the different layers within a given class. Having once arisen, the leadership invariably arises above its class and thereby becomes predisposed to the pressure and influence of other classes.

The proletariat may "tolerate" for a long time a leadership that has already suffered a complete inner degeneration but has not as yet had the opportunity to express this degeneration amid great events. A great historic shock is necessary to reveal sharply the contradiction between the leadership and the class. The mightiest historical shocks are wars and revolutions. Precisely for this reason the working class is often caught unawares by war and revolution.

But even in cases where the old leadership has revealed its internal corruption, the class cannot improvise immediately a new leadership, especially if it has not inherited from the previous period strong revolutionary cadres capable of utilising the collapse of the old leading party. The Marxist, i.e. dialectic and not scholastic interpretation of the inter-relationship between a class and its leadership does not leave a single stone unturned of our author's legalistic sophistry.

How the Russian Workers Matured

He conceives of the proletariat's maturity as something purely static. Yet during a revolution the consciousness of a class is the most dynamic process directly determining the course of the revolution. Was it possible in January 1917 or even in March, after the overthrow of Czarism, to give an answer to the question whether the Russian proletariat had sufficiently "matured" for the conquest of power in eight to nine months? The working class was at that time extremely heterogeneous socially and politically. During the years of the war it had been renewed by 30-40 percent from the ranks of the petty bourgeoisie, often reactionary, from backward peasants, from women, and from youth. The Bolshevik party in March 1917 was followed by an insignificant minority of the working class and furthermore there was discord within the party itself. The overwhelming majority of the workers supported the Mensheviks and the "Socialists-Revolutionists" i.e., conservative social-patriots. The situation was even less favourable with regard to the army and the peasantry. We must add to this: the general low level of culture in the country, the lack of political experience among the broadest layers of the proletariat, especially in the provinces, let alone the peasants and soldiers.

What was the "active"[1] of Bolshevism? A clear and thoroughly thought-out revolutionary conception at the beginning of the revolution was held only by Lenin. The Russian cadres of the party were scattered and to a considerable degree bewildered. But the party had authority among the advanced workers. Lenin had great authority with the party cadres. Lenin's political conception corresponded to the actual development of the revolution and was reinformed by each new event. These elements of the "active" worked wonders in a revolutionary situation, that is, in conditions of bitter class struggle. The party quickly aligned its policy to correspond with Lenin's conception, to correspond that is with the actual course of the revolution. Thanks to this it met with firm support among tens of thousands of advanced workers. Within a few months, by basing itself upon the development of the revolution the party was able to convince the majority of the workers of the correctness of its slogans. This majority organised into Soviets was able in its turn to attract the soldiers and peasants.

How can this dynamic, dialectic process be exhausted by a formula of the maturity or immaturity of the proletariat? A colossal factor in the maturity of the Russian proletariat in February or March 1917 was Lenin. He did not fall from the skies. He personified the revolutionary tradition of the working class.

For Lenin's slogans to find their way to the masses there had to exist cadres, even though numerically small at the beginning; there had to exist the confidence of the cadres in the leadership, a confidence based on the entire experience of the past. To cancel these elements from one's calculations is simply to ignore the living revolution, to substitute for it an abstraction, the "relationship of forces," because the development of the revolution precisely consists of this, that the relationship of forces keeps incessantly and rapidly changing under the impact of the changes in the consciousness of the proletariat, the attraction of backward layers to the advanced, the growing assurance of the class in its own strength. The vital mainspring in this process is the party, just as the vital mainspring in the mechanism of the party is its leadership. The role and the responsibility of the leadership in a revolutionary epoch is colossal.

Relativity of "Maturity"

The October victory is a serious testimonial of the "maturity" of the proletariat. But this maturity is relative. A few years later the very same proletariat permitted the revolution to be strangled by a bureaucracy which rose from its ranks. Victory is not at all the ripe fruit of the proletariat's "maturity." Victory is a strategical task. It is necessary to utilise the favourable conditions of a revolutionary crisis in order to mobilise the masses; taking as a starting point the given level of their "maturity" it is necessary to propel them forward, teach them to understand that the enemy is by no means om-

---

1. Translator's note: Untranslatable term, which means in part "liquid assets".

nipotent, that it is torn asunder with contradictions, that behind the imposing facade panic prevails. Had the Bolshevik party failed to carry out this work, there couldn't even be talk of the victory of the proletarian revolution. The Soviets would have been crushed by the counter-revolution and the little sages of all countries would have written articles and books on the keynote that only uprooted visionaries could dream in Russia of the dictatorship of the proletariat, so small numerically and so immature.

Auxiliary Role of Peasants

Equally abstract, pedantic and false is the reference to the "lack of independence" of the peasantry. When and where did our sage ever observe in capitalist society a peasantry with an independent revolutionary program or a capacity for independent revolutionary initiative? The peasantry can play a very great role in the revolution, but only an auxiliary role.

In many instances the Spanish peasants acted boldly and fought courageously. But to rouse the entire mass of the peasantry, the proletariat had to set an example of a decisive uprising against the bourgeoisie and inspire the peasants with faith in the possibility of victory. In the meantime the revolutionary initiative of the proletariat itself was paralysed at every step by its own organisations.

The "immaturity" of the proletariat, the "lack of independence" of the peasantry are neither final nor basic factors in historical events. Underlying the consciousness of the classes are the classes themselves, their numerical strength, their role in economic life. Underlying the classes is a specific system of production which is determined in its turn by the level of the development of productive forces. Why not then say that the defeat of the Spanish proletariat was determined by the low level of technology?

The Role of Personality

Our author substitutes mechanistic determinism for the dialectic conditioning of the historical process. Hence the cheap jibes about the role of individuals, good and bad. History is a process of the class struggle. But classes do not bring their full weight to bear automatically and simultaneously. In the process of struggle the classes create various organs which play an important and independent role and are subject to deformations. This also provides the basis for the role of personalities in history. There are naturally great objective causes which created the autocratic rule of Hitler but only dull-witted pedants of "determinism" could deny today the enormous historic role of Hitler. The arrival of Lenin in Petrograd on April 3, 1917 turned the Bolshevik party in time and enabled the party to lead the revolution to victory.

Our sages might say that had Lenin died abroad at the beginning of 1917, the October revolution would have taken place "just the same." But that is not so. Lenin represented one of the living elements of the historical process. He personified the experience and the perspicacity of the most active section of the proletariat. His timely appearance on the arena of the revolution was necessary in order to mobilise the vanguard and provide it with an opportunity to rally the working class and the peasant masses. Political leadership in the crucial moments of historical turns can become just as decisive a factor as is the role of the chief command during the critical moments of war. History is not an automatic process. Otherwise, why leaders? why parties? why programs? why theoretical struggles?

Stalinism in Spain

"But why, in the devil's name," asks the author as we have already heard, "did the revolutionary masses who left their former leaders, rally to the banner of the Communist Party?" The question is falsely posed. It is not true that the revolutionary masses left all of their former leaders. The workers who were previously connected with specific organisations continued to cling to them, while they observed and checked. Workers in general do not easily break with the party that awakens them to conscious life. Moreover the existence of mutual protection within the People's Front lulled them: Since everybody agreed, everything must be all right. The new and fresh masses naturally turned to the Comintern as the party which had accomplished the only victorious proletarian revolution and which, it was hoped, was capable of assuring arms to Spain. Furthermore the Comintern was the most zealous champion of the idea of the People's Front; this inspired confidence among the inexperienced layers of workers. Within the People's Front the Comintern was the most zealous champion of the bourgeois character of the revolution; this inspired the confidence of the petty and in part the middle bourgeoisie. That is why the masses "rallied to the banner of the Communist Party."

Our author depicts the matter as if the proletariat were in a well-stocked shoe store, selecting a new pair of boots. Even this simple operation, as is well known, does not always prove successful. As regards new leadership, the choice is very limited. Only gradually, only on the basis of their own experience through several stages can the broad layers of the masses become convinced that a new leadership is firmer, more reliable, more loyal than the old. To be sure, during a revolution, i.e., when events move swiftly, a weak party can quickly grow into a mighty one provided it lucidly understands the course of the revolution and possesses staunch cadres that do not become intoxicated with phrases and are not terrorised by persecution. But such a party must be available prior to the revolution inasmuch as the process of educating the cadres requires a considerable period of time and the revolution does not afford this time.

Treachery of the POUM

To the left of all the other parties in Spain stood the POUM, which undoubtedly embraced revolutionary proletarian elements not previously firmly tied to anarchism. But it was precisely this party that played a fatal role in the development of the Spanish revolution. It could not become a mass party because in order to do so it was first necessary to overthrow the old parties and it was possible to overthrow them only by an irreconcilable struggle, by a merciless exposure of their bourgeois character. Yet the POUM while criticising the old parties subordinated itself to them on all fundamental questions. It participated in the "People's" election bloc; entered the government which liquidated workers' committees; engaged in a struggle to reconstitute this governmental coalition; capitulated time and again to the anarchist leadership; conducted, in connection with this, a false trade union policy; took a vacillating and non-revolutionary attitude toward the May 1937 uprising.

From the standpoint of determinism in general it is possible of course to recognise that the policy of the POUM was not accidental. Everything in this world has its cause. However, the series of causes engendering the centrism of the POUM are by no means a mere reflection of the condition of the Spanish or Catalonian proletariat. Two causalities moved toward each other at an angle and at a certain moment they came into hostile conflict. It is possible by taking into account previous international experience, Moscow's influence, the influence of a number of defeats, etc. to explain politically and psychologically why the POUM unfolded as a centrist party. But this does not alter its centrist character, nor does it alter the fact that a centrist party invariably acts as a brake upon the revolution, must each time smash its own head, and may bring about the collapse of the revolution. It does not alter the fact that the Catalonian masses were far more revolutionary than the POUM, which in turn was more revolutionary than its leadership. In these conditions to unload the responsibility for false policies on the "immaturity" of the masses is to engage in sheer charlatanism frequently resorted to by political bankrupts.

Responsibility of Leadership

The historical falsification consists in this, that the responsibility for the defeat of the Spanish masses is unloaded on the working masses and not those parties which paralysed or simply crushed the revolutionary movement of the masses. The attorneys of the POUM simply deny the responsibility of the leaders, in order thus to escape shouldering their own responsibility. This impotent philosophy, which seeks to reconcile defeats as a necessary link in the chain of cosmic developments, is completely incapable of posing and refuses to pose the question of such concrete factors as programs, parties, personalities that were the organisers of defeat. This philosophy of fatalism and prostration is diametrically opposed to Marxism as the theory of revolutionary action.

Civil war is a process wherein political tasks are solved by military means. Were the outcome of this war determined by the "condition of class forces," the war itself would not be necessary. War has its own organisation, its own policies, its own methods, its own leadership by which its fate is directly determined. Naturally, the "condition of class forces" supplies the foundation for all other political factors; but just as the foundation of a building does not reduce the importance of walls, windows, doors, roofs, so the "condition of classes" does not invalidate the importance of parties, their strategy, their leadership. By dissolving the concrete in the abstract, our sages really halted midway. The most "profound" solution of the problem would have been to declare the defeat of the Spanish proletariat as due to the inadequate development of productive forces. Such a key is accessible to any fool.

By reducing to zero the significance of the party and of the leadership these sages deny in general the possibility of revolutionary victory. Because there are not the least grounds for expecting conditions more favourable. Capitalism has ceased to advance, the proletariat does not grow numerically, on the contrary it is the army of unemployed that grows, which does not increase but reduces the fighting force of the proletariat and has a negative effect also upon its consciousness. There are similarly no grounds for believing that under the regime of capitalism the peasantry is capable of attaining a higher revolutionary consciousness. The conclusion from the analysis of our author is thus complete pessimism, a sliding away from revolutionary perspectives. It must be said — to do them justice — that they do not themselves understand what they say.

As a matter of fact, the demands they make upon the consciousness of the masses are utterly fantastic. The Spanish workers, as well as the Spanish peasants gave the maximum of what these classes are able to give in a revolutionary situation. We have in mind precisely the class of millions and tens of millions.

Que Faire represents merely one of these little schools, or churches or chapels who frightened by the course of the class struggle and the onset of reaction publish their little journals and their theoretical études in a corner, on the sidelines away from the actual developments of revolutionary thought, let alone the movement of the masses.

Repression of Spanish Revolution

The Spanish proletariat fell the victim of a coalition composed of imperialists, Spanish republicans, socialists, anarchists, Stalinists, and on the left flank, the POUM. They all paralysed the socialist revolution which the Spanish proletariat had actually begun to realise.

It is not easy to dispose of the socialist revolution. No one has yet devised other methods than ruthless repressions, massacre of the vanguard, execution of the leaders, etc. The POUM of course did not want this. It wanted on the one hand to participate in the Republican government and to enter as a loyal peace-loving opposition into the general bloc of ruling parties; and on the other hand to achieve peaceful comradely relations at a time when it was a question of implacable civil war. For this very reason the POUM fell victim to the contradictions of its own policy.

The most consistent policy in the ruling bloc was pursued by the Stalinists. They were the fighting vanguard of the bourgeois-republican counter-revolution. They wanted to eliminate the need of Fascism by proving to the Spanish and world bourgeoisie that they were themselves capable of strangling the proletarian revolution under the banner of "democracy." This was the gist of their policies. The bankrupts of the Spanish People's Front are today trying to unload the blame on the GPU. I trust that we cannot be suspected of leniency toward the crimes of the GPU. But we see clearly and we tell the workers that the GPU acted in this instance only as the most resolute detachment in the service of the People's Front. Therein was the strength of the GPU, therein was the historic role of Stalin. Only ignorant philistines can wave this aside with stupid little jokes about the Chief Devil.

These gentlemen do not even bother with the question of the social character of the revolution. Moscow's lackeys, for the benefit of England and France, proclaimed the Spanish revolution as bourgeois. Upon this fraud were erected the perfidious policies of the People's Front, policies which would have been completely false even if the Spanish revolution had really been bourgeois. But from the very beginning the revolution expressed much more graphically the proletarian character than did the revolution of 1917 in Russia.

In the leadership of the POUM gentlemen sit today who consider that the policy of Andres Nin was too "leftist," that the really correct thing was to have remained the left flank of the People's Front. The real misfortune was that Nin, covering himself with the authority of Lenin and the October revolution, could not make up his mind to break with the People's Front. Victor Serge, who is in a hurry to compromise himself by a frivolous attitude toward serious questions, writes that Nin did not wish to submit to commands from Oslo or Coyoacan. Can a serious man really be capable of reducing to petty gossip the problem of the class content of a revolution?

The sages of *Que Faire* have no answer whatever to this question. They do not understand the question itself. Of what significance indeed is the fact that the "immature" proletariat founded its own organs of power, seized enterprises, sought to regulate production while the POUM tried with all its might to keep from breaking with bourgeois anarchists who in an alliance with the bourgeois republicans and the no less bourgeois socialists and Stalinists assaulted and strangled the proletarian revolution! Such "trifles" are obviously of interest only to representatives of "ossified orthodoxy." The sages of *Que Faire* possess instead a special apparatus which measures the maturity of the proletariat and the relationship of forces independently of all questions of revolutionary class strategy...

Tradition and revolutionary policy

Leon Trotsky

The question of the relationship of tradition and party policy is far from simple, especially in our epoch. More than once, recently, we have had occasion to speak of the immense importance of the theoretical and practical tradition of our party and have declared that we could in no case permit the breaking of our ideological lineage. It is only necessary to come to an agreement on what is meant by the tradition of the party. To do that, we must begin largely by the inverse method and take some historical examples in order to base our conclusions upon them.

Let us take the "classic" party of the Second International, the German social democracy. Its half century of "traditional" policy was based upon an adaptation to parliamentarism and to the unbroken growth of the organisation, the press, and the treasury. This tradition, which is profoundly alien to us, bore a semiautomatic character: each day flowed "naturally" from the day before and just as "naturally" prepared the day to follow. The organisation grew, the press developed, the cash box swelled.

It is in this automatism that the whole generation following Bebel took shape: a generation of bureaucrats, of philistines, of dullards whose political character was completely revealed in the first hours of the imperialist war. Every congress of the social democracy spoke invariably of the party's old tactics, consecrated by tradition. And the tradition was indeed powerful. It was an automatic tradition, uncritical, conservative, and it ended by stifling the revolutionary will of the party...

It is clear that, as a conservative element, as the automatic pressure of yesterday upon today, tradition represents an extremely important force at the service of the conservative parties and deeply inimical to the revolutionary party. The whole strength of the latter lies precisely in its freedom from conservative traditionalism. Does this mean that it is free with regard to tradition in general? Not at all. But the tradition of a revolutionary party is of an entirely different nature.

If we now take our Bolshevik Party in its revolutionary past and in the period following October, it will be recognised that its most precious fundamental tactical quality is its unequalled ability to orient itself rapidly, to change tactics quickly, to renew its armament and to apply new methods, in a word, to carry out abrupt turns. Tempestuous historical conditions have made this tactic necessary. Lenin's genius gave it a superior form. This is not to say, naturally, that our party is completely free of a certain conservative traditionalism: a mass party cannot be ideally free. But its strength and potency have manifested themselves in the fact that inertia, traditionalism, routinism, were reduced to a minimum by a farsighted, profoundly revolutionary tactical initiative, at once audacious and realistic.

It is in this that the genuine tradition of the party consists and should consist. The relatively strong bureaucratisation of the party apparatus is inevitably accompanied by the development of conservative traditionalism with all its effects. It is better to exaggerate this danger than to underrate it. The undeniable fact that the most conservative elements of the apparatus are inclined to identify their opinions, their methods, and their mistakes with the "Old Bolshevism," and seek to identify the criticism of bureaucratism with the destruction of tradition, this fact, I say, is already by itself the incontestable expression of a certain ideological petrifaction.

Marxism is a method of historical analysis, of political orientation, and not a mass of decisions prepared in advance. Leninism is the application of this method in the conditions of an exceptional historical epoch. It is precisely this union of the peculiarities of the epoch and the method that determines that courageous, self-assured policy of brusque turns of which Lenin gave us the finest models, and which he illuminated theoretically and generalized on more than one occasion.

Marx said that the advanced countries, to a certain extent, show the backward countries the image of their future. Out of this conditional proposition an effort was made to set up an absolute law which was at the root of the "philosophy" of Russian Menshevism. By means of it, limits were fixed for the proletariat, flowing not from the course of the revolutionary struggle but from a mechanical pattern; Menshevik Marxism was and remains solely the expression of the needs of bourgeois society, an expression adapted to a belated "democracy." In reality, it turned out that Russia, joining in its economy and its politics extremely contradictory phenomena, was the first to be pushed onto the road of the proletarian revolution.

Neither October, nor Brest-Litovsk, nor the creation of a regular peasant army, nor the system of requisitioning food products, nor the NEP, nor the State Planning Commission, were or could have been foreseen or predetermined by pre-October Marxism or Bolshevism. All these facts and turns were the result of the independent, critical application of the methods of Bolshevism, marked by the spirit of initiative, in situations that differed in each case.

Every one of these decisions, before being adopted, provoked struggles. The simple appeal to tradition never decided anything. As a matter of fact, with each new task and at each new turn, it is not a question of searching in tradition and discovering there a nonexistent reply, but of profiting from all the experience of the party to find by oneself a new solution suitable to the situation and, by doing so, enriching tradition. It may even be put more sharply: Leninism consists of being courageously free of conservative retrospection, of being bound by precedent, purely formal references, and quotations.

Lenin himself not so long ago expressed this thought in Napoleon's words: "On s'engage et puis on voit" (start fighting and then see). To put it differently, once engaged in the struggle, don't be excessively pre-occupied with canon and precedent, but plunge into reality as it is and seek there the forces necessary for victory and the roads leading to it. It is by following this line that Lenin, not once but dozens of times, was accused in his own party of violating tradition and repudiating "Old Bolshevism"...

The more ingrown the party apparatus, the more imbued it is with the feeling of its own intrinsic importance, the slower it reacts to needs emanating from the ranks and the more inclined it is to set formal tradition against new needs and tasks. And if there is one thing likely to strike a mortal blow to the spiritual life of the party and the doctrinal training of the youth, it is certainly the transformation of Leninism from a method demanding for its application initiative, critical thinking, and ideological courage, into a canon which demands nothing more than interpreters appointed for good and all. Leninism cannot be conceived of without theoretical breadth, without a critical analysis of the material bases of the political process. The weapon of Marxist investigation must be constantly sharpened and applied. It is precisely in this that tradition consists, and not in the substitution of a formal reference or an accidental quotation. Least of all can Leninism be reconciled with ideological superficiality and theoretical slovenliness.

Lenin cannot be chopped up into quotations suited for every possible case, because for Lenin the formula never stands higher than the reality; it is always the tool that makes it possible to grasp the reality and to dominate it. It would not be hard to find in Lenin dozens and hundreds of passages which, formally speaking, seem to be contradictory. But what must be seen is not the formal relationship of one passage to another, but the real relationship of each of them to the concrete reality in which the formula was introduced as a lever. The Leninist truth is always concrete!

As a system of revolutionary action, Leninism presupposes a revolutionary sense sharpened by reflection and experience, which, in the social realm, is equivalent to the muscular sensation in physical labour. But revolutionary sense cannot be confused with demagogical flair. The latter may yield ephemeral successes, sometimes even sensational ones. But it is a political instinct of an inferior type. It always leans toward the line of least resistance. Leninism, on the other hand, seeks to pose and resolve the fundamental revolutionary problems, to overcome the principal obstacles; its demagogical counterpart consists in evading the problems, in creating an illusory appeasement, in lulling critical thought to sleep.

Leninism is, first of all, realism, the highest qualitative and quantitative appreciation of reality, from the standpoint of revolutionary action. Precisely because of this it is irreconcilable with flying from reality behind the screen of hollow agitationalism, with passive loss of time, with haughty justification of yesterday's mistakes on the pretext of saving the tradition of the party.

Leninism is genuine freedom from formalistic prejudices, from moralising doctrinairism, from all forms of intellectual conservatism attempting to stifle the will to revolutionary action. But to believe that Leninism signifies that "anything goes" would be an irremediable mistake. Leninism includes the morality, not formal but genuinely revolutionary, of mass action and the mass party. Nothing is so alien to it as functionary arrogance and bureaucratic cynicism. A mass party has its own morality, which is the bond of fighters in and for action. Demagogy is irreconcilable with the spirit of a revolutionary party because it is deceitful: by presenting one or another simplified solution for the difficulties of the hour, it inevitably undermines the future and weakens the party's self-confidence.

Swept by the wind and gripped by a serious danger, demagogy easily dissolves into panic. It is hard to juxtapose, even on paper, panic and Leninism.

Leninism is warlike from head to foot. War is impossible without cunning, without subterfuge, without deception of the enemy. Victorious war cunning is a constituent element of Leninist politics. But at the same time, Leninism is a supreme revolutionary honesty toward the party and the working class. It admits of no fiction, no bubble-blowing, no pseudo-grandeur!

Leninism is orthodox, obdurate, irreducible, but it does not contain so much as a hint of formalism, canon, or bureaucratism. In the struggle, it takes the bull by the horns. To make out of the traditions of Leninism a supra-theoretical guarantee of the infallibility of all the words and thoughts of the interpreters of these traditions, is to scoff at a genuine revolutionary tradition and transform it into official bureaucratism. It is ridiculous and pathetic to try to hypnotise a great revolutionary party by the repetition of the same formulas, according to which the right line should be sought not in the essence of each question, not in the methods of posing and solving this question, but in information... of a biographical character...

Whatever the difficulties and the differences of opinion may be in the future, they can be victoriously overcome only by the party's collective thinking, checking up on itself each time and thereby maintaining the continuity of development. This character of the revolutionary tradition is bound up with the peculiar character of revolutionary discipline. Where tradition is conservative, discipline is passive and is violated at the first moment of crisis. Where, as in our party, tradition consists of the highest revolutionary activity, discipline attains its maximum point, for its decisive importance is constantly checked in action. That is the source of the indestructible alliance of revolutionary initiative, of critical, bold elaboration of questions, with iron discipline in action, and it is only by this superior activity that the youth can receive from the old tradition of discipline and carry it on.

We cherish the traditions of Bolshevism as much as anybody. But let no one dare identify bureaucratism with Bolshevism, tradition with officious routine.

Index

Sources for Part 2 and Part 4

Part 2

The party that won the victory: *New International* November 1944

Did Bolshevism produce Stalinism? *New International* July-August 1950

The working class and socialism: *Labor Action*, 30 November 1953

The "mistakes" of the Bolsheviks: *New International*, November 1943

Bolshevism and gossip: *New International*, November 1938

October was a true working class revolution: ISL Internal Bulletin, September 1957

Part 4

Shachtman confronts Kerensky: *Labor Action*, 19 February 1951 (unsigned)

An anti-Bolshevik eye-witness: *Labor Action*, 14 and 21 June 1954

What is Trotskyism? From *The Struggle for the New Course*, preface to an edition of Trotsky's *The New Course*, 1943

Blanquism and social democracy: *Czerwony Sztander* [Red Flag], Cracow, 27 June 1906

Hands off Rosa Luxemburg! *The Militant*, 6 and 13 August 1932

Trotskyism and the PSOP: *New International*, October 1939

The class, the party and the leadership: *Fourth International*, December 1940 (an unfinished article found in Trotsky's papers after his death).

Tradition and revolutionary policy: from *The New Course*